Aquaculture and the Environment: An Integrated Approach

Aquaculture and the Environment: An Integrated Approach

Edited by Douglas Rodriquez

Syrawood
PUBLISHING HOUSE

New York

Published by Syrawood Publishing House,
750 Third Avenue, 9th Floor,
New York, NY 10017, USA
www.syrawoodpublishinghouse.com

Aquaculture and the Environment: An Integrated Approach
Edited by Douglas Rodriquez

International Standard Book Number: 978-1-68286-733-4(Hardback)

Cataloging-in-Publication Data

Aquaculture and the environment : an integrated approach / edited by Douglas Rodriquez.
    p. cm.
Includes bibliographical references and index.
ISBN 978-1-68286-733-4
1. Aquaculture. 2. Aquatic ecology. 3. Aquaculture--Environmental aspects.
I. Rodriquez, Douglas.
SH135 .A68 2019
639.8--dc23

TABLE OF CONTENTS

Permissions

List of Contributors

Index

PREFACE

This book aims to highlight the current researches and provides a platform to further the scope of innovations in this area. This book is a product of the combined efforts of many researchers and scientists, after going through thorough studies and analysis from different parts of the world. The objective of this book is to provide the readers with the latest information of the field.

Aquaculture is the rearing and breeding of freshwater and saltwater aquatic organisms in controlled conditions. Aquatic organisms include fish, molluscs, crustaceans and aquatic plants. To enhance production, various intervention strategies like regular stocking, feeding, protection from predators, etc. are implemented. Various environmental concerns can arise due to unregulated aquaculture practices. Some of these potential damages include the introduction of invasive organisms in an aquatic ecosystem, the competition between wild and farmed species, antibiotic side-effects, etc. This book provides comprehensive insights into the field of aquaculture. Most of the topics introduced herein cover new techniques and their applications in a comprehensive manner. It is appropriate for students, researchers and experts seeking detailed information in this area.

I would like to express my sincere thanks to the authors for their dedicated efforts in the completion of this book. I acknowledge the efforts of the publisher for providing constant support. Lastly, I would like to thank my family for their support in all academic endeavors.

Editor

Carbon and nitrogen flow, and trophic relationships, among the cultured species in an integrated multi-trophic aquaculture (IMTA) bay

Tariq Mahmood[1,4,*], **Jianguang Fang**[2], **Zengjie Jiang**[2], **Jing Zhang**[3]

[1]School of Resources and Environmental Science, East China Normal University, 3663 North Zhongshan Road, Shanghai 200062, PR China

[2]Key Laboratory of Sustainable Utilization of Marine Fisheries Resources, Ministry of Agriculture, Yellow Sea Fisheries Research Institute, Chinese Academy of Fishery Sciences, Qingdao 266071, PR China

[3]State Key Laboratory of Estuarine and Coastal Research, East China Normal University, 3663 North Zhongshan Road, Shanghai 200062, PR China

[4]*Present address:* National Institute of Oceanography, St 47, Block 1, Clifton, Karachi 75600, Pakistan

ABSTRACT: Stable isotopic signatures of organic carbon (δ^{13}C) and total nitrogen (δ^{15}N) were measured on suspended particulates and sediments in order to understand the sources of organic matter (OM), water quality and flow of organic carbon and nitrogen among integrated multi-trophic aquaculture (IMTA) species, as well as to evaluate the role of IMTA practice in accumulation and assimilation of OM during wet and dry seasons. OM distribution and composition were studied during 2011 in Sanggou Bay (SGB) of northern China, a system that receives terrestrial and oceanic inputs, and which is used for IMTA ventures. Results showed that higher terrestrial input of OM occurs during the wet compared to the dry season in the SGB. OM in suspended particulates (POM) showed marine- and terrestrial-derived signatures during the wet season, as revealed from their ranges in δ^{13}C (−27.4 to −20.7‰) and δ^{15}N (4.7 to 9.4‰). Sedimentary organic matter (SOM) showed signatures of marine-derived OM during both seasons, with ranges in δ^{13}C and δ^{15}N of −22.4 to −21.4‰ and 1.7 to 6.4‰, respectively. Shellfish and combined (shellfish, seaweed) cultures in SGB have the potential to reduce OM received from the fish cages as well as from the seasonal inputs from rivers. Mixing with Yellow Sea water, combined with prevailing circulation, favours the dispersal, dilution and transformation of OM and maintains and improves water quality. Based on our results, and compared with previous studies, the water quality of the SGB is likely to be sustained by IMTA activities.

KEY WORDS: IMTA · POM · SOM · Carbon isotope · Nitrogen isotope · Trophic levels · Sanggou Bay

INTRODUCTION

Excess amounts of carbon and nitrogen produced either from land-based or offshore aquaculture activities are considered to be one of the main sources of pollution in coastal environments. Increasing coastal area development as well as aquaculture activities have been of particular concern to the health of coastal ecosystems. Land-based aquaculture waste is often discharged directly into shallow coastal areas, causing excessive organic and nutrient loads (Alabaster 1982). Offshore cage culture is considered to be a direct source of organic matter (OM) to the surrounding waters in the form of suspended detritus (Karakassis et al. 2000, Mazzola & Sarà 2001), which mainly consists of uneaten feed and excretion products from the cultured fish (Holby & Hall 1991, Hall et al. 1992). Furthermore, anthropogenic input provides

*Corresponding author: tariqnio@gmail.com

additional nutrient and OM enrichment in the coastal marine system (Evgenidou & Valiela 2002). This waste affects not only the area in close proximity to the sources but can alter a wider coastal zone at various ecosystem levels; reducing the biomass, density and diversity of the benthos, plankton and nekton, and modifying natural food webs and stimulating eutrophication (Gowen et al. 1991, Pillay 1991, Vollenweider 1992, Duarte 1995). However, the offshore cultivation of shellfish together with seaweed could reduce the impact of OM waste and nutrients on the environment, as substantiated by land-based integrated aquaculture practice (Shpigel et al. 1991, Shpigel & Neori 1996). The aquaculture-derived nutrients can be removed by seaweed biofilters (Buschmann et al. 2008). Such a combined species cultivation method, so-called integrated multi-trophic aquaculture (IMTA), is practiced in Chinese coastal zones. Besides the feasible ventures in mariculture schemes, the combination of trophic levels among cultured species in IMTA systems is also important in improving water quality. The IMTA of shellfish, seaweed and fish is common on the coast of northern China and has been in practice over 3 decades (Fang et al. 1996a,b, 2009).

Sanggou Bay (SGB) receives OM from both natural and anthropogenic sources, which subsequently impact the water quality of the bay. SGB is surrounded by a population of ca. 0.6 million in Rongcheng City of Shandong Peninsula. River runoff from Rongcheng City is considered the main source of nutrients into SGB and is composed on average of 65% crop land waste and 35% urban waste (Project SPEAR; Ferreira et al. 2007). Stable isotope analysis has been used successfully in determining sources of nutrition for consumers, evaluation of trophic relationship among organisms, understanding different sources of OM (terrestrial and marine) and environmental impact assessment (Wada et al. 1987, Risk & Erdmann 2000, Costanzo et al. 2001). Stable isotope ratios of organic carbon ($\delta^{13}C$) and total nitrogen ($\delta^{15}N$) have also been used to determine the impact of aquaculture waste on the environment (Ye et al. 1991, Vizzini & Mazzola 2004, Yokoyama et al. 2006, Jiang et al. 2012). Aquaculture waste enters the food web and alters the natural isotopic composition of OM sources at both the base and upper trophic levels. Nitrogen-rich fish waste mainly affects $\delta^{15}N$ values without or little alteration of $\delta^{13}C$ (Vizzini & Mazzola 2004). Aquaculture and human waste can affect at different levels of the ecosystem—reducing the biomass, density and diversity of the benthos, plankton and

nekton—and modify natural food webs in coastal areas (Gowen et al. 1991, Pillay 1991).

In the present study, our first goal was to investigate the carbon and nitrogen flow from (1) phytoplankton, particulate OM (POM), sediment OM (SOM) or seaweed to filter feeders and (2) trash fish (feed provided to fish in fish cages) or plankton to omnivorous fish in an IMTA system in SGB using dual isotopic technique. A second objective was to study the isotopic profile ($\delta^{13}C$ and $\delta^{15}N$) of SOM and POM to understand the sources of carbon and nitrogen in SGB. Our study focused on understanding the role of lower trophic levels in the reduction of OM and clarifying whether aquaculture- and land-derived OM impact the water quality of the bay.

MATERIALS AND METHODS

Study area

The SGB (37° 01' to 37° 09' N and 122° 24' to 122° 35' E) is located in Rongcheng Town, in Weihai City, on the Shandong Peninsula in northeastern China (Fig. 1). The bay is semi-enclosed and opens into the Yellow Sea (YS) in the east, covering an area of 144 km². Freshwater inputs to the bay are mainly from one large river (the Gu River) and some small rivers (Ba, Sanggan, Yetao and Xiaolou Rivers). The bay experiences seasonal terrigenous inputs, with freshwater inflow being maximum in summer and with an average discharge of 1.7×10^8 m³ to 2.3×10^8 m³ (Rongcheng River Report 2012, www.rcsl.gov.cn). Water in the bay is well mixed and depth varies between 7.5 and 21 m (Zhao et al. 1996). IMTA is an important commercial activity in SGB. On the basis of culturing activities, the bay is divided into 4 culture areas. The southwest is used for shellfish and fish culture (hereafter, SF+F), the central part is dominated by polyculture of shellfish and seaweed (SF+SW), and the outer bay is cultivated with seaweed (SW) monoculture along the eastern boundary that opens into the YS (Fig. 1). Fish is cultured between May and October, while bivalve culture lasts between 1 and 2 yr. Red seaweed and kelp are cultivated from June–October and November–April, respectively (Zhao et al. 1996, SPEAR 2007). Shellfish and seaweed are cultivated in long lines around fish cages. Bivalve production includes the Chinese scallop *Chlamys farreri* (~60 × 10³ t yr⁻¹) and the Pacific oyster *Crassostrea gigas* (~15 × 10³ t yr⁻¹). Seaweed production includes kelp *Saccharina japonica* (~84 × 10³ t yr⁻¹) and red alga *Gracilaria lemaneiformis*

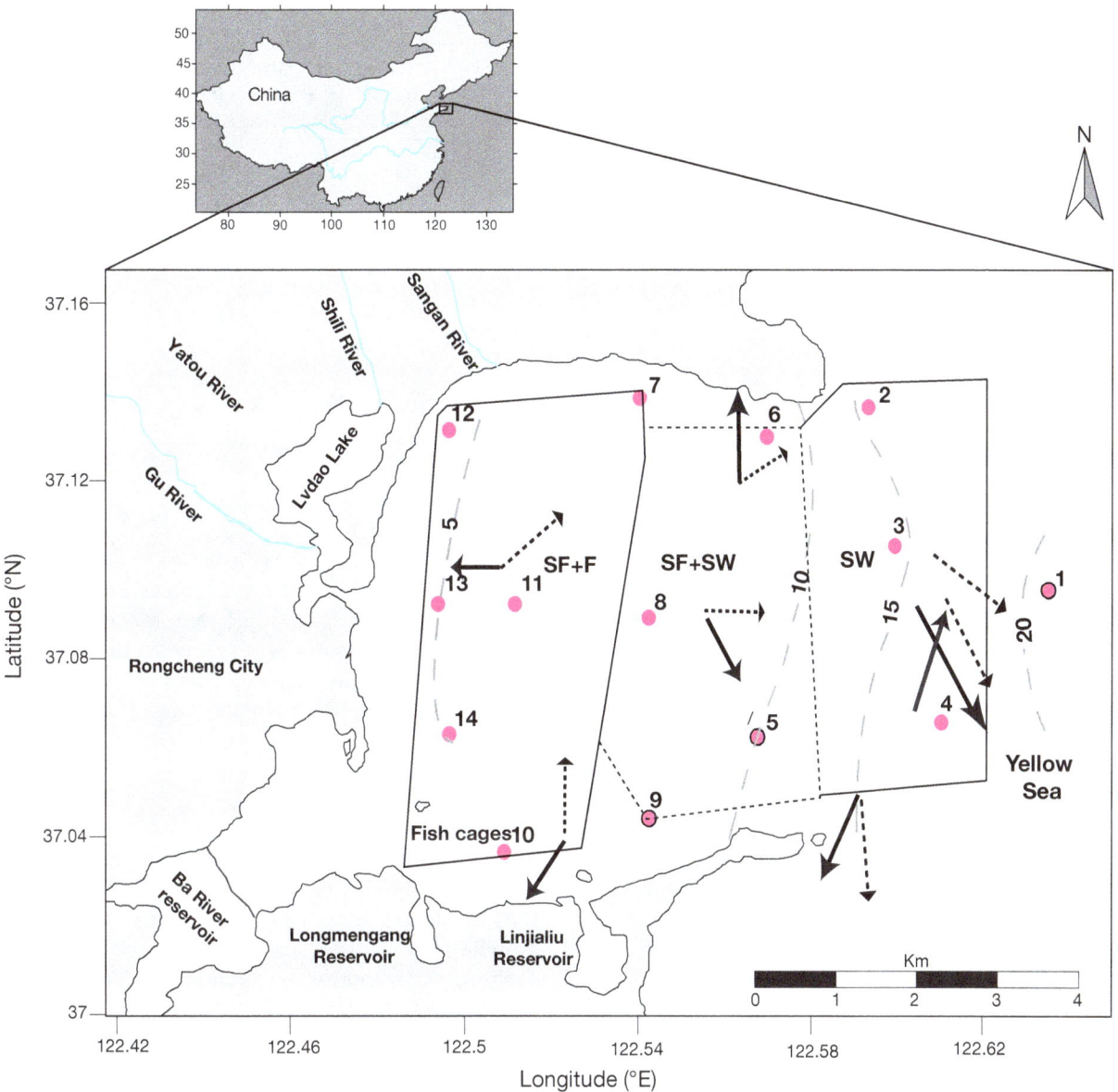

Fig. 1. Map of Sanggou Bay, showing culture areas (polygons with solid and dotted lines) and 14 stations (red dots) of cruises in August 2011 (wet season, summer) and January 2012 (dry season, winter). Culture areas include combined culture of shellfish and fish (SF+F), shellfish and seaweed (SF+SW) and monoculture of seaweed (SW). Solid arrows denote surface water current and dashed arrows bottom flow (source: Ferreira et al. 2007). Grey dashed lines denote isobaths (m)

(~25 × 10³ t yr⁻¹). The production of Japanese flounder *Paralichthys olivaceus* is ~24 × 10³ t yr⁻¹ (Rongcheng Fisheries Technology Extension Station 2012 statistics [www.rchy.gov.cn], summarized in Table 1).

Sampling and analysis

Samples for hydrographic parameters, POM, SOM, phytoplankton, zooplankton, shellfish (oyster and scallop), seaweed, cultured fish and trash fish were collected in August 2011 (wet season, i.e. summer) and January 2012 (dry season, i.e. winter). Surface water samples were collected using a Niskin water sampler at 14 stations covering all 3 culture areas in SGB (Fig. 1). The water samples were immediately screened through a 200 μm mesh net to remove larger zooplankton and debris. They were filtered under vacuum onto prewashed, pre-combusted (450°C, 4h) and pre-weighed Whatman GF/F filter papers (0.7 μm pore size). The samples were subsequently stored at −40 °C in a freezer until laboratory analysis.

Table 1. Summary of aquaculture in Sanggou Bay, where species are cultured in combination (SF+F, SF+SW) and monoculture (SW) in integrated multi-trophic aquaculture (IMTA). Additional details on the cultured area, annual production, and stocking, harvesting and culture periods for the different groups are also given (data from Zhao et al. 1996, Ferreira et al. 2007, Rongcheng Fisheries Technology Extension Station 2012 statistics [www.rchy.gov.cn])

Cultured species	Cultured area (km²), total per group	Stocking period	Harvesting period	Culture period	Production (t yr⁻¹)
Shellfish (SF)					
Chlamys farreri (Chinese scallop)	32	May	March	1–2 yr	~60 × 10³
Crassostrea gigas (Pacific oyster)		May	March	1–2 yr	~15 × 10³
Seaweed (SW)					
Saccharina japonica (kelp)	40	November	April	6 mo	~84 × 10³
Gracilaria lemaneiformis (Gracilaria)		June	October	5 mo	~25 × 10³
Fish (F)					
Paralichthys olivaceus (Japanese flounder)	0.36	May	October	6 mo	~24 × 10³

Bottom sediment samples were collected with a Van Veen grab (Hydro-bios) from a few stations and then frozen at –20°C until analysis. Salinity and chlorophyll *a* (chl *a*) were measured *in situ* with a multi-parameter instrument (Model: YSI Professional plus USA) and an ACLW-RS chlorophyll sensor, respectively. Cultured fish, shellfish, seaweed and trash fish samples were collected by local fishermen at some sampling sites. Phytoplankton (60 µm) and zooplankton (200 µm) nets were used to collect plankton samples. Plankton samples were filtered through Whatman GF/F filter papers, then frozen at –40 °C until analysis. All samples of fish, shellfish and trash fish were rinsed carefully with filtered seawater and guts were removed to reduce bias. Muscle of cultured fish, trash fish and shellfish, as well as sediments and particulate samples, were dried at 60°C for at least 24 h prior to stable isotope analysis. Cultured fish, trash fish and bivalve samples were soaked in 1.2 N HCl for 30 min, rinsed with distilled water, dried at 60°C and ground to a powder. The bottom sediment samples were ground and sieved through a 0.2 µm mesh, and then both the sediments and particulate samples were digested with 1 M HCl to remove carbonates and dried at 60°C for 12 h. Samples for total nitrogen concentration and isotopes were directly measured without the acid treatment (Cui et al. 2012).

Organic carbon, total nitrogen content and isotopes of carbon and nitrogen were measured using a Finnigan EA-1112 elemental analyzer interfaced with a Finnigan Delta plus XP continuous flow isotope ratio mass spectrometer. Carbon and nitrogen isotope ratios are expressed in the delta notation $\delta^{13}C$ and $\delta^{15}N$ relative to Vienna Pee Dee Belemnite and atmospheric nitrogen, respectively, and expressed as (Hayes 2004):

$$\delta X = [(R_{sample}/R_{standard}) - 1] \times 1000 \ (\text{‰}) \quad (1)$$

where, $X = {}^{13}C$ or ${}^{15}N$, and $R = {}^{13}C{:}^{12}C$ for $\delta^{13}C$ or ${}^{15}N{:}^{14}N$ for $\delta^{15}N$.

Internal standards of caffeine and cellulose were used for calibration during the measurements. The average precision for organic carbon and total nitrogen measurements during this study was ±0.1 %.

Trophic levels among the cultured species were calculated using the following formula (Wan et al. 2010):

$$\text{Trophic level} = [(\text{consumer } \delta^{15}N - \text{phyto } \delta^{15}N)/3.2] + 1 \quad (2)$$

where 3.2 represents the average enrichment of $\delta^{15}N$ among trophic levels in the present study, obtained by calculating the average value of $\delta^{15}N$ of each trophic level. This value is close to the enrichment factor of 3.1 reported by Wan et al. (2010) in a YS trophic level study.

Statistical analysis

SPSS 17.0 and Golden Software Grapher 9 were used to perform data analysis. Seasonal variation in $\delta^{13}C$ and $\delta^{15}N$ of POM and SOM were examined using 1-way ANOVA. Difference of $\delta^{13}C$ and $\delta^{15}N$ values of POM and SOM were analyzed by a paired *t*-test (Cui et al. 2012).

RESULTS

Hydrographic parameters

A negative correlation between salinity and chl *a* was observed in the wet season ($r^2 = -0.82$; $p < 0.05$). The coastal region was dominated by low salinity

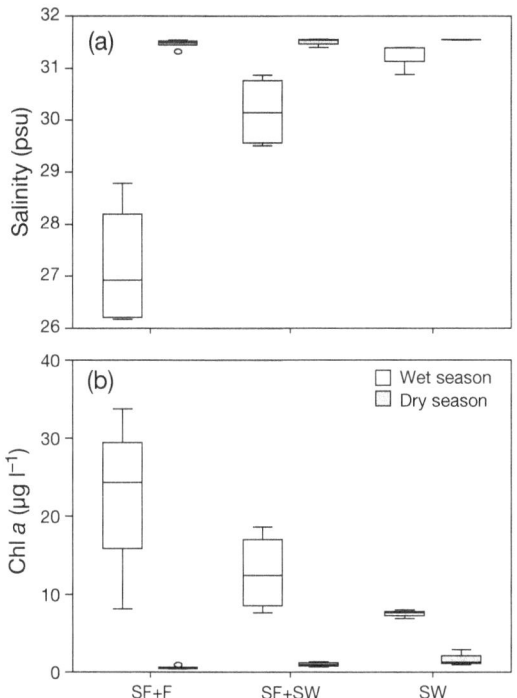

Fig. 2. Surface distribution of (a) salinity and (b) chlorophyll *a* in the 3 culture areas (see Fig. 1) in Sanggou Bay during the wet and dry seasons. Box plots show the median value (line), 25 and 75% quantiles (box), 5 and 95% quantiles (whiskers), and outliers (circles)

and high chl *a* concentration. Slightly lower salinity and higher chl *a* concentrations were found in the SF+F culture area of the bay compared to SF+SW and SW culture areas. The other 2 culture areas showed high salinity and low chl *a* concentrations. The maximum salinity and minimum chl *a* values were observed in the SW culture region (Fig. 2). The average values of salinity and chl *a* during the wet season in SGB were 29.4 ± 2.0 psu and 15.5 ± 10.9 µg l⁻¹ (Fig. 2), respectively. There was no significant variation in salinity among the aquaculture areas during the dry season (Fig. 2), due to low freshwater input into the bay. Considering all culture areas of SGB in the dry season, salinity ranged between 31 and 32 psu, with an average (±SD) of 31.5 ± 0.07 psu. During the dry season, the average (±SD) chl *a* concentration was 1.0 ± 0.63 µg l⁻¹. Chl *a* was significantly higher in the SW culture area in the offshore region than in the SF+F area in the coastal region of the bay (Fig. 2).

Stable isotope analysis of biological samples

The weight percentages of organic carbon in cultured fish and shellfish were higher than in plankton

and seaweed. The maximum values of nitrogen (% dry wt) were found in cultured fish and minimum values in plankton (Fig. 3). The C/N ratios of cultured fish, oysters, scallops and trash fish were in the range of 2.7–2.8, 3.6–4.0, 5.2–5.4 and 4.4–4.5, respectively, which were lower than the C/N ratios of phytoplankton (9.9), zooplankton (11.6) and *Gracilaria* spp. (hereafter simply *Gracilaria*) (10.0). δ¹³C versus δ¹⁵N values of SOM, POM, biological samples and the trophic level of the cultured species are shown in Fig. 4. The respective average values (±SD) of δ¹³C and δ¹⁵N were −21.1 ± 0.1‰ and 9.2 ± 0.4‰ for scallops, −21.1 ± 0.2‰ and 11.2 ± 0.3‰ for oysters, −20.9 ± 0.1‰ and 6.7‰ for *Gracilaria*, −19.0 ± 0.2‰ and −21.0 ± 0.6‰ for cultured fish, and 11.1 ± 0.3‰ and 9.6 ± 1.2‰ for trash fish.

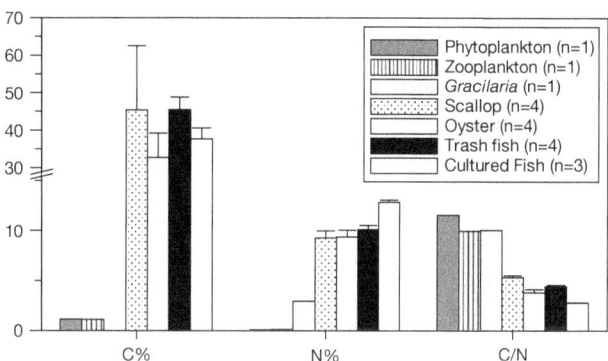

Fig. 3. Carbon and nitrogen contents (% dry wt) and C/N ratios of cultured species (seaweed, shellfish and fish; see Table 1) and of phyto- and zooplankton and input feed (i.e trash fish) in Sanggou Bay during the wet season. Means ± SD

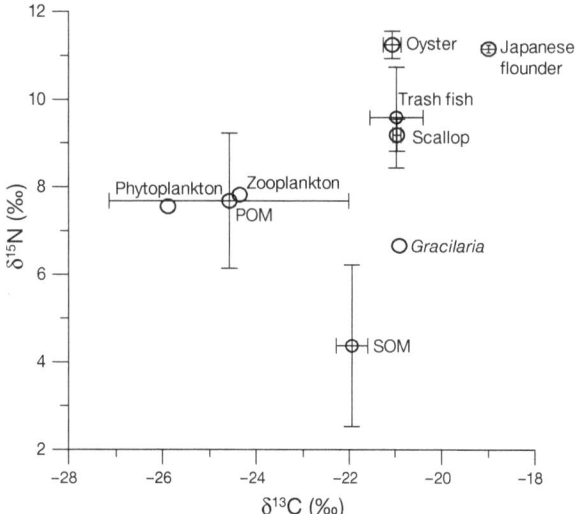

Fig. 4. δ¹⁵N‰ versus δ¹³C (‰) isotopic signatures of plankton, cultured species, trash fish, and particulate and sediment organic matter from Sanggou Bay during the wet season. Means given ± SD, if n > 1

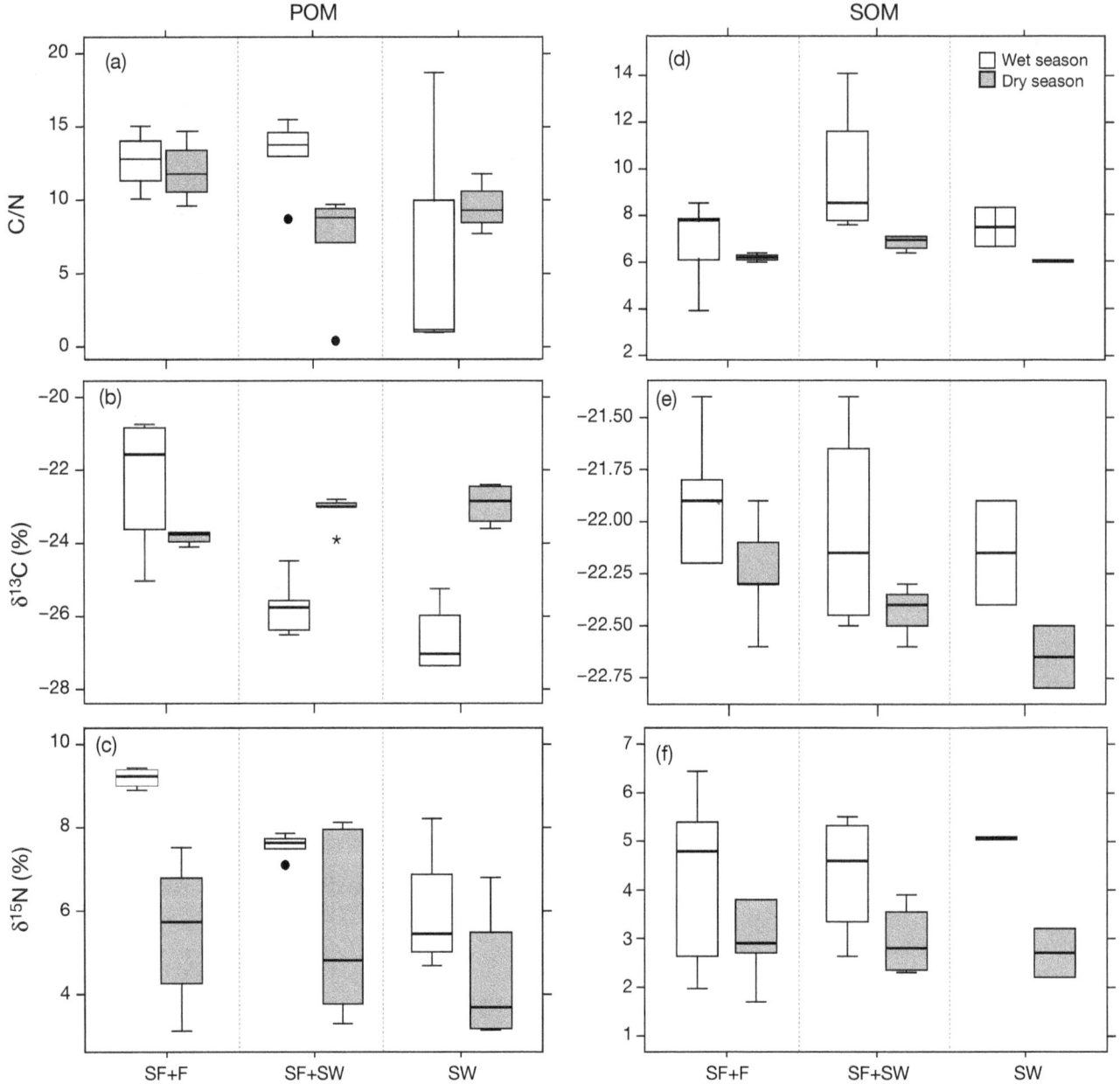

Fig. 5. Distribution of C/N, $\delta^{13}C$ (‰) and $\delta^{15}N$ (‰) of (a–c) particulate (POM) and (d–f) sediment organic matter (SOM) in the 3 culture areas (see Fig. 1) of Sanggou Bay during the wet and dry seasons. Box plots show the median value (line), 25 and 75 % quantiles (box), 5 and 95 % quantiles (whiskers), outliers (black dots) and extremes (stars)

Stable isotope analysis of SOM and POM in culture areas

The distribution of C/N, $\delta^{13}C$ and $\delta^{15}N$ of SOM (n = 26) and POM (n = 28) in the 3 culture areas of SGB during the wet and dry seasons is shown in Fig. 5. The fish cage culture and long-line culture of *Gracilaria* in SGB are performed during the wet season. Mixing of the bay water with the YS is higher in the SW culture area compared to the central (SF+SW) area. In the wet season, the lowest (1.16) and highest (18.68) C/N values of POM were observed in the SW culture area (Fig. 5a). For SOM, the lowest C/N value (3.93) was found in the SF+F culture area and the highest (14.09) in the SF+SW area (Fig. 5d). Highest values of $\delta^{13}C$ and $\delta^{15}N$ of POM were found in the SF+F culture area (−20.74‰ and 9.43‰, respectively) and the lowest in the SW culture area (−27.35‰ and 4.68‰, respectively) (Fig. 5b,c). The $\delta^{15}N$ values of POM showed a decreasing trend from SF+F to seaward (Fig. 5c). In contrast, no significant difference was found in the distribution of $\delta^{13}C$ of SOM among the 3 culture areas in the wet season (Fig. 5e). $\delta^{15}N$ values of SOM also showed no significant difference among the 3 culture areas in the wet season (Fig. 5f).

In the dry season, POM maximum and minimum C/N ratios were observed in SF+F (14.77) and SF+SW (0.39) culture areas, respectively, whereas for SOM no significance difference was found in C/N ratios among the 3 culture areas (Fig. 5a,d). The $\delta^{13}C$ values of POM and SOM were in the range of $-24.06‰$ to $-21.88‰$, with only minor variations being observed between POM and SOM (Fig. 5b,e). The lowest value (3.12‰) of $\delta^{15}N$ of POM was found in the SF+F culture area and the highest (8.12‰) in SF+SW (Fig. 5c). A slight, though non-significant, decrease in SOM $\delta^{15}N$ was observed from SF+F to SW culture areas (Fig. 5f).

Within SGB overall, significant differences in $\delta^{13}C$, $\delta^{15}N$ and C/N values of SOM and POM between wet and dry seasons (p < 0.05) were found. In both seasons, SOM had slightly higher values of $\delta^{13}C$ than POM. In contrast, SOM had lower values of $\delta^{15}N$ and C/N compared to POM in both seasons.

DISCUSSION

Trophic relationships among the cultured species

In the present study, the C/N ratio (>11) of phytoplankton (being a major fraction of POM) indicates that terrestrial material from the rivers is a major source of carbon, since these values are higher than those previously reported for marine phytoplankton (range: 6.7–10) and closer to vascular plants (>12) (Redfield et al. 1963, Holligan et al. 1984, Meyers 1994, Hedges & Oades 1997, Bale & Morris 1998, Bates et al. 2005, Lamb et al. 2006). The $\delta^{13}C$ values of plankton (range: $-25.4‰$ to $-25.9‰$) in this study were lower than those reported for Narragansett Bay, USA (mean ± SD: $-22 ± 0.6‰$) and Osaka Bay in Japan (range: $-18.0‰$ to $-24.0‰$) (Gearing et al. 1984, Mishima et al. 1996). The $\delta^{15}N$ values of phyto- and zooplankton (range: 7.6–7.8‰) were within the range reported for marine phytoplankton (3.0–10‰) (Wada et al. 1991). For oysters, we determined relatively lower values of $\delta^{13}C$ (mean ± SD: $-20.03 ± 0.18‰$) and higher values of $\delta^{15}N$ (8.27 ± 0.13‰) compared to values reported from oysters around a fish cage area in Ailian Bay, China (mean ± SD: $-20.03 ± 0.18‰$; Jiang et al. 2012), indicating that river runoff has been a source of carbon and nitrogen

in oysters of the present study. The $\delta^{13}C$ and $\delta^{15}N$ values ($-19.0‰$ and 11.1‰) we determined in cultured fish were lower than the average values ($-17‰$ and 13‰, respectively) observed for marine fishes (Mays 2000).

The wet season in SGB is characterized by peak IMTA activities, when fish cage culture occurs in conjunction with shellfish and seaweed. In addition, maximum freshwater inputs influence the sources and flow of OM (carbon and nitrogen) among cultured species and other organisms at various trophic levels. In the wet season, along with integrated aquaculture, primary production is a large carbon source for higher trophic levels. In the summer months, SGB usally experiences comparatively high light intensity and water temperature, which promote phytoplankton growth. This is reflected by the high chl a concentrations we observed in this season and the positive correlation between chl a and the $\delta^{13}C$ of POM that is dominated by phytoplankton (Fig. 6). Similar findings were reported by Lehmann et al. (2004), who showed that an increase in $\delta^{13}C$ values of POC is associated with increasing primary productivity due to the seasonal environmental conditions, including water temperature and light intensity. The enhanced primary production is connected to the high input of nutrients by freshwater inflow, as indicated by depleted $\delta^{13}C$ values of POM in the present study. It is possible that zooplankton in this bay feed on terrestrial detritus, which has $\delta^{13}C$ and $\delta^{15}N$ values similar to POM. Shellfish are usually considered to derive a large proportion of organic carbon from phytoplankton (Xu & Yang 2007). By identifying the relative con-

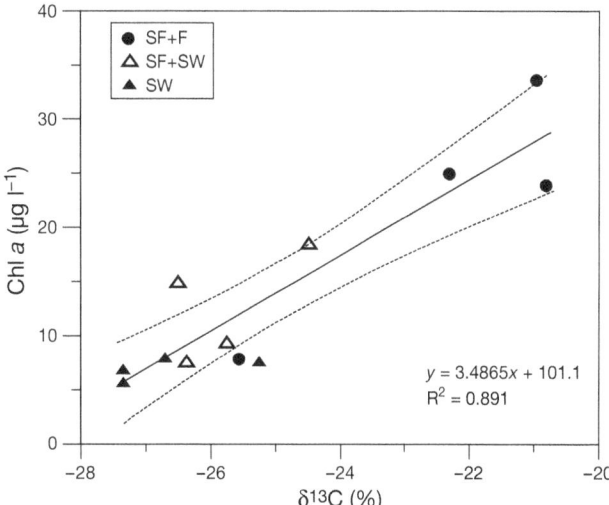

Fig. 6. Relationship between $\delta^{13}C$ (‰) of particulate organic matter and chlorophyll a concentration in 3 different aquaculture areas (see Fig. 1) of Sanggou Bay during the wet season

tribution of aquaculture-derived OM and its impact on water quality, the present study shows that shellfish can be considered to function as biological filters in coastal integrated aquaculture, as was reported previously for land-based integrated aquaculture (Shpigel et al. 1991, Shpigel & Neori 1996). In the coastal area different kinds of POM are present that may serve as a food source for shellfish (oyster and scallop) (Dame 1996). The observed increase in $\delta^{15}N$ from phytoplankton and POM to omnivorous fish was indicative of the trophic position of the cultured species in SGB: $\delta^{15}N$ ranged from 6.7‰ for autotrophs to up to 11.2‰ for heterotrophs, reflecting the enrichment in $\delta^{15}N$ with increasing trophic level. The $\delta^{15}N$ signatures of primary producers (phytoplankton and seaweed) clearly separated the filter feeders (shellfish) from omnivorous fish (Japanese flounder) (Fig. 4). Some species shared the same trophic level, such as cultured fish and oyster (2.16), but differed in $\delta^{13}C$ values (fish: −19.0 ± 0.2‰, oyster: −21.1 ± 0.2‰; mean ± SD), indicating that these species are up-taking carbon from different sources. In spite of this, cultured fish showed 2% enrichment in $\delta^{13}C$ from its primary input source of feeding, i.e. trash fish, while oysters also showed a $\delta^{13}C$ signature similar to trash fish with 0% enrichment. In contrast, some species, such as scallop and oyster, showed similar $\delta^{13}C$ values (−21.0 ± 0.1‰ and −21.1 ± 0.2‰ [mean ± SD], respectively) indicating the same carbon source, but the difference in their $\delta^{15}N$ values revealed that they belong to different trophic levels (1.52 and 2.16, respectively). Similar findings of the same trophic relationship (i.e. different carbon sources with same trophic level and similar carbon sources with different trophic levels) were reported in Jinghai Bay, China (Feng et al. 2014).

In the present study, scallop showed low $\delta^{15}N$ isotopic fractionation compared to the average fractionation factor reported elsewhere (3.4; Minagawa & Wada 1984). Several studies have reported low nitrogen fractionation values for shellfish (Raikow & Hamilton 2001, Post 2002, Marin-Leal et al. 2008), suggesting that low $\delta^{15}N$ enrichment may be due to the specific physiological characteristics of scallops. Moreover, the $\delta^{13}C$ values of trash fish, seaweed and shellfish were close to each other. We did not collect faeces samples but used an average ($\delta^{13}C = -21.8‰$) of respective values from the literature (Table 2). This average value was close to the $\delta^{13}C$ value of shellfish, suggesting that shellfish in SGB may also use carbon

Table 2. Carbon and nitrogen isotopes signatures of fish faecal material reported in previous literature and resulting average value that was adopted as reference value for this study

Study area	$\delta^{13}C$ (‰)	$\delta^{15}N$ (‰)	References
Gokasho Bay, Japan	−24.3	6.3	Yokoyama et al. (2006)
Gokasho Bay, Japan	−24.7	5.6	Yokoyama et al. (2010)
Gokasho Bay, Japan	−20.6	6.2	Yokoyama (2013)
Kat O Bay, Hong Kong	−21.6	4.4	Wai et al. (2011)
Nansha Bay, China	−17.5	7.5	Jiang et al. (2012)
Average	−21.8	6.0	

sources from faecal material released from fish cages, uneaten particles of trash fish and rotten seaweed. Therefore, shellfish cultured in SGB possibly not only help in reducing OM but may also be able to increase the economic benefit and production and survival rate of other species in the IMTA system by maintaining water quality. Based upon stable isotope analysis, a conceptual model of OM flow among the integrated aquaculture species in SGB was established (Fig. 7). The trophic level efficiency was calculated by dividing the $\delta^{13}C$ and $\delta^{15}N$ values of one trophic level to the next. POM integrated both phytoplankton and zooplankton and acted as a large source of OM that could be transferred to all the upper trophic levels in the integrated food web structure of SGB. Stable isotope results indicate that scallop and oyster are taking up >80% of the OM from these sources in SGB. Bivalves accumulated approx. 90% of their carbon and 60% of the nitrogen from fish faeces and uneaten particles of trash fish, but during the wet season only; as opposed to the dry season, when shellfish mostly relied on POM, phytoplankton and zooplankton. Feeding on faeces and trash fish remains during the warm wet season probably helped to meet the high metabolic demand of the shellfish in warmer water temperatures. Alternative sources of OM in the dry season at low temperature may be provided through large-scale cultivation of kelp. Kelp culture produces a considerable amount of rotten kelp particles that can serve as a source of OM to shellfish, whereas shellfish would only be provided a minute amount (1%) of OM from *Gracilaria* culture. Omnivorous cultured fish obtained most of their carbon (90%) and nitrogen (60%) from trash fish, while other sources were OM from producers and herbivores. In the food web structure of the cultured species of the present study, shellfish played a crucial role in OM accumulation from various sources. In summary, the water quality of SGB is not impacted by OM generated by caged fish and shellfish culture activities; on the contrary, shellfish

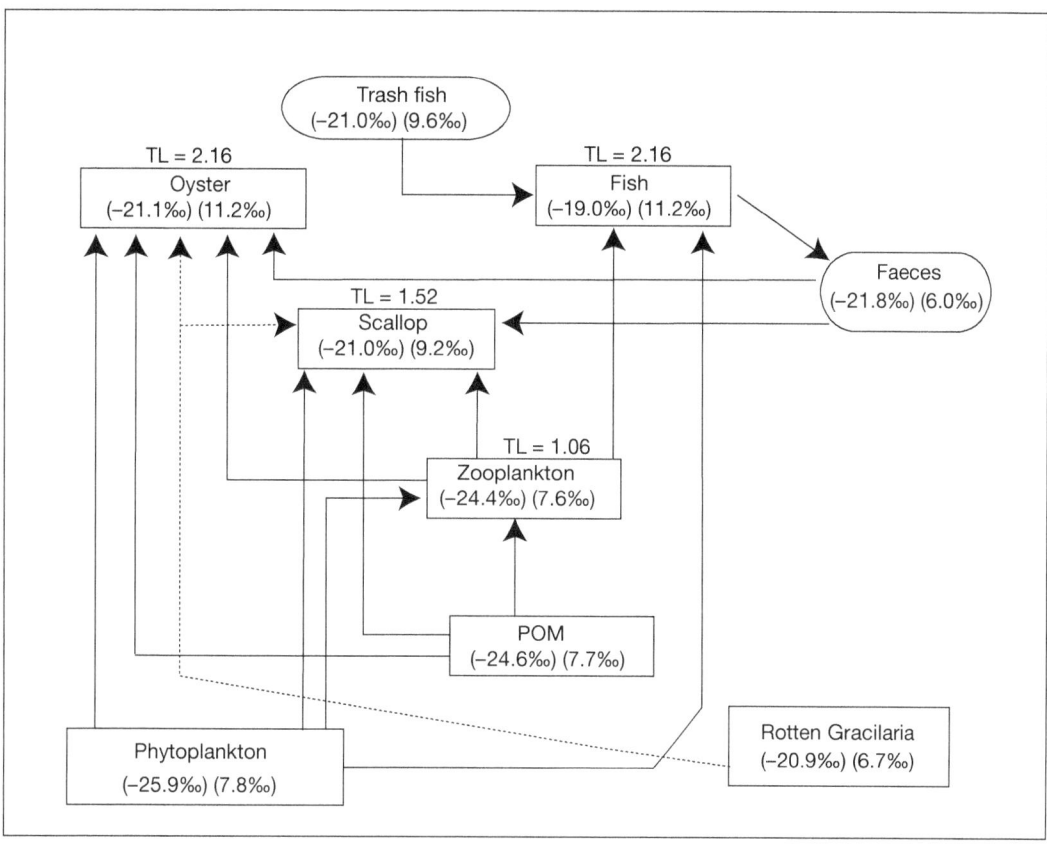

Fig. 7. Conceptual model of carbon flow among the aquaculture species in Sanggou Bay on the basis of stable isotope analysis. Values in parentheses are $\delta^{13}C$ (first) and $\delta^{15}N$ (second) isotopic signatures (‰) of various species and represent the flow (arrows; dotted arrows originate from rotten *Gracilaria*) from one trophic level (TL) to the next, showing the trophic efficiency. The TL is given for each cultured species

co-culture combined with proximity to the YS to allow for water mixing may be helpful in maintaining the water quality of the bay.

Sources of suspended and sedimentary OM across the bay

In the wet season, higher C/N values (>10) of POM in the SF+F (near coast) and SF+SW (central bay) culture areas indicate the influence of terrestrial OM. The lower C/N ratio in the *Gracilaria* monoculture area (near YS) may indicate the high consumption of nitrogen in this area or mixing with YS water. In the wet season, POM in the SF+F culture area showed higher values of $\delta^{13}C$ with a decreasing trend towards offshore, indicating OM load in the SF+F near-coast area compared to the other 2 areas. The lower values of $\delta^{13}C$ towards offshore may have resulted from the presence of degraded OM (Khodse et al. 2007). Another reason could be that high freshwater discharge during the wet season may have resulted

in the rapid distribution of OM to offshore waters, preventing utilization and deposition of OM in the bay. By contrast, the higher values of $\delta^{15}N$ of POM in the SF+F culture area (near-shore area) compared to the central SF+SW and outer SW culture areas may be attributed to nitrate derived from human activities coupled with increased denitrification (Michener & Schell 1994, McClelland et al. 1997, Chanton and Lewis 1999, Miller et al. 2010). The decreasing trend of $\delta^{15}N$ in POM towards the sea suggests an offshore source of nitrogen (Miller et al. 2011). In the wet season, higher values of $\delta^{13}C$ (−22.4 ‰ to −21.4‰) in SOM of 3 culture areas displayed the isotopic signature of marine-derived OM (Wada et al. 1987, Tan et al. 1991, Mishima et al. 1996, Barros et al. 2010). Similar results for SOM $\delta^{13}C$ were found by Meksumpun et al. (2005) (avg. $\delta^{13}C$ = −21.0‰) in the Gulf of Thailand, as well as in an earlier study by Gearing et al. (1984), who reported $\delta^{13}C$ values indicative of a plankton source in SOM, ranging from −22.2 ± 0.6‰ to −20.3 ± 0.6‰ in Narragansett Bay, USA and an average value of −21.0‰ in Malaysian waters. Relatively low values of $\delta^{15}N$ in SOM of all culture areas in the present study indicate a marine source of the deposited OM. This is supported by an increasing trend of $\delta^{15}N$ in SOM of the 3 aquaculture areas from shellfish to polyculture to seaweed, suggesting the import of OM from the sea.

In the present study, $\delta^{13}C$, $\delta^{15}N$ and C/N of POM are applied to describe OM sources The $\delta^{13}C$ of POM in the wet season has either lower or higher values than SOM in the 3 culture areas. Therefore, in the wet season, due to maximum freshwater discharge into the bay, as indicated by a decreasing inshore salinity trend, fluctuations in $\delta^{13}C$ and $\delta^{15}N$ values of POM among the stations imply different sources of OM. The results of the present study suggest that during the wet season, OM in SGB originates from 2 sources; marine and terrestrial. Hence to quantify the relative contribution of each source, a 2 end-member mixing model has been applied to the wet season data, using terrestrial and marine end-members values based on the model by Calder & Parker (1968).

The equation used in this model is given as:

$$TC\ (\%) = \delta^{13}C_{mar} - \delta^{13}C_{sam} / \delta^{13}C_{mar} - \delta^{13}C_{ters} \times 100 \quad (3)$$

where TC is the terrestrial carbon, $\delta^{13}C_{mar}$ is the marine end-member, $\delta^{13}C_{ters}$ is the terrestrial end-member, and $\delta^{13}C_{sam}$ is the measured value of the samples at each station. Generally, terrestrial OM has relatively low values of $\delta^{13}C$ and $\delta^{15}N$. Therefore, in our study, $\delta^{13}C$ (−27.4‰) and $\delta^{15}N$ (4.7‰) values of POM were selected as terrestrial end-members, which are closer to terrestrial end-member values of $\delta^{13}C$ and $\delta^{15}N$ identified in a number of previous studies (Peters et al. 1978, Wada et al. 1987, Middleburg & Nieuwenhuize 1998, Barros et al. 2010). In the present study, mean $\delta^{13}C$ (−19.0‰) and $\delta^{15}N$ (9.4‰) values of cultured fish and oyster, respectively, have been selected as marine end-members and are close to the values of Middleburg & Nieuwenhuize (1998). Model results indicated that during the wet season in

SGB, an average of ~72% of OM in POM is derived from the land.

In contrast to the wet season, during the dry season the range and average values of $\delta^{13}C$ and $\delta^{15}N$ of POM in all culture areas were within the range of marine-derived OM reported in previous studies (Gearing et al. 1984, Wada & Hattori 1991, Meyers 1997, Lamb et al. 2006). The high C/N values observed among SF+F culture stations might have resulted from the presence of degraded OM (Khodse et al. 2007) due to limited river inflow during the dry season, while in the other 2 culture areas, C/N values were in the range of marine-derived OM (Meyers 1994). SOM of all culture areas was assumed to be derived from suspended matter during the dry season, as indicated by their mean values of $\delta^{13}C$ (SOM = −22.4 ± 0.3‰ and POM = −23.2 ± 0.6‰; ANOVA, $p < 0.05$), revealing material exchange between the 2 different OM pools (Meksumpun et al. 2005).

Comparing both seasons, significant differences were found between $\delta^{13}C$ and $\delta^{15}N$ values of SOM and POM (ANOVA, $p < 0.05$). The relatively high values of $\delta^{13}C$ in SOM showed that SOM in SGB was derived from the same marine source in both seasons. The reason for this could be that sediments were receiving OM from autochthonous sources originating from diatoms, bacteria, and green macroalgae (Gao et al. 2012). The significant difference between SOM and POM in the wet season shows less exchange between the 2 OM pools, the reason being either high freshwater inflow or assimilation of terrestrial-derived OM in the upper water column. In both seasons, the $\delta^{15}N$ values were also close to those reported for marine-derived OM in previous studies (Gearing et al. 1984, Wada et al. 1991). The comparison of our carbon and nitrogen isotopic signatures of the POM in SGB with that of other bays (Table 3) suggests that the water quality of SGB is not significantly impacted by land-based sources of OM. The

Table 3. Ranges of carbon and nitrogen isotope values of the present study compared to previous values reported in the literature from different coastal areas having aquaculture activities or being impacted by various sources of organic matter. nd: not determined

Study area	Activity / source of impact	$\delta^{13}C$ (‰)	$\delta^{15}N$ (‰)	References
Southwestern Thailand	Land-based aquaculture	−27.3 to −20.6	3.1–8.4	Kuramoto & Minagawa (2001)
Gaeta Gulf (Mediterranean)	Bivalve and cage culture	−25.0 to −19.8	nd	Mazzola & Sarà (2001)
Kat O Bay, Hong Kong	Land-based aquaculture	−21.2 to −20.1	8.5–10.2	Wai et al. (2011)
Simon Bay, South Africa	Anthropogenic	−24.8 to −19.3	nd	Filgueira & Castro (2011)
Kosirina Bay, Croatia	Anthropogenic	nd	4.3–8.3	Dolenec et al. (2011)
Sanggou Bay	Bivalve and cage culture	−27.4 to −19.0	4.7–9.4	Present study

high production of phytoplankton and the $\delta^{13}C$ values in all cultured species indicate that the bay acts as source of carbon, and that this carbon is utilized by cultured species and removed from the bay at their harvest. However, the high C/N values indicate that SGB may act as a sink for anthropogenic material (river input).

CONCLUSIONS

In SGB, phytoplankton production is one of the main sources of OM to higher trophic levels during the wet season, as indicated by a positive correlation between $\delta^{13}C$ and POM, the latter of which containing a large proportion of phytoplankton. Trophic relationships showed that cultured fish and oyster take up carbon from different sources while sharing the same trophic level (2.16). On the other hand, oyster and scallop used the same carbon sources in spite of different trophic levels (2.16 and 1.52 respectively). Based on the results of the stable isotope analysis, our conceptual model for the wet season suggested that ~80% of the OM including faecal material and riverine OM in the form of POM is extracted by oyster and scallop. In the dry season, these species still mainly rely on POM but to some extent also use rotten kelp. C/N values >11 for POM indicate the partly terrestrial origin of OM in SGB; however, in the wet season the bay also functions as a source of carbon due to the high phytoplankton production and aquaculture activities, while high C/N values indicate that SBG may also be a sink of anthropogenic material (river input). Therefore, both culture areas SF+F (avg. C/N = 12.69) and SF+SW (avg. C/N = 13.11) in SGB are highly impacted by OM from river inflow and human activities in the wet season, as indicated by average C/N ratios in POM. In the dry season, POM in the near-shore SF+F culture area showed high C/N values of 11.97 ± 2.08 (mean ± SD) relative to the other 2 areas with C/N values (<10) indicative of more marine-derived OM. The outer SW culture area (near YS) is highly impacted by YS water. However, C/N values (<10, typical of a marine source) indicate the influence of YS, but the $\delta^{13}C$ values show the signature of terrestrial OM that may result from river input and degraded OM. Results from the 2 end-member mixing model revealed that for POM an average of 72% OM is derived from land during the wet season. $\delta^{13}C$ and $\delta^{15}N$ signatures show that OM in SOM during both the wet and dry seasons is mostly of marine origin. However, a detailed study on terrestrial organic input from rivers into the SGB is required to better understand the sources of OM and its influence on the water quality of the bay. In addition, studies investigating the role of benthic, non-aquaculture organisms and seagrass could further the understanding of the detailed food web structure of the bay.

Acknowledgements. This study was supported by the Ministry of Science and Technology of China (MoST-China) through 'Sustainability of Marine Ecosystem Production under Multi-stressors and Adaptive Management (MEco-PAM)' Project 973-3 (No. 2011CB409801) in the period of 2011–2015. The authors are very grateful to colleagues from the Ocean University of China and East China Normal University for their help with the fieldwork and laboratory experiment. T.M. thanks the China Scholarship Council (CSC) for providing a PhD scholarship and the National Institute of Oceanography Pakistan for granting study leave and moral support.

LITERATURE CITED

Alabaster JS (1982) Report of the EIFAC workshop on fish-farm effluents. Silkeborg, Denmark, 26–28 May 1981. EIFAC Tech Rep 41

Bale A, Morris A (1998) Organic carbon in suspended particulate material in the North Sea: effect of mixing resuspended and background particles. Cont Shelf Res 18: 1333–1345

Barros GV, Martinelli LA, Novais TMO, Ometto JPHB, Zuppi GM (2010) Stable isotopes of bulk organic matter to trace carbon and nitrogen dynamics in an estuarine ecosystem in Babitonga Bay (Santa Catarina, Brazil). Sci Total Environ 408:2226–2232

Bates NR, Dennis A, Hansell DA, Moran SB, Codispoti LA (2005) Seasonal and spatial distribution of particulate organic matter (POM) in the Chukchi and Beaufort Seas. Deep-Sea Res II 52:3324–3343

Buschmann AH, Varela DA, Hernández-González MC, Huovinen P (2008) Opportunities and challenges for the development of an integrated seaweed-based aquaculture activity in Chile: determining the physiological capabilities of *Macrocystis* and *Gracilaria* as biofilters. J Appl Phycol 20:571–577

Calder JA, Parker PL (1968) Stable carbon isotope ratios as indices of petrochemical pollution of aquatic systems. Environ Sci Technol 2:535–539

Chanton JP, Lewis FG (1999) Plankton and dissolved inorganic carbon isotopic composition in a river-dominated estuary: Apalachicola Bay, Florida. Estuaries 22:575–583

Costanzo SD, O'Donohue MJ, Dennison WC, Loneragan NR, Thomas M (2001) A new approach for detecting and mapping sewage impacts. Mar Pollut Bull 42:149–156

Cui Y, Wu Y, Zhang J, Wang N (2012) Potential dietary influence on the stable isotopes and fatty acid compositions of jellyfishes in the Yellow Sea. J Mar Biol Assoc UK 92: 1325–1333

Dame R (1996) Ecology of marine shellfish: an ecosystem approach. CRC Press, Boca Raton, FL

Dolenec M, Žvab P, Mihelčić G, Lambaša Belak Ž and others (2011) Use of stable nitrogen isotope signatures of anthropogenic organic matter in the coastal environment: a case study of the Kosirina Bay (Murter Island, Croatia). Geol Croatica 64:143–152

Duarte C (1995) Submerged aquatic vegetation in relation to different nutrient regimes. Ophelia 41:87–112

Evgenidou A, Valiela I (2002) Response of growth and density of a population of *Geukensia demissa* to land-derived nitrogen loading, in Waquoit Bay, Massachusetts. Estuar Coast Shelf Sci 55:125–138

Fang JG, Funderud J, Qi ZH, Zhang JH, Jiang ZJ, Wang W (2009) Sea cucumbers enhance IMTA system with abalone, kelp in China. Global Aquacult Advocate July/August 2009:49–51

Fang JG, Kuang SH, Sun HL, Sun Y (1996a) Study on the carrying capacity of Sanggou Bay for the culture of scallop *Chlamys farreri*. Mar Fish Res 17:18–31 (in Chinese with English Abstract)

Fang JG, Sun HL, Kuang SH, Sun Y (1996b) Assessing the carrying capacity of Sanggou Bay for culture of kelp *Laminaria japonica*. Mar Fish Res 17:7–17 (in Chinese with English Abstract)

Feng JX, Gao QF, Dong SL, Sun ZL, Zhang K (2014) Trophic relationships in a polyculture pond based on carbon and nitrogen stable isotope analyses: a case study in Jinghai Bay, China. Aquaculture 428–429:258–264

Ferreira G, Andersson HC, Corner RA, Desmit X and others (eds) (2007) SPEAR: sustainable options for people, catchment and aquatic resources. EU 6th framework programme FP6-2002-INCO-DEV-1, IMAR—Institute of Marine Research, Coimbra

Filgueira R, Castro BG (2011) Study of the trophic web of San Simón Bay (Ría de Vigo) by using stable isotopes. Cont Shelf Res 31:476–487

Gao X, Yang Y, Wang C (2012) Geochemistry of organic carbon and nitrogen in surface sediments of coastal Bohai Bay inferred from their ratios and stable isotopic signatures. Mar Pollut Bull 64:1148–1155

Gearing GN, Gearing PL, Rudnick DT, Requejo AG, Hutchins MJ (1984) Isotope variability of organic carbon in a phytoplankton based temperate estuary. Geochim Cosmochim Acta 48:1089–1098

Gowen RJ, Weston DP, Ervik A (1991) Aquaculture and the benthic environment: a review. In: Cowey CB, Cho CY (eds) Nutritional strategies and aquaculture waste. Fish Nutrition Research Laboratory, University of Guelph, Guelph, p 187–205

Hall POJ, Holby O, Kollberg S, Samuelsson MO (1992) Chemical fluxes and mass balances in a marine fish cage farm. IV. Nitrogen. Mar Ecol Prog Ser 89:81–91

Hayes JM (2004) An introduction to isotopic calculations. Woods Hole Oceanographic Institution, Woods Hole, MA, p 1–10

Hedges JI, Oades JM (1997) Comparative organic geochemistries of soils and marine sediments. Org Geochem 27:319–361

Holby O, Hall POJ (1991) Chemical fluxes and mass balances in a marine fish cage farm. II. Phosphorus. Mar Ecol Prog Ser 70:263–272

Holligan PM, Harris RP, Newell RC, Harbour DS and others (1984) Vertical distribution and partitioning of organic carbon in mixed, frontal and stratified waters of the English Channel. Mar Ecol Prog Ser 14:111–127

Hu J, Peng P, Jia G, Mai B, Zhang G (2006) Distribution and sources of organic carbon, nitrogen and their isotopes in sediments of the subtropical Pearl River estuary and adjacent shelf, Southern China. Mar Chem 98:274–285

Jiang ZJ, Wang GH, Fang JG, Mao YZ (2012) Growth and food sources of Pacific oyster *Crassostrea gigas* integrated culture with sea bass *Lateolabrax japonicus* in Ailian Bay, China. Aquacult Int 21:45–52

Karakassis I, Tsapakis M, Hatziyanni E, Papadopoulou KN, Plaiti W (2000) Impact of cage farming of fish on the seabed in three Mediterranean coastal areas. ICES J Mar Sci 57:1462–1471

Khodse VB, Fernandes L, Gopalkrishna VV, Bhosle NB, Fernandes V, Matondkar SGP, Bhushan R (2007) Distribution and seasonal variation of concentrations of particulate carbohydrates and uronic acids in the northern Indian Ocean. Mar Chem 103:327–346

Kuramoto T, Minagawa M (2001) Stable carbon and nitrogen isotopic characterization of organic matter in a mangrove ecosystem on the southwestern coast of Thailand. J Oceanogr 57:421–431

Lamb AL, Wilson GP, Leng MJ (2006) A review of coastal palaeoclimate and relative sea-level reconstructions using $\delta^{13}C$ and C/N ratios in organic material. Earth Sci 75: 29–57

Lehmann M, Bernasconi SM, Mckenzie JA, Barbieri A, Simona M, Veronesi M (2004) Seasonal variation of the $\delta^{13}C$ and $\delta^{15}N$ of particulate and dissolved carbon and nitrogen in Lake Lugano: constraints on biogeochemical cycling in a eutrophic lake. Limnol Oceanogr 49:415–429

Marin Leal JC, Dubois S, Orvain F, Galois R, Blin JL and others (2008) Stable isotopes ($\delta^{13}C$, $\delta^{15}N$) and modeling as tools to estimate the trophic ecology of cultivated oysters in two contrasting environments. Mar Biol 153:673–688

Mays S (2000) New directions in the analysis of stable isotopes in excavated bones and teeth. In: Cox M, Mays S (eds) Human osteology in archaeology and forensic science. Greenwich Medical Media, London, p 425–438

Mazzola A, Sarà G (2001) The effect of fish farming organic waste on food availability for bivalve molluscs (Gaeta Gulf, Central Tyrrhenian, MED): stable carbon isotopic analysis. Aquaculture 192:361–379

McClelland JW, Valiela I, Michener RH (1997) Nitrogen stable isotope signatures in estuarine food webs: a record of increasing urbanization in coastal watersheds. Limnol Oceanogr 42:930–937

Meksumpun S, Meksumpun C, Hoshika A, Mishima Y, Tanimoto T (2005) Stable carbon and nitrogen isotope ratios of sediment in the gulf of Thailand: evidence for understanding of marine environment. Cont Shelf Res 25: 1905–1915

Meyers PA (1994) Preservation of elemental and isotopic source identification of sedimentary organic matter. Chem Geol 114:289–302

Meyers PA (1997) Organic geochemical proxies of paleoceanographic, paleolimnologic, and paleoclimatic processes. Org Geochem 27:213–250

Michener RH, Schell DM (1994) Stable isotope ratios as tracers in marine aquatic food webs. In: Lajtha K, Michener RH (eds) Stable isotopes in ecology. Blackwell Scientific Publications, Oxford, p 138–157

Middelburg JJ, Nieuwenhuize J (1998) Carbon and nitrogen stable isotopes in suspended matter and sediments from the Schelde Estuary. Mar Chem 60:217–225

Miller TW, Omori K, Hamaoka H, Shibata JY, Hidejiro O (2010) Tracing anthropogenic inputs to production in the Seto Inland Sea, Japan—a stable isotope approach. Mar Pollut Bull 60:1803–1809

Miller TW, Jaquinto G, McGlone M, Isobe A, Shibata JY,

Hamaoka H, Omori K (2011) Tracing dynamics of organic material flow in coastal marine ecosystems: results from Manila Bay (Philippines) and Kyucho Intrusion (Japan). In: Omori K, Guo X, Yoshie N, Fujii N, Handoh IC, Isobe A, Tanabe S (eds) Interdisciplinary studies on environmental chemistry: marine environmental modeling and analysis. Terrapub, Tokyo, p 95–104

▶ Minagawa M, Wada E (1984) Stepwise enrichment of ^{15}N along food chains: further evidence and the relation between δ^{15}N and animal age. Geochim Cosmochim Acta 48:1135–1140

Mishima Y, Hoshika A, Tanimoto T (1996) Movement of terrestrial organic matter in the Osaka Bay, Japan. In: Proc 3rd Int Symp ETERNET-APR: conservation of the hydrospheric environment, Bangkok, Thailand, 3–4 December 1996. Environmental Research Institute, Chulalongkorn University, Bangkok, p 20–25

▶ Peters KE, Sweeney RE, Kaplan IR (1978) Correlation of carbon and nitrogen stable isotopes in sedimentary organic matter. Limnol Oceanogr 23:598–604

Pillay TVR (1991) Aquaculture and the environment. Blackwell Scientific, London

▶ Post DM (2002) Using stable isotopes to estimate trophic position: models, methods, and assumptions. Ecology 83:703–718

▶ Raikow DF, Hamilton SK (2001) Bivalve diets in a midwestern U.S. stream: a stable isotope enrichment study. Limnol Oceanogr 46:514–522

Redfield AC, Ketchum BH, Richards FA (1963) The influence of organisms on the composition of seawater. In: Hill MN (ed) The sea, Vol 2. Wiley Interscience, New York, NY, p 26–77

▶ Risk MJ, Erdmann MV (2000) Isotopic composition of nitrogen in stomatopod (Crustacea) tissues as an indicator of human sewage impacts on Indonesian coral reefs. Mar Pollut Bull 40:50–58

▶ Shpigel M, Neori A (1996) The integrated culture of seaweed, abalone, fish and clams in modular intensive land-based systems: I. Proportions of size and projected revenues. Aquacult Eng 15:313–326

Shpigel M, Neori A, Gordin H (1991) Oyster and clam production in the outflow water of marine fish pond in Israel. EAS Spec Publ 14:295

▶ Tan FC, Cai DL, Edmond JM (1991) Carbon isotope geochemistry of the Changjiang Estuary. Estuar Coast Shelf Sci 32:395–403

▶ Vizzini S, Mazzola A (2004) Stable isotope evidence for the environmental impact of a land-based fish farm in the western Mediterranean. Mar Pollut Bull 49:61–70

Vollenweider RA (1992) Coastal marine eutrophication: principles and control. In: Vollenweider RA, Marchetti R, Viviani R (eds) Marine coastal eutrophication. Elsevier, Amsterdam, p 1–20

Wada E, Hattori A (1991) Nitrogen in the sea: forms, abundances and rate processes. CRC Press, Boca Raton, FL

▶ Wada E, Minagawa M, Mizutani H, Tsuji T, Imaizumi R, Karasawa K (1987) Biogeochemical studies on the transport of organic matter along the Otsuchi River watershed, Japan. Estuar Coast Shelf Sci 25:321–336

▶ Wada E, Mizutani H, Minagawa M (1991) The use of stable isotopes for food web analysis. Crit Rev Food Sci Nutr 30:361–371

▶ Wai TC, Leung KMY, Wu RSS, Shin PKS, Cheung SG, Li XY, Lee JHW (2011) Stable isotopes as a useful tool for revealing the environmental fate and trophic effect of open-sea-cage fish farm wastes on marine benthic organisms with different feeding guilds. Mar Poll Bull 63:77–85

▶ Wan RJ, Wu Y, Huang L, Zhang J, Gao L, Wang N (2010) Fatty acids and stable isotopes of a marine ecosystem: study on the Japanese anchovy (*Engraulis japonicus*) food web in the Yellow Sea. Deep-Sea Res II 57:1047–1057

Xu Q, Yang H (2007) Food sources of three shellfish living in two habitats of Jiaozhou Bay (Qingdao, China): indicated by lipid biomarkers and stable isotope analysis. J Shellfish Res 26:561–567

▶ Ye L, Ritz DA, Fenton GE, Lewis ME (1991) Tracing the influence on sediments of organic waste from a salmonid farm using stable isotope analysis. J Exp Mar Biol Ecol 145:161–174

▶ Yokoyama H (2013) Growth and food source of the sea cucumber Apostichopus japonicus cultured below fish cages—potential for integrated multi-trophic aquaculture. Aquaculture 372–375:28–38

Yokoyama H, Abo K, Ikuta K, Kamiyama T, Higano J, Toda S (2006) Country scenarios for ecosystem approaches for aquaculture. 4. Japan. In: McVey JP, Lee CS, O'Bryen PJ (eds) Aquaculture and ecosystems: an integrated coastal and ocean management approach. The World Aquaculture Society, Baton Rouge, LA, p 71–89

Yokoyama H, Ishihi Y, Abo K, Takashi T (2010) Quantification of waste feed and fish feces using stable carbon and nitrogen isotopes. Bull Fish Res Agen 31:71–76

Zhao J, Zhou S, Sun Y, Fang J (1996) Research on Sanggou Bay aquaculture hydro-environment. Mar Fish Res 17:68–79

2

Effects of temperature and ocean acidification on shell characteristics of *Argopecten purpuratus*: implications for scallop aquaculture in an upwelling-influenced area

Nelson A. Lagos[1,*], Samanta Benítez[1], Cristian Duarte[2], Marco A. Lardies[3], Bernardo R. Broitman[4], Christian Tapia[5], Pamela Tapia[5], Steve Widdicombe[6], Cristian A. Vargas[7]

[1]Centro de Investigación e Innovación para el Cambio Climático (CiiCC), Facultad de Ciencias, Universidad Santo Tomás, 8370003 Santiago, Chile
[2]Departamento de Ecología y Biodiversidad, Facultad de Ecología y Recursos Naturales, Universidad Andrés Bello, 8370251 Santiago, Chile
[3]Facultad de Ingeniería & Ciencias y Facultad de Artes Liberales, Universidad Adolfo Ibáñez, 7941169 Santiago, Chile
[4]Centro de Estudios Avanzados en Zonas Áridas (CEAZA), Universidad Católica del Norte, Larrondo 1281, 1781421 Coquimbo, Chile
[5]Cultivos Invertec Ostimar S.A., Tongoy, 1780000 Coquimbo, Chile
[6]Plymouth Marine Laboratory, Prospect Place, West Hoe, PL1 3DH Plymouth, UK
[7]Laboratorio de Funcionamiento de Ecosistemas Acuáticos (LAFE), Departamento de Sistemas Acuáticos, Facultad de Ciencias Ambientales, Universidad de Concepción, 4070386 Concepción, Chile

ABSTRACT: Coastal upwelling regions already constitute hot spots of ocean acidification as naturally acidified waters are brought to the surface. This effect could be exacerbated by ocean acidification and warming, both caused by rising concentrations of atmospheric CO_2. Along the Chilean coast, upwelling supports highly productive fisheries and aquaculture activities. However, during recent years, there has been a documented decline in the national production of the native scallop *Argopecten purpuratus*. We assessed the combined effects of temperature and pCO_2-driven ocean acidification on the growth rates and shell characteristics of this species farmed under the natural influence of upwelling waters occurring in northern Chile (30°S, Tongoy Bay). The experimental scenario representing current conditions (14°C, pH ~8.0) were typical of natural values recorded in Tongoy Bay, whilst conditions representing the low pH scenario were typical of an adjacent upwelling area (pH ~7.6). Shell thickness, weight, and biomass were reduced under low pH (pH ~7.7) and increased temperature (18°C) conditions. At ambient temperature (14°C) and low pH, scallops showed increased shell dissolution and low growth rates. However, elevated temperatures ameliorated the impacts of low pH, as evidenced by growth rates in both pH treatments at the higher temperature treatment that were not significantly different from the control treatment. The impact of low pH at current temperature on scallop growth suggests that the upwelling could increase the time required for scallops to reach marketable size. Mortality of farmed scallops is discussed in relation to our observations of multiple environmental stressors in this upwelling-influenced area.

KEY WORDS: Calcification · Shell growth · Scallop farming · Upwelling · Chile

INTRODUCTION

The increasing concentration of carbon dioxide (CO_2) in the atmosphere due to anthropogenic activities has triggered major changes in the global climate system, leading to increased global mean temperatures (i.e. global warming) of approximately 0.7°C, a trend that will continue and fluctuate over a broad range (1–4°C) towards the end of the century (IPCC 2014). These changes in the thermal environ-

*Corresponding author: nlagoss@santotomas.cl

ment may influence the distribution and abundances of the marine biota through challenges to the eco-physiological capabilities of different organisms (Helmuth et al. 2006). However, environmental changes due to warming alone are only one of a set of possible interacting climatic variables that will drive ecological and evolutionary responses in marine ecosystems (Harley et al. 2006). Rising atmospheric CO_2 levels are also driving an increased uptake of CO_2 into the ocean, which in turn produces a series of changes to the carbonate chemistry of seawater, a process now well established as ocean acidification (Caldeira & Wickett 2003, Orr et al. 2005, Gattuso et al. 2015). Ocean acidification (OA) negatively affects a variety of marine calcifying organisms by reducing calcification rates or by increasing the dissolution of calcareous structures (Fabry et al. 2008, Doney et al. 2009, Ries et al. 2009, Gazeau et al. 2013, Kroeker et al. 2013). Studies of mollusks specifically have suggested that the negative impacts of OA include reduced shell mass along with compromised shell structural integrity and strength (Buschbaum et al. 2007, McClintock et al. 2009, Welladsen et al. 2010, Bressan et al. 2014, Duarte et al. 2014, Fitzer et al. 2014, Mackenzie et al. 2014). However, the responses of different species and populations are highly variable when the impacts of OA are assessed in combination with warming and other stressors such as salinity and oxygen (e.g. Duarte et al. 2014, 2015, Ko et al. 2014). Thus, the combined effects of OA and warming on commercial shellfish species are of growing concern given that mollusks represent an important source of global seafood production (Branch et al. 2013). In addition, as global warming and OA are occurring concomitantly, affecting many physiological processes of marine organisms, their combined effects must be evaluated (Byrne & Przeslawski 2013).

Shellfish aquaculture is reliant on growing individuals at high densities, meaning that shell robustness and size are important individual attributes with important implications for levels of production. Shell integrity is particularly important during the thinning process in which bivalves are physically handled as their densities are manipulated to reduce the interference in feeding so the scallops can grow faster to market size (e.g. Fréchette et al. 2010, Cubillo et al. 2012). OA and warming could also have implications for decision-making processes associated with the adequate timing of release of hatchery-reared mollusks into nature (Jory & Iversen 1988, Grefsrud & Strand 2006). Thus, in addition to knowing how OA and warming may affect the ability of the shell to

provide physical support for soft internal organs and defense against predators and/or other environmental stressors (Waldbusser et al. 2013), a knowledge of how shell characteristics may change under future environmental conditions is crucial information for underpinning the future management strategies employed by the aquaculture industry (e.g. Grefsrud & Strand 2006, Welladsen et al. 2010, Mackenzie et al. 2014).

The aquaculture industry relies on key ecosystem services provided by the coastal ocean, where most shellfish production takes place. Larval production in onshore hatcheries provides some control over the natural variability observed in coastal waters, such as food content and water temperature, but they are transferred as juveniles to the natural environment, where they are reared until harvest. However, the carbon chemistry of seawater is also subject to natural variability, especially in coastal upwelling areas, where it may negatively impact aquaculture production (Barton et al. 2012). In northern Chile, the fishery of the native scallop *Argopecten purpuratus* (Mollusca; Bivalvia; Pectinidae) collapsed during the mid-1980s following overexploitation of natural stocks. The fishery is still banned and was replaced by production from suspended culture activities, although current production only represents a fraction (10–15%) of the historical national landings (Stotz 2000). During the last decade, the Chilean landings of mollusks in general have ranged from 300 000 to 550 000 t. Over the same period, the production of scallops showed a decreasing trend, with the production in 2014 being equivalent to only 20% of the landings recorded 10 yr before (SERNAPESCA 2014; Fig. 1). This reduction in scallop production is challenging

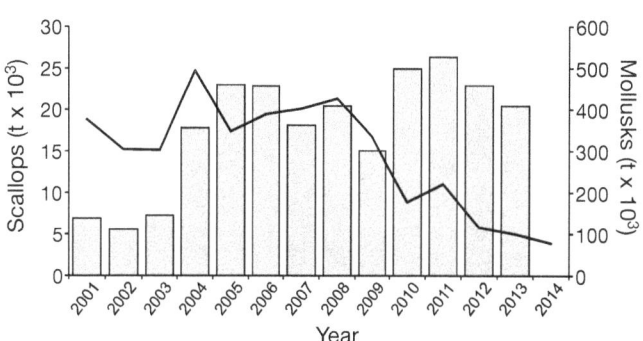

Fig. 1. Landings of *Argopecten purpuratus* (black line) and total mollusks harvested in Chile for the period 2001 to 2014 (source: Servicio Nacional de Pesca, www.sernapesca.cl; accessed 4 June 2015)

the sustainability of this socio-economic sector in northern Chile, thus highlighting the need to perform studies aimed at understanding causes of this reduction and to provide information to support mitigation and adaptation strategies. For example, studies carried out with the mussel *Mytilus chilensis*, the most widely cultivated species in Chile, showed that OA, but not temperature increase, will decrease the biomass production of this highly valuable bivalve by 30 % (Navarro et al. 2013, Duarte et al. 2014, 2015).

The commercial production of *A. purpuratus* is currently restricted to aquaculture activities which use suspended cultures in natural embayments, with northern Chile and Tongoy Bay (30° 12′ S, 71° 34′ W, Fig. 2) being among the most important locations for scallop aquaculture along the Chilean coast (Thiel et al. 2007). Tongoy Bay is located ca. 10 km north of the Pt. Lengua de Vaca upwelling center (30° S, Fig. 2). The bay is part of the highly productive Coquimbo upwelling system, which is characterized by the presence of cold, CO_2-enriched, and poorly oxygenated waters, particularly during the spring and summer months (Thiel et al. 2007). More recent evidence suggests that this upwelling area is naturally acidified and characterized by low pH levels and a

low carbonate saturation state (Torres & Ampuero 2009). These conditions are similar to projections of progressive acidification in other eastern boundary coastal upwelling areas, such as off the coasts of Oregon and California, USA (Feely et al. 2008, Gruber et al. 2012). This environmental pattern implies that the sustainability of the shellfish aquaculture industry within Tongoy Bay could be vulnerable to the penetration of corrosive waters upwelled at Pt. Lengua de Vaca.

Upwelling is intensifying in eastern boundary current systems (Sydeman et al. 2014), particularly in the Oregon–California and Humboldt systems, which may further increase the intrusion of corrosive waters in the coastal domain. This ongoing global trend emphasizes the challenge of establishing a coherent scientific basis with which to predict the response of specific aquaculture species living in regions subject to upwelling disturbances (Barton et al. 2012, Ekstrom et al. 2015). The current study represents the first effort specifically aimed at experimentally evaluating the combined effects of temperature and OA upon the native scallop *A. purpuratus* cultured under the influence of upwelling waters. A range of biological responses were measured, including calcification rates, shell dissolution, shell thickness, dry shell weight, and tissue biomass. These results are discussed in relation to updated information on the physical–chemical variability inside Tongoy Bay and events of *in situ* mortality of *A. purpuratus* observed in the commercial aquaculture operations in the bay.

MATERIALS AND METHODS

Scallop collection

Juvenile individuals of *Argopecten purpuratus* (~41 mm ± 1.3 SD in shell length) were obtained on 20 August 2014, from suspended culture facilities (ca. 2 m depth) belonging to the at-sea commercial aquaculture production operations from Invertec–Ostimar Co. located in Tongoy Bay (Fig. 1). After collection, scallops were transported in a thermobox (ca. 14°C, emersed but wet conditions) to the Calfuco Coastal Laboratory (Valdivia, Chile) and aquaria (9 l, 15–20 scallops per aquarium). Before the experimental period, scallops were kept for 3 d in running seawater (ca. 13–14°C), with natural photoperiod, and fed daily with microalgae (*Tetraselmis* spp., ~65 × 10^6 cells ml^{-1}) to allow the scallops to acclimate to laboratory conditions.

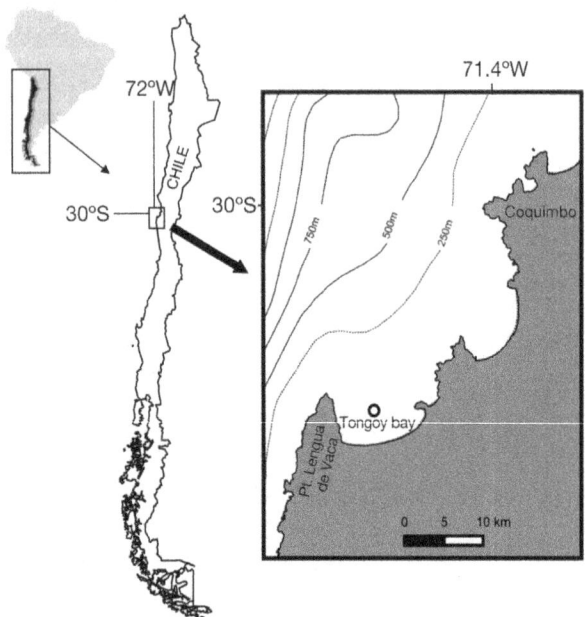

Fig. 2. Geographic location of Tongoy Bay and Pt. Lengua de Vaca upwelling center. The circle indicates the location of the CEAZAMET buoy, which is moored adjacent to the aquaculture facilities of the Invertec–Ostimar aquaculture company

Experimental setup

To examine the effects of pH and temperature, 4 experimental treatments were selected: (1) present-day conditions (control): 14°C and pH ~8.0; (2) control temperature (14°C) and low pH (~7.7); (3) high temperature (18°C) and control pH; (4) high temperature and low pH. Control conditions in temperature represent annual average sea surface temperature (14°C, a value currently recorded during the scallop collection, see 'Results'), whereas high temperature, in addition to the projected increase in 4°C (IPCC 2014), represents the maximum surface temperature (18°C) reported for Tongoy Bay (Aravena et al. 2014). The temperature of each treatment was stabilized using external chillers (±0.1°C). To obtain the different pH scenarios, the methodology previously described by Torres et al. (2013) (see Navarro et al. 2013, Duarte et al. 2014, 2015) was followed. Briefly, for present-day conditions (i.e. ~390 µatm pCO_2 in seawater), atmospheric air was bubbled into experimental aquaria and head tanks; for a low pH scenario, blended dry air was generated by compressing atmospheric air (117 psi) using an oil-free compressor with pure CO_2 using mass flow controllers (MFCs, Aalborg™); this blend was then bubbled into experimental aquaria and head tanks reaching ~900 µatm pCO_2 in seawater. The seawater of each aquarium was replaced every 2 d using the pre-equilibrated seawater from the head tank. In the current experimental system, this increase in pCO_2 in seawater resulted in a corresponding drop in pH (~0.3 units) yielding a target pH level of ~7.7 for the low pH scenario, while the present-day pH level remained at ~8.0 units (Table 1). These pCO_2 levels in seawater were selected taking into account the rate of change projected for the atmospheric CO_2 by the year 2100, consistent with the IPCC A2 emission scenario (e.g. Meinshausen et al. 2011, IPCC 2014).

Four scallops were randomly assigned to each of 20 aquaria (9 l), and each aquarium was then randomly assigned to a pH/temperature treatment following a systematic design, with each treatment being replicated 5 times. All animals were labeled using bee tags glued to the shell. In addition, 1 empty shell of *A. purpuratus* of similar size to those in the experimental aquaria, previously tagged and weighed, was placed in each container to provide an estimate of net shell dissolution in each treatment. Beforehand, dead shells were cleaned with distilled water and then dried for 24 h (60°C) until constant dry weight. Before assigning individuals to experimental treatments, live animals were characterized in terms of their shell length, shell height, and buoyant weight. There were no statistical differences across treatments in these shell characteristics at the beginning of the experiments (1-way ANOVA, p > 0.05 in all cases). The exposure period lasted for 18 d, and scallops were fed daily with *Tetraselmis* spp. and maintained under natural photoperiod conditions, as in the previous acclimation period.

Carbonate system parameter monitoring

The seawater in each aquarium was gently replaced every day, with the corresponding seawater pre-equilibrated at the target pCO_2 levels in the head tank. Over the experimental period, pH and total alkalinity (A_T) were monitored on Days 2, 6, 10, 14, and 18 to estimate carbonate system parameters. Samples for pH were collected in 50 ml syringes, avoiding formation of bubbles during collection and handling of the sample, and immediately transferred

Table 1. Carbonate system parameters (mean ± SE) registered at each experimental treatment combining present-day (low pCO_2/pH ~8.0) and future acidification conditions (high pCO_2/pH ~7.7) with temperatures dominating in Tongoy Bay, Chile, during the experimental period (14°C) and under warmer conditions (18°C). NBS: National Bureau of Standards

Carbonate system parameter	14°C		18°C	
	pH ~8.0	pH ~7.7	pH ~8.0	pH ~7.7
Temperature (°C)	14.13 ± 0.19	14.08 ± 0.14	18.13 ± 0.11	18.14 ± 0.13
Salinity (PSU)	34.48 ± 1.78	35.36 ± 0.50	33.28 ± 1.97	34.00 ± 0.68
pH$_{NBS}$	8.058 ± 0.017	7.754 ± 0.027	8.032 ± 0.041	7.720 ± 0.021
Total alkalinity (µmol kg^{-1})	1580.41 ± 206.40	1778.14 ± 127.82	1666.75 ± 117.55	1695.96 ± 133.49
CO_3^{-2} (µmol kg^{-1})	86.54 ± 10.15	53.76 ± 5.58	94.76 ± 3.92	52.13 ± 5.07
pCO_2 (µatm)	367.77 ± 62.17	891.26 ± 77.76	435.96 ± 74.43	969.48 ± 74.27
$\Omega_{calcite}$	2.07 ± 0.24	1.28 ± 0.13	2.29 ± 0.09	1.25 ± 0.13
$\Omega_{aragonite}$	1.32 ± 015	0.82 ± 0.09	1.48 ± 0.06	0.81 ± 0.08

to a 25 ml thermostated closed cell at 25.0 ± 0.1°C for standardization (DOE 1994, Torres et al. 2013), using a Metrohm® pH–meter with a glass combined double junction Ag/AgCl electrode (Metrohm model 6.0258.600) calibrated with standard buffer of pH 4 (Metrohm® 6.2307.200), pH 7 (6.2307.210), and pH 9 (6.2307.220). pH values are reported on the National Bureau of Standards scale. Samples for A_T were poisoned with 50 µl of saturated $HgCl_2$ solution and stored in 500 ml borosilicate bottles (Pyrex, Corning®) with ground-glass stoppers lightly coated with Apiezon L® grease and stored in the dark at room temperature. Additionally, temperature and salinity were monitored during incubations by using a portable Salinometer (Salt6+, Oakton®, accuracy: ±1% and ±0.5°C, respectively). A_T was determined using the open-cell titration method (Dickson et al. 2007) using an automatic alkalinity titrator (Model AS-ALK2 Apollo SciTech) equipped with a combination pH electrode (8102BNUWP, Thermo Scientific) and temperature probe (Star ATC, Thermo Scientific) connected to a pH meter (Orion Star A211, Thermo Scientific). All samples were analyzed at 25°C (±0.1°C) with temperature regulated using a water bath (Lab Companion CW-05G). Accuracy was controlled against a certified reference material (CRM; batch no. 140 supplied by A. Dickson, University of California, San Diego, CA). Every sample was analyzed with 2 or 3 replicates, and an accuracy of 2 to 3 µmol kg^{-1} was observed with respect to CRM. Temperature and salinity data were used to calculate the rest of the carbonate system parameters (e.g. pCO_2, CO_3^{2-}) and the saturation stage of aragonite ($\Omega_{aragonite}$) and calcite ($\Omega_{calcite}$). Analyses were performed using CO2SYS software for MS Excel (Pierrot et al. 2006) set with Mehrbach solubility constants (Mehrbach et al. 1973) refitted by Dickson & Millero (1987). The $KHSO_4$ equilibrium constant determined by Dickson (1990) was used for all calculations (Table 1).

Biological responses

At the end of the experimental period, shell thickness (mm) was measured and averaged over 3 measurements taken between the ribs located at the posterior region of scallop shells (i.e. the edge of the newly deposited shell; Bibby et al. 2007, Thomsen et al. 2010, Bressan et al. 2014). The tissue was removed and the final dry shell weight (mg) was measured after drying the shell at 60°C overnight (Binder) and then weighed in an analytical balance (to the nearest 0.01 mg, Mettler Toledo). Because potential problems of water held in the shell and soft tissues, the final wet biomass was described using a relative index estimated as the difference between total and buoyant weight (see below), and expressed as a percentage of the total weight. Relative indexes are traditionally used to describe bivalve quality under aquaculture conditions (e.g. Maguire et al. 1999, Filgueira et al. 2013). Individual growth rates based on shell height (length perpendicular to the hinge) and length were estimated based on measurements performed (to the nearest 0.01 m using a Mitutoyo® caliper) on Days 1 and 18 of the experiment. Measurements of both shell height and length are used routinely during the thinning process of scallops in commercial aquaculture operations (Hennen & Hart 2012). In addition, calcification and dissolution rates (see also Gazeau et al. 2015) were estimated from changes in the buoyant weight of individual scallops and changes in empty shell weight recorded on Days 1 and 18 of the experiment using an analytical balance. Briefly, the buoyant weight is a non-destructive technique useful to estimate the shell weight in gastropods, where the whole animal is immersed in seawater (Palmer 1982). Due to differences in specific gravity of shell and tissue, when immersed in seawater the mass of the animal is mainly accounted for by the shell. In addition, buoyant weight increment is an important proxy for growth because it is equivalent to the calcification rate and is not affected by the amount of seawater and tissue weight (Palmer 1982). However, the changes in empty shell weight used as a proxy of shell dissolution must be interpreted with caution (Nienhuis et al. 2010), given the role of organic layers such as the shell periostracum in protecting live animals from dissolution (Tunnicliffe et al. 2009). To avoid errors from air trapped inside the animals during the measurements of buoyant weight, each specimen was gently moved from the rearing container to the analytical balance. Following this procedure, only seawater remained within the valves and yielded consistent buoyant weights for the experimental specimens. The relationship between dry shell weight (SW) and buoyant weight (BW) was verified by constructing the scaling relationship between both variables using an additional batch of scallops encompassing all sizes available from the Invertec-Ostimar farm (Log_{10} SW = 0.1995 + 0.9950 Log_{10} BW; n = 71; p < 0.001; r^2 = 0.99). This methodology was successfully validated previously for the mussel *Mytilus chilensis* (Duarte et al. 2014). No mortality was recorded in any treatment during the experimental period.

In situ biological and environmental monitoring

During July 2014, the Center for Advanced Research in Arid Zones (CEAZA) in collaboration with the Invertec–Ostimar Co. monitored water quality parameters in Tongoy Bay in close proximity to the area of scallop farming. This monitoring involved the deployment (10 m depth) of a CTD SBE 16plus v2 RS232 (Sea-Bird Electronics) measuring conductivity (S m^{-1}), salinity (PSU), temperature (°C), and dissolved oxygen (mg l^{-1}, ml l^{-1}, and % saturation) every 15 min (www.ceazamet.cl/index.php?pag=mod_estacion&e_cod=BTG). In this study, the environmental data from July to December 2014 are summarized, jointly with estimates of the cumulative mortality of scallops for several size ranges recorded during the thinning process performed by the Invertec–Ostimar Company. The thinning process consists of a mechanical separation of scallops, reducing density to make room for those remaining in order to ensure better growth and productivity of the individuals (Maguire et al. 1999, Filgueira et al. 2013). During the thinning process, individuals that have died during the growing period are also removed. In this study, we report the cumulative mortalities recorded after 4 to 5 mo of growing period for 4 shell range sizes (i.e. 7–15 to 45–50 mm in shell height), and their corresponding densities (i.e. from 200 to 25 scallops per lantern level). The adult fraction of the scallops size range corresponds to individuals between 45 and 50 mm, which are cultivated at densities of 25 individuals on each level of the lanterns. The harvest is performed when the scallop shell sizes reach 65 to 70 mm. During December 2014, the first-ever survey of carbonate system parameters (pH$_T$, A$_T$, and estimates of pCO$_2$ and Ω) within Tongoy Bay was conducted, using similar methodology as described above. However, in contrast to the monitoring techniques employed within the laboratory, the field monitoring campaign used seawater buffer supplied by CRM, and pH was computed on the 'total' hydrogen scale (pH$_T$). This study only presents the pH$_T$ and temperature data, and more environmental field data will be published elsewhere.

Statistical analyses

In each analysis, the replicate was the average response of the 4 scallops in each tank. Differences between treatments in scallop shell thickness and dry shell weight recorded at the end of the experiment were tested using ANCOVA models, including the final measurement of shell length and buoyant weight as covariates, respectively, in order to control for the inherent positive scaling relationship between shell length and both shell weight and thickness. ANCOVA results evidenced no significant influence of the shell length upon scallop shell thickness (covariate, $F_{1,15} = 1.74$; $p = 0.207$), but a significant influence of buoyant weight upon dry shell weight (covariate, $F_{1,15} = 82.69$; $p < 0.001$). In this case, further comparisons among treatments were done estimating the predicted value (least square mean) of the regression between dry shell weight with buoyant weight, and comparing them at the mean value of this covariate (3.681 g ± 0.37 SD; see also Watson et al. 2012). Two-way ANOVAs were used to evaluate whether temperature, pH levels, or the interaction between factors affected the measured biological responses (biomass, shell growth, calcification, and shell dissolution rates) during the experiment. Tukey's HSD was used as an *a posteriori* test when the main factors indicated significant differences between levels of the corresponding factor (Underwood 1997). These ANOVAs were implemented in a generalized linear model and the coefficient of the interaction term was estimated. Positive or negative coefficients for significant interaction terms were interpreted as synergistic or antagonistic effects between temperature and pH upon the corresponding biological response (Kroeker et al. 2013). Only growth rates based on shell length and height required log$_{10}$ transformation to meet ANOVA assumptions. In both ANOVA and ANCOVA, normality was assessed over model residuals using the Kolmogorov–Smirnov test, and homoscedasticity was evaluated using Bartlett tests (Sokal & Rohlf 1995). Finally, a χ^2 test on scallop cumulative mortality data recorded in the sea farm was used to test whether, over time (month), mortalities were dependent on the shell range sizes used in the scallop thinning process. All analyses were carried out using Minitab v14.

RESULTS

The experimental setup and average environmental and carbonate chemistry parameters recorded during the experiments with *Argopecten purpuratus* are shown in Table 1. Low A$_T$ values were observed for all treatments (<2000 µmol kg^{-1}), due to the influence of riverine discharges (Valdivia River Estuary) close to the coastal laboratory (Torres et al. 2013).

pCO_2 values ranged from 367 to 435 µatm for the low pCO_2 treatment at 14 and 18°C, respectively, and from 891 to 969 µatm for the high pCO_2 at 14 and 18°C, respectively. $\Omega_{aragonite}$ varied less between temperatures, but undersaturation was reached under the high pCO_2 treatment ($\Omega_{aragonite}$ from 0.81 to 0.82; Table 1).

At the end of the experimental period, shell thickness, dry shell weight, and wet biomass of the scallops were significantly lower in the treatments representing combinations of low pH levels at increased temperature when compared with current environmental conditions (i.e. 14°C and pH ~8.0; Fig. 3). These patterns of variability were dominated by differences in temperature among treatments and to a lesser extent to pH variability (Table 2). Thus, thicker and heavier shells and increased biomass were observed on scallops growing under current environmental conditions when compared with all other combinations of temperatures and pH level (Tukey HSD, p < 0.05).

Control or current temperature (~14°C) and low pH (~7.7) significantly restricted the shell growth in length and height (Table 2, Tukey HSD, p < 0.05, Fig. 4), but elevated temperatures appeared to ame-

Table 2. Summary results comparing *Argopecten purpuratus* shell thickness and shell weight (ANCOVA), biomass and growth rates in shell height and length, and calcification and dissolution rates (2-way ANOVA) between combinations of pH (7.7 and 8.0) and temperature (14 and 18°C) used in experimental treatments. Significant p-values (p < 0.05) are shown in **bold**

Biological response	Environmental variable	df (source, error)	F	p
Shell thickness (mm)	Temperature	1, 15	18.07	**0.001**
	pH	1, 15	1.24	0.283
	Temperature × pH	1, 15	0.50	0.488
Shell dry weight (g)	Temperature	1, 15	6.08	**0.026**
	pH	1, 15	3.89	0.067
	Temperature × pH	1, 15	2.19	0.159
Wet biomass (%)	Temperature	1, 16	8.36	**0.011**
	pH	1, 16	2.52	0.132
	Temperature × pH	1, 16	2.61	0.126
Growth in height (mm d^{-1})	Temperature	1, 16	7.4	**0.015**
	pH	1, 16	6.11	**0.025**
	Temperature × pH	1, 16	6.52	**0.021**[a]
Growth in length (mm d^{-1})	Temperature	1, 16	14.1	**0.002**
	pH	1, 16	9.00	**0.008**
	Temperature × pH	1, 16	4.28	0.055
Calcification rate (g d^{-1})	Temperature	1, 16	24.39	**<0.001**
	pH	1, 16	0.97	0.340
	Temperature × pH	1, 16	3.39	0.066
Dissolution rate (mg d^{-1})	Temperature	1, 16	2.21	**0.015**
	pH	1, 16	12.90	**0.002**
	Temperature × pH	1, 16	0.72	0.408

[a]Significant synergistic effects of both variables

liorate the negative impacts of reduced pH (i.e. combination pH ~7.7/18°C; Fig. 4). Temperature and pH showed a significant and positive interaction term (coefficient = 0.0104 ± 0.0041 SE; Table 2), thus indi-

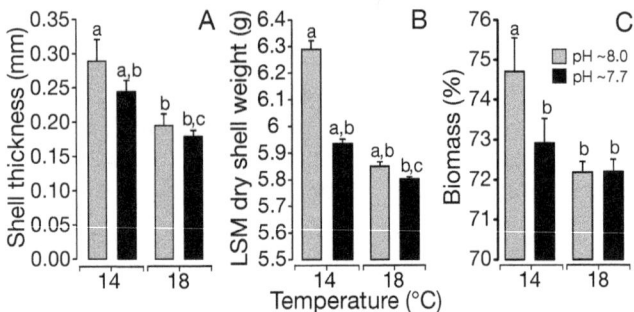

Fig. 3. Mean ± SE (A) shell thickness, (B) dry shell weight, and (C) wet biomass of *Argopecten purpuratus* reared at nominal pH levels in seawater and at 2 temperatures. All measurements were recorded at the end of the study period (18 d). LSM: least square mean, which corresponds to the predicted value of dry shell weight at the mean value of buoyant weight, used as covariate in an ANCOVA model (see 'Materials and methods'). Different letters indicate significant differences among treatments (n = 5) using a post hoc Tukey HSD test

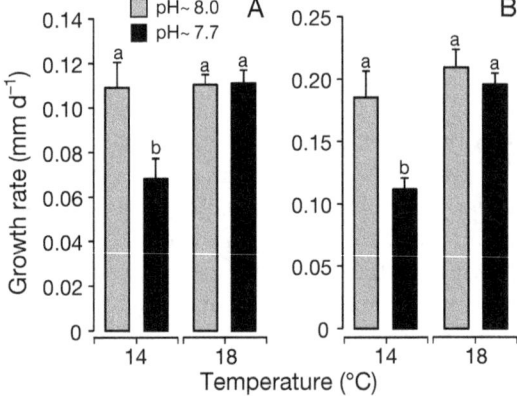

Fig. 4. Growth rates measured as increments in the shell height and length (mean ± SE) of *Argopecten purpuratus* individuals reared at 2 temperatures and 2 nominal pH levels in seawater. These rates were estimated based on the difference between initial and final measurements after the 18 d experimental period. Different letters indicate significant differences among treatments (n = 5) using a post hoc Tukey HSD test

cating a synergistic effect of temperature and pH on the growth rate in shell height of the scallops. For the remaining biological responses, we found non-significant interactive effects of pH and temperature (Table 2).

The net calcification rate over the study period was not affected by low pH, but temperature had a significant positive effect (Table 2, Fig. 5A). Shell dissolution, estimated from changes in shell weight of dead scallops, was significantly higher at the combination of low pH and at control temperature (i.e. pH ~7.7/14°C; Fig. 5B).

The monitoring program that started during 2014 provided updated information on the seawater properties of Tongoy Bay. In particular, during the experimental period (mid-August to mid-September), the seawater temperature measured at 10 m depth inside the bay ranged between 12 and 15°C, with low levels of oxygen concentration (3–6 ml l^{-1}) and low variability in salinity (ca. 34.4 psu; Fig. 6A–C). On 2 September, cooling events were observed that coincided with periods of oxygen depletion, with the second event displaying oxygen levels below 2 ml l^{-1}, values that are consistent with a hypoxia event (Fig. 6B). During the rest of the year (October to December), the temperatures regularly fluctuated below 15°C, with drops in temperature being concurrent with decreased levels of dissolved oxygen. From July to December 2014, the cumulative mortalities (integrating the last 4–5 mo) of scallops were dependent on shell size (χ^2 = 142, df = 15, p < 0.001, Fig. 6D). In particular, scallop mortality was reduced during September, but an important

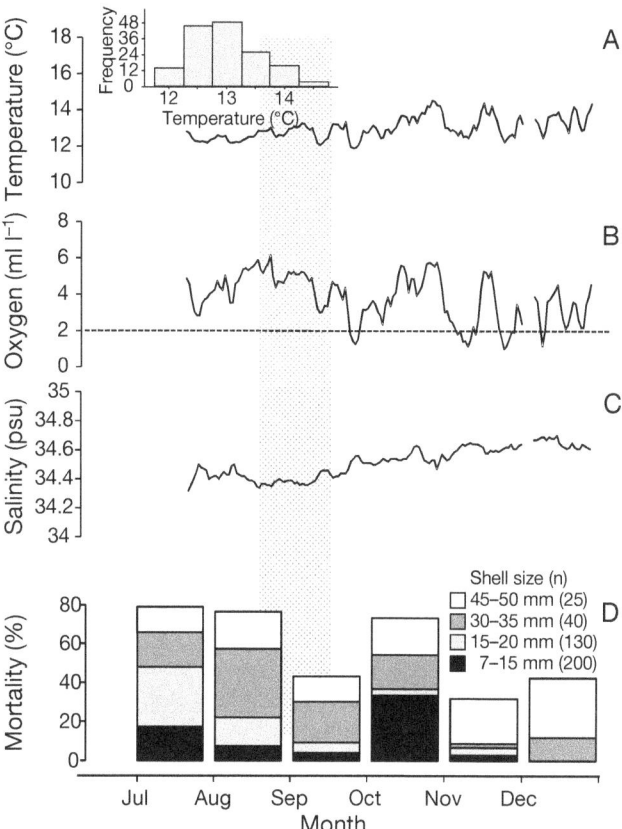

Fig. 6. Temporal fluctuations in (A) temperature, (B) oxygen concentration, and (C) salinity measured from June to December 2014 by the CEAZAMET buoy at 10 m depth in Tongoy Bay. (D) Cumulative mortalities in 4 shell range sizes of *Argopecten purpuratus* recorded during the scallop thinning procedure performed from July to December 2014 at the Invertec–Ostimar Co. scallop farm; each growing period lasted for ca. 4 to 5 mo. The initial number of scallops at each level of the lantern nets and in each size range (see 'Materials and methods') is shown in brackets. Inset in (A) shows the histogram for temperature observations recorded during the study period; the horizontal dashed line in (B) indicates hypoxia levels, and the stippled area represents the timing of the experimental period

increase in mortality was observed in October mostly in small scallops (7–15 mm). The mortality of scallops with larger shell sizes (>45 mm) remained fairly constant (ca. 20–30 %) throughout the monitoring period (Fig. 6D).

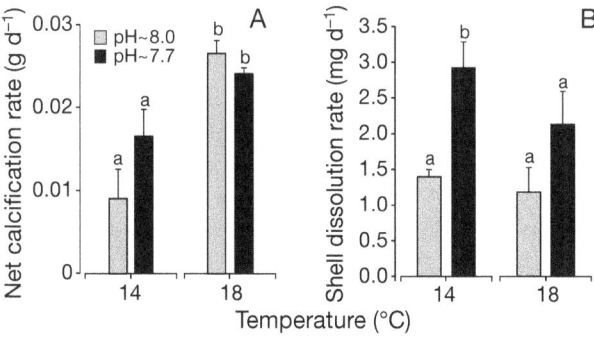

Fig. 5. (A) Net calcification rate (mean ± SE) of *Argopecten purpuratus* reared at 2 temperatures and 2 nominal pH levels in seawater. (B) Dissolution rate (mean ± SE) of dead shells of the scallops exposed to the same experimental conditions. These rates were estimated based on the difference between initial and final measurement after the 18 d experimental period. Different letters indicate significant differences among treatments (n = 5) using a post hoc Tukey HSD test

DISCUSSION

Despite the short period of exposure used in our study (18 d), we detected significant effects of OA and temperature on juvenile individuals of *Argopecten purpuratus*; our results were similar to a previous study, in which a short exposure time was sufficient to detect OA and warming effects on this marine mollusk (Nienhuis et al. 2010, Duarte et al. 2014). These results suggest that OA and warming might have significant effects on the fitness of these individuals in a short period of time. This information is relevant to understand the effects of upwelling conditions on these animals, because this phenomenon is of short duration and is not constant. In this study, no mortality effects were found on juveniles in any of the experimental treatments, which indicates that the projected increase of CO_2 levels and temperature had chronic, but not lethal, effects on these animals. This lack of mortality may result from an underlying natural tolerance of *A. purpuratus* to the changes in temperature and pH used in the current study, all of which fall within the extremes of environmental variability observed inside Tongoy Bay (see discussion below and Fig. 7).

Despite the lack of lethal effects, significant physiological responses to both temperature and pH changes were observed. Shell thickness, shell weight, and live biomass of *A. purpuratus* individuals were significantly reduced at low pH (pH ~7.7) and increased temperature (18°C) compared to control individuals (pH = 8.0 at 14°C). Reduced shell growth and increased dissolution rates were observed at

ambient temperature (14°C) in low pH treatments. These dissolution rates fluctuated between 3 and 10% of net calcification, but did not preclude animal growth, as illustrated by increased shell length and weight across all treatments. Impacts of lowered pH were ameliorated at warmer temperatures (18°C), through a significant increase in calcification rates. Thus, the results of this study agree with previous studies where shell characteristics, such as shell dissolution, shell size, and calcification rates, were used to examine the impacts of temperature and OA on marine calcifying organisms, and that warming could reverse the negative effects of OA (e.g. Beniash et al. 2010, Welladsen et al. 2010, Byrne 2012, Chan et al. 2013, Thomsen et al. 2013, Bressan et al. 2014, Fitzer et al. 2014). Although we did not assess additional biological sources of stress in the present study, they may also affect shell formation and structure. For instance, food limitation could be crucial if feeding rates increased beyond the food supply provided in this experiment (Mackenzie et al. 2014).

Significant reductions of shell thickness and weight (ca. −36% and −7%, respectively) were observed under experimental conditions of low pH at 18°C with respect to control scallops (i.e. present-day conditions, 14°C/pH ~8.0). Such a response could have implications for the long-term survival of individuals, as shell strength in scallops is a function of shell height, thickness, corrugation, and convexity (Pennington & Currey 1984). Lafrance et al. (2003) demonstrated that the shells of wild scallops were significantly heavier and stronger than those of cultured scallops, suggesting that cultured juvenile scallops have an increased vulnerability to shell-cracking predators. This reduction in shell strength was associated with factors related to the suspended culture (e.g. density, suspension depth, fouling, handling), leading to the study of mechanisms enhancing shell strength and strategies involving the timing of release of scallops into bottom culture (Grefsrud & Strand 2006). Our study suggests that, in addition to factors related to suspended culture conditions, other stressors such as low seawater pH also influence the shell strength of *A. purpuratus*. Similar reductions in shell thickness and associated robustness have been described by several studies in other mollusks exposed to low pH conditions (Bibby et al. 2007, Gazeau et al. 2013). In particular, the larval shell of the scallop *A. irradians* decreased in thickness when exposed to the low pH (high pCO_2) seawater conditions projected for the end of this century (Talmage & Gobler 2010). However, other studies reported that mussels grew thicker shells at warmer temperatures

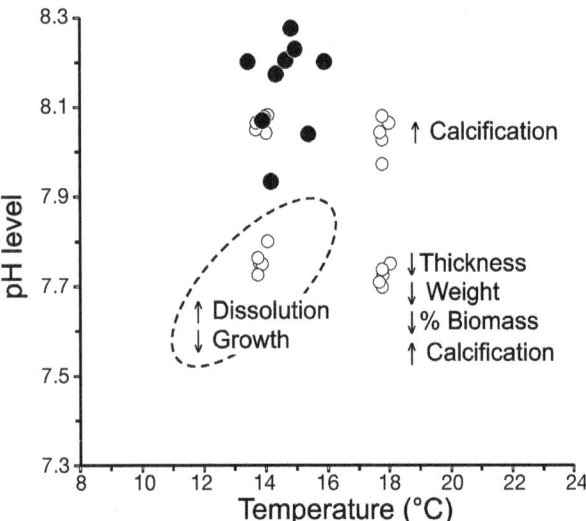

Fig. 7. Graphical summary for the variability of each temperature/pH combination in the experimental treatments (open circles); the natural variability in temperature/pH recorded inside Tongoy Bay in December 2014 (black dots) during a field survey, and the environmental range of pH and temperature reported for Pt. Lengua de Vaca upwelling area (dashed ellipse, see 'Discussion'). The significant effects recorded in each shell characteristic of *Argopecten purpuratus* are indicated by up or down arrows and presented beside each pH/temperature combination

independent of pH conditions (*Mytilus galloprovincialis*, Kroeker et al. 2013), or showed no reduction in shell thickness (*M. edulis*, Thomsen et al. 2013). More recently, Bressan et al. (2014) pointed out that a reduction in shell thickness could be species specific and could take place after 3 and 6 mo of incubation under acidification conditions in clams *Chamelea gallina* and mussels *M. galloprovincialis*, respectively. Since low pH conditions can reduce the shell thickness in scallops, it is realistic to expect effects in areas, such as Tongoy Bay, that are under the influence of low pH upwelling waters (Torres & Ampuero 2009, see below and Fig. 6). Thus, studies that aim to improve our understanding of the mechanisms that govern scallop shell strength and related mineralogical properties are urgently needed.

Reduced shell growth under acidified or low pH conditions have been reported specifically for scallops (Talmage & Gobler 2009, White et al. 2013) and other mollusks (Melzner et al. 2011, Hiebenthal et al. 2013, Kroeker et al. 2013, Bressan et al. 2014). As in those studies, we found that a significant reduction in shell length and height of *A. purpuratus* occurs even though seawater remained above the saturation state for carbonate precipitation ($\Omega_{calcite} > 1$). In addition, our experimental scenarios are within the environmental conditions described as setting a threshold for negative effects of pH (7.5–7.4; Berge et al. 2006, Bressan et al. 2014) and temperature (25°C) on shell growth of bivalves (Hiebenthal et al. 2013, Kroeker et al. 2013). Thus, to a lesser extent but also important, the reduction in shell growth of *A. purpuratus* may be ascribed to dissolution as recorded in dead shells exposed to low pH and low temperature conditions (Nienhuis et al. 2010, Duarte et al. 2014). In contrast to other materials, the kinetic dissolution/solubility of carbonate shells increases at low temperatures (Morse et al. 2007). This property of carbonates may help to explain the increase in shell calcification of *A. purpuratus* at warmer temperatures. However, saturation states were roughly similar across temperature treatments at respective pH levels. This suggests a major role for warmer temperatures, which could potentially offset the reduction in calcification caused by OA (Byrne & Przeslawski 2013, Kroeker et al. 2013, Duarte et al. 2014). However, calcification must be regarded as part of the whole biology of the individuals and thus the reduction in shell increments under low pH could be because energy allocation is being diverted away to other key physiological processes such as acid–base balance, reproduction, and immune function (Wood et al. 2008, Findlay et al. 2009). Finally, the increment in the calcification rates

at 18°C registered in our study (see Fig. 7) reinforces the notion that *A. purpuratus* can benefit from warmer temperatures and agrees with previous observations relating increased developmental rates and population abundances of *A. purpuratus* during El Niño – Southern Oscillation (ENSO) events which regularly impact upwelling ecosystems in northern Chile (Thiel et al. 2007). This finding indicates that scallop aquaculture in northern Chile will remain exclusively associated with embayment areas, where increased temperatures may provide a 'refuge' for scallop production.

Currently, there is no consensus of how the effects of warming and acidification will interact (e.g. additive, synergistic, or antagonistic) to influence the fitness of mollusks (e.g. Berge et al. 2006, Findlay et al. 2009, Gooding et al. 2009, Comeau et al. 2010, Byrne 2011, Kroeker et al. 2013, Duarte et al. 2014). These discrepancies have led to generalizations that temperature may have a stronger effect on the overall survival of invertebrates (Findlay et al. 2009, Lischka et al. 2011), while OA will have its greatest influence primarily on calcification and growth rates (Byrne & Przeslawski 2013). In our study, temperature increase and OA operated synergistically only on growth rates of *A. purpuratus*, estimated through changes in shell length and height, a result which is in agreement with previous evidence for synergistic effects of temperature and acidification on mollusks (Rodolfo-Metalpa et al. 2011). Thus, our study suggests that changes in shell growth can result from increased dissolution at low pH and temperature conditions and from increased calcification rates at warmer temperatures, which may actually overcome the negative impact of acidification. However, we also record a moderate but significant reduction in biomass (−2 %) under low pH, regardless of seawater temperature. This would imply that low pH was increasing the biological cost of living for the scallops, which could have additional implications for the commercial production of *A. purpuratus* under the influence of upwelling waters (e.g. Talmage & Gobler 2010, Barton et al. 2012).

The increased calcification rates of *A. purpuratus* observed at warmer temperatures appeared to be occurring at the expense of reducing the shell thickness, weight, and live biomass. For instance, at the end of the experiment, the mean shell length was positively correlated with mean dry weight (Pearson's r = 0.99; p = 0.003, n = 4). That is, across treatments, larger shells imply heavier shells. However, calcification rate was negatively associated with shell thickness (Pearson's r = −0.97; p = 0.033, n = 4), which suggests that those shell size/weight increments promoted by increased temperature at both pH conditions could occur at the cost of reducing shell thickness. This potential trade-off highlights some strategies that may be implemented by this cultured species to cope with variability in both climate stressors. In particular, our study suggests that scallops may increase their calcification rate and reach larger shell size during warmer periods (e.g. ENSO) but would be thinner and frailer, and thus more vulnerable to predators and parasites (e.g. perforating polychaetes), as well as increasing the risks of shell breakage and mortality during the manipulation and mechanical sorting symptomatic of the commercial thinning process. In addition, under projected global acidification scenarios (which are already occurring in our study area due to upwelling events, see below and Fig. 7), the observed reduction in growth rates would suggest that it will take longer for scallops to reach marketable size and will require a modification to the timing of the thinning process performed by the aquaculture industry. Future studies should examine strategies that facilitate the survival of scallops when subjected to the thinning and sorting process.

Coastal upwelling modulates the physical and chemical properties of seawater over large areas of the coastal ocean. In relation to the current study, the Pt. Lengua de Vaca upwelling increases CO_2-fluxes (fCO_2 ~ 1000 µatm) and promotes strong reductions in pH (pH ~7.6) in Tongoy Bay and the surrounding area (Torres & Ampuero 2009). These pH values have previously been proposed as a threshold for negative effects of OA on shell growth in bivalves (e.g. Berge et al. 2006, Bressan et al. 2014). Thus, the significant net shell dissolution and reduced growth rates of *A. purpuratus* registered at acidified (pH ~7.7) and current temperature (14°C) conditions suggest that the shell characteristics of the farmed scallops in Tongoy Bay are already compromised during upwelling periods. More generally, it has been suggested that changes in ocean chemistry, which have been driven by rising anthropogenic CO_2 emissions, may be inhibiting the development and survival of larval shellfish, and are probably contributing to a global decline in bivalve populations of particular scallop species (Talmage & Gobler 2010). Some upwelling areas of the Humboldt Current system are showing cooling trends (Gutierrez et al. 2011), which may represent an additional stressor due to the increased kinetic dissolution of carbonate at lower temperatures (Morse et al. 2007). Thus, further environmental monitoring is required to better

understand the extent, timing, and persistence of upwelling on the physical-chemical properties of the waters masses circulating inside Tongoy Bay, and their role in reducing shell growth and biomass production of *A. purpuratus* farming operations.

Initial physical-chemical monitoring and water sampling related to the scallop aquaculture was started during 2014. This monitoring has already indicated that during the austral spring (December), the seawater inside Tongoy Bay had pH values between 7.9 and 8.3, and temperatures between 14 and 16°C. This variability compares well with the present-day experimental scenario used in the current study (Fig. 7). This implies that, in absence of upwelling influences, Tongoy Bay provides suitable conditions to support scallops with appropriate shell thickness, shell weight, and live tissue biomass as required by scallop aquaculture. In addition, previous studies in the area have suggested the role of upwelling influences in providing an adequate food supply into Tongoy Bay (Gonzalez et al. 1999). However, Tongoy Bay also showed low levels of oxygenation at 10 m depth (Fig. 6B), which adds a third climatic stressor associated with upwelling ecosystems (e.g. Gruber 2011). Occasional occurrences of hypoxic conditions may also affect the development of *A. purpuratus* and decrease their physiological capacity to withstand other stressors often encountered by cultured scallops (Brokordt et al. 2013). The mortality data gathered at Invertec-Ostimar farms suggest that, under current temperatures, drops in oxygen concentration below 2 mg l^{-1} (i.e. hypoxia events, Díaz & Rosenberg 2008) may be related to the increased mortality of *A. purpuratus* juveniles (7–15 mm) in September and October, while the mortality of the adult fraction (>45 mm) remains almost stable from July to October. However, hypoxic events are more common from November to January and correspond with an important reduction in mortality of juveniles. This suggests that, compared to the adult fraction, juveniles are more resilient to sources of mortality potentially induced by the underlying environmental variability occurring during the Nov–Jan monitoring period inside Tongoy Bay. Ramajo et al. (2016) recently reported that juvenile *A. purpuratus* are tolerant to OA conditions and showed increased metabolism, shell growth, net calcification, and ingestion rates under OA conditions (pH ~7.6). However, as in our study, these positive responses to OA also occurred at increased temperatures (i.e. at 18°C). These observations reflect the occurrence of multiple stressors that may interfere with temperature and acidification upwelling-influenced areas with potential impacts on growth and production of the native scallop. Overall, the results of the current study indicate that a review of existing data for scallop production and environmental variability in the area would be valuable, as would be additional experimentation using a multiple-stressor approach.

Acknowledgements. We thank Pedro Alcayaga at Invertec–Ostimar Hatchery facilities, Jorge López, Sebastian Osores, and Paulina Contreras for their valuable assistance during the experiments and fieldwork. This study was supported by the Millennium Nucleus Center for the Study of Multiple-drivers on Marine Socio-Ecological Systems (MUSELS) funded by MINECON NC120086 and FONDECYT grant nos. 1140938 and 1140092 to N.A.L. and M.A.L. C.A.V. was also supported by MINECON IC120019 and Red 14 Doctoral REDOC.CTA, MINEDUC project UCO1202 at the Universidad de Concepción.

LITERATURE CITED

ä Aravena G, Broitman B, Stenseth NC (2014) Twelve years of change in coastal upwelling along the central-northern coast of Chile: spatially heterogeneous responses to climatic variability. PLoS ONE 9:e90276

ä Barton A, Hales B, Waldbusser GG, Langdon C, Feely RA (2012) The Pacific oyster, *Crassostrea gigas*, shows negative correlation to naturally elevated carbon dioxide levels: implications for near-term ocean acidification effects. Limnol Oceanogr 57:698–710

ä Beniash E, Ivanina A, Lieb NS, Kurochkin I, Sokolova IM (2010) Elevated level of carbon dioxide affects metabolism and shell formation in oysters *Crassostrea virginica*. Mar Ecol Prog Ser 419:95–108

ä Berge JA, Bjerkeng B, Pettersen O, Schaanning MT, Øxnevad S (2006) Effects of increased sea water concentrations of CO_2 on growth of the bivalve *Mytilus edulis* L. Chemosphere 62:681–687

ä Bibby R, Cleall-Harding P, Rundle S, Widdicombe S, Spicer J (2007) Ocean acidification disrupts induced defences in the intertidal gastropod *Littorina littorea*. Biol Lett 3: 699–701

ä Branch TA, Dejoseph BM, Ray LJ, Wagner CA (2013) Impacts of ocean acidification on marine seafood. Trends Ecol Evol 28:178–186

ä Bressan M, Chinellato A, Munari M, Matozzo V and others (2014) Does seawater acidification affect survival, growth and shell integrity in bivalve juveniles? Mar Environ Res 99:136–148

Brokordt K, Pérez H, Campos F (2013) Environmental hypoxia reduces the escape response capacity of juvenile and adult scallops *Argopecten purpuratus*. J Shellfish Res 32:369–376

ä Buschbaum C, Buschbaum G, Schrey I, Thieltges DW (2007) Shell-boring polychaetes affect gastropod shell strength and crab predation. Mar Ecol Prog Ser 329:123–130

Byrne M (2011) Impact of ocean warming and ocean acidification on marine invertebrate life history stages: vulnerabilities and potential for persistence in a changing ocean. Oceanogr Mar Biol 49:1–42

ä Byrne M (2012) Global change ecotoxicology: identification of early life history bottlenecks in marine invertebrates, variable species responses and variable experimental approaches. Mar Environ Res 76:3–15

ä Byrne M, Przeslawski R (2013) Multistressor impacts of warming and acidification of the ocean on marine invertebrates' life histories. Integr Comp Biol 53:582–596

ä Caldeira K, Wickett ME (2003) Oceanography: anthropogenic carbon and ocean pH. Nature 425:365

ä Chan VBS, Thiyagarajan V, Lu XW, Zhang T, Shih K (2013) Temperature dependent effects of elevated CO_2 on shell composition and mechanical properties of *Hydroides elegans*: insights from a multiple stressor experiment. PLoS ONE 8:e78945

ä Comeau S, Gorsky G, Alliouane S, Gattuso JP (2010) Larvae of the pteropod *Cavolinia inflexa* exposed to aragonite undersaturation are viable but shell-less. Mar Biol 157: 2341–2345

ä Cubillo AM, Peteiro LG, Fernández-Reiriz MJ, Labarta U (2012) Influence of stocking density on growth of mussels (*Mytilus galloprovincialis*) in suspended culture. Aquaculture 342-343:103–111

ä Díaz RJ, Rosenberg R (2008) Spreading dead zones and consequences for marine ecosystems. Science 321: 926–929

Dickson AG (1990) Standard potential of the reaction $AgCl(s)+1/2\ H_2(g) = Ag(s)+HCl(aq)$ and the standard acidity constant of the ion HSO_4^- in synthetic sea water from 273.15 to 318.15 K. J Chem Thermodyn 22:113–127

ä Dickson AG, Millero FJ (1987) A comparison of the equilibrium constants for the dissociation of carbonic acid in seawater media. Deep-Sea Res Part A 34:1733–1743

Dickson AG, Sabine CL, Christian JR (eds) (2007) Guide to best practices for ocean CO_2 measurements. PICES Special Publication 3, North Pacific Marine Sciences Organization, Sidney

DOE (US Department of Energy) (1994) Handbook of methods for the analysis of the various parameters of the carbon dioxide system in seawater; version 2.1. http://cdiac.ornl.gov/oceans/DOE_94.pdf (accessed 29 August 2013)

ä Doney SC, Fabry VJ, Feely RA, Kleypas JA (2009) Ocean acidification: the other CO_2 problem. Ann Rev Mar Sci 1: 169–192

ä Duarte C, Navarro JM, Acuña K, Torres R and others (2014) Combined effects of temperature and ocean acidification on the juvenile individuals of the mussel *Mytilus chilensis*. J Sea Res 85:308–314

ä Duarte C, Navarro J, Acuña K, Torres R and others (2015) Intraspecific variability in the response of the edible mussel *Mytilus chilensis* (Hupe) to ocean acidification. Estuar Coast 38:590–598

ä Ekstrom JA, Suatoni L, Cooley SR, Pendleton LH and others (2015) Vulnerability and adaptation of US shellfisheries to ocean acidification. Nat Clim Change 5:207–214

ä Fabry VJ, Seibel BA, Feely RA, Orr JC (2008) Impacts of ocean acidification on marine fauna and ecosystem processes. ICES J Mar Sci 65:414–432

ä Feely RA, Sabine CL, Hernandez-Ayon JM, Ianson D, Hales B (2008) Evidence for upwelling of corrosive 'acidified' water onto the continental shelf. Science 320:1490–1492

ä Filgueira R, Comeau LA, Landry T, Grant J, Guyondet T, Mallet A (2013) Bivalve condition index as an indicator of aquaculture intensity: a meta-analysis. Ecol Indic 25: 215–229

Findlay HS, Wood HL, Kendall MA, Spicer JI, Twitchett RJ,

Widdicombe S (2009) Calcification, a physiological process to be considered in the context of the whole organism. Biogeosci Discuss 6:2267–2284

Fitzer SC, Cusack M, Phoenix VR, Kamenos NA (2014) Ocean acidification reduces the crystallographic control in juvenile mussel shells. J Struct Biol 188:39–45

Fréchette M, Lachance-Bernard M, Daigle D (2010) Body size, population density and factors regulating suspension-cultured blue mussel (*Mytilus* spp.) populations. Aquat Living Resour 23:247–254

Gattuso JP, Magnan A, Billé R, Cheung WWL and others (2015) Contrasting futures for ocean and society from different anthropogenic CO_2 emissions scenarios. Science 349:aac4722, doi:10.1126/science.aac4722

Gazeau F, Parker LM, Comeau S, Gattuso JP and others (2013) Impacts of ocean acidification on marine shelled mollusks. Mar Biol 160:2207–2245

Gazeau F, Urbini L, Cox TE, Alliouane S, Gattuso JP (2015) Comparison of the alkalinity and calcium anomaly techniques to estimate rates of net calcification. Mar Ecol Prog Ser 527:1–12

González L, López A, Pérez C, Riquelme A and others (1999) Growth of the scallop, *Argopecten purpuratus*, in southern Chile (Lamarck, 1819). Aquaculture 175:307–316

Gooding RA, Harley CDG, Tang E (2009) Elevated water temperature and carbon dioxide concentration increase the growth of a keystone echinoderm. Proc Natl Acad Sci USA 106:9316–9321

Grefsrud SE, Strand Ø (2006) Comparison of shell strength in wild and cultured scallops (*Pecten maximus*). Aquaculture 251:306–313

Gruber N (2011) Warming up, turning sour, losing breath: ocean biogeochemistry under global change. Philos Trans A Math Phys Eng Sci 369:1980–1996

Gruber N, Hauri C, Lachkar Z, Loher D, Fröhlicher TL, Plattner GK (2012) Rapid progression of ocean acidification in the California Current System. Science 337:220–223

Gutiérrez D, Bouloubassi I, Sifeddine A, Purca S and others (2011) Coastal cooling and increased productivity in the main upwelling zone off Peru since the mid-twentieth century. Geophys Res Lett 38:L07603, doi:10.1029/2010 GL046324

Harley CDG, Hughes AR, Hultgren KM (2006) The impacts of climate change in coastal marine systems. Ecol Lett 9: 228–241

Helmuth B, Mieszkowska N, Moore P, Hawkins S (2006) Living on the edge of two changing worlds: forecasting the responses of rocky intertidal ecosystems to climate change. Annu Rev Ecol Evol Syst 37:373–404

Hennen DR, Hart DR (2012) Shell height to weight relationships for Atlantic sea scallops (*Placopecten magellanicus*) in offshore U.S. waters. J Shellfish Res 31:1133–1144

Hiebenthal C, Philipp EER, Eisenhauer A, Wahl M (2013) Effects of seawater pCO_2 and temperature on shell growth, shell stability, condition and cellular stress of Western Baltic Sea *Mytilus edulis* (L.) and *Arctica islandica* (L.). Mar Biol 160:2073–2087

IPCC (Intergovernmental Panel on Climate Change) (2014) Climate change 2014: impacts, adaptation, and vulnerability. Contribution of Working Group II to the 5th Assessment Report of the Intergovernmental Panel on Climate Change. Cambridge University Press, Cambridge

Jory DE, Iversen ES (1988) Shell strength of queen conch, *Strombus gigas* L: aquaculture implications. Aquacult Fish Manag 19:45–51

Ko GW, Dineshram R, Campanati C, Chan VB, Havenhand J, Thiyagarajan V (2014) Interactive effects of ocean acidification, elevated temperature, and reduced salinity on early-life stages of the Pacific oyster. Environ Sci Technol 48:10079–10088

Kroeker KJ, Kordas R, Crim R, Hendriks I and others (2013) Impacts of ocean acidification on marine organisms: quantifying sensitivities and interaction with warming. Glob Change Biol 19:1884–1896

Lafrance M, Cliche G, Haugum GA, Guderley H (2003) Comparison of cultured and wild sea scallops *Placopecten magellanicus*, using behavioural responses and morphometric and biochemical indices. Mar Ecol Prog Ser 250:183–195

Lischka S, Büdenbender J, Boxhammer T, Riebesell U (2011) Impact of ocean acidification and elevated temperatures on early juveniles of the polar shelled pteropod *Limacina helicina*: mortality, shell degradation, and shell growth. Biogeosciences 8:919–932

Mackenzie CL, Ormondroyd GA, Curling SF, Ball RJ, Whitely NM, Malham SK (2014) Ocean warming, more than acidification, reduces shell strength in a commercial shellfish species during food limitation. PLoS ONE 9: e86764

Maguire J, Fleury P, Burnell G (1999) Some methods for quantifying quality in scallops *Pecten maximus* (L.). J Shellfish Res 18:59–66

McClintock JB, Angus RA, McDonald MR, Amsler CD, Catledge SA, Vohra YK (2009) Rapid dissolution of shells of weakly calcified Antarctic benthic macroorganisms indicates high vulnerability to ocean acidification. Antarct Sci 21:449–456

Mehrbach C, Culberson CH, Hawley JE, Pytkowicz RN (1973) Measurement of the apparent dissociation constants of carbonic acid in seawater at atmospheric pressure. Limnol Oceanogr 18:897–907

Meinshausen M, Smith SJ, Calvin K, Daniel JS and others (2011) The RPC greenhouse gas concentrations and their extensions from 1765 to 2300. Clim Change 109:213–241

Melzner F, Stange P, Trübenbach K, Thomsen J and others (2011) Food supply and seawater pCO_2 impact calcification and internal shell dissolution in the blue mussel *Mytilus edulis*. PLoS ONE 6:e24223

Morse JW, Arvidson RS, Luttge A (2007) Calcium carbonate formation and dissolution. Chem Rev 107:342–381

Navarro JM, Torres R, Acuña K, Duarte C and others (2013) Impact of medium-term exposure to elevated pCO_2 levels on the physiological energetics of the mussel *Mytilus chilensis*. Chemosphere 90:1242–1248

Nienhuis S, Palmer R, Harley C (2010) Elevated CO_2 affects shell dissolution rate but not calcification rate in a marine snail. Proc R Soc Lond B Biol Sci 277:2553–2558

Orr JC, Fabry VJ, Aumont O, Bopp L and others (2005) Anthropogenic ocean acidification over the twenty-first century and its impact on calcifying organisms. Nature 437:681–686

Palmer AR (1982) Growth in marine gastropods: a non-destructive technique for independently measuring shell and body weight. Malacologia 23:63–73

Pennington BJ, Currey JD (1984) A mathematical model for the mechanical properties of scallop shells. J Zool 202: 239–263

Pierrot D, Lewis E, Wallace DWR (2006) MS Excel program

developed for CO_2 system calculations, ORNL/CDIAC-105. Carbon Dioxide Information Analysis Center, Oak Ridge National Laboratory, US Department of Energy, Oak Ridge, TN

➤ Ramajo L, Marbà N, Prado L, Peron S and others (2016) Biomineralization changes with food supply confer juvenile scallops (*Argopecten purpuratus*) resistance to ocean acidification. Glob Change Biol 22:2025–2037

Ries JB, Cohen AL, McCorkle DC (2009) Marine calcifiers exhibit mixed responses to CO_2-induced ocean acidification. Geology 37:1131–1134

➤ Rodolfo-Metalpa R, Houlbreque F, Tambutte E and others (2011) Coral and mollusc resistance to ocean acidification adversely affected by warming. Nat Clim Change 1: 308–312

SERNAPESCA (Servicio Nacional de Pesca y Acuicultura) (2014) Anuarios estadísticos del Servicio Nacional de Pesca y Acuicultura. www.sernapesca.cl

Sokal RR, Rohlf FJ (1995) Biometry: the principles and practice of statistics in biological research. WH Freeman, New York, NY

➤ Stotz WB (2000) When aquaculture restores and replaces a overfished stock, is the conservation of the species assured? The case of the scallop *Argopecten purpuratus* (Lamarck, 1819) in northern Chile. Aquacult Int 8:237–247

➤ Sydeman WJ, García-Reyes M, Schoeman DS, Rykaczewski RR, Thompson SA, Black BA, Bograd SA (2014) Climate change and wind intensification in coastal upwelling ecosystems. Science 345:77–80

➤ Talmage S, Gobler C (2009) The effects of elevated carbon dioxide concentrations on the metamorphosis, size, and survival of larval hard clams (*Mercenaria mercenaria*), bay scallops (*Argopecten irradians*), and eastern oysters (*Crassostrea virginica*). Limnol Oceanogr 54:2072–2080

➤ Talmage SC, Gobler C (2010) Effects of past, present, and future ocean carbon dioxide concentrations on the growth and survival of larval shellfish. Proc Natl Acad Sci USA 107:17246–17251

➤ Thiel M, Macaya EC, Acuña E, Arntz WE and others (2007) The Humboldt Current system of northern and central Chile. Oceanographic processes, ecological interactions and socioeconomic feedback. Oceanogr Mar Biol Annu Rev 45:195–344

➤ Thomsen J, Gutowska MA, Saphörster J, Heinemann A and others (2010) Calcifying invertebrates succeed in a naturally CO_2-rich coastal habitat but are threatened by high levels of future acidification. Biogeosciences 7:3879–3891

➤ Thomsen J, Casties I, Pansch C, Körtzinger A, Melzner F (2013) Food availability outweighs ocean acidification effects in juvenile *Mytilus edulis*: laboratory and field experiments. Glob Chang Biol 19:1017–1027

➤ Torres R, Ampuero P (2009) Strong CO_2 outgassing from high nutrient low chlorophyll coastal waters off central Chile (30°S): the role of dissolved iron. Estuar Coast Shelf Sci 83:126–132

➤ Torres R, Manriquez PH, Duarte C, Navarro JM, Lagos NA, Vargas CA, Lardies MA (2013) Evaluation of a semi-automatic system for long-term seawater carbonate chemistry manipulation. Rev Chil Hist Nat 86:443–451

➤ Tunnicliffe V, Davies KT, Butterfield DA, Embley RW, Rose JM, Chadwick Jr WW (2009) Survival of mussels in extremely acidic waters on a submarine volcano. Nat Geosci 2:344–348

Underwood AJ (1997) Experiments in ecology: their logical design and interpretation using analysis of variance. Cambridge University Press, Cambridge

➤ Waldbusser GG, Brunner EL, Haley BA, Hales B, Langdon CJ, Prahl FG (2013) A developmental and energetic basis linking larval oyster shell formation to acidification sensitivity. Geophys Res Lett 40:2171–2176

➤ Watson SA, Peck LS, Tyler PA, Southgate PC, Tan KS, Day RW, Morley SA (2012) Marine invertebrate skeleton size varies with latitude, temperature and carbonate saturation: implications for global change and ocean acidification. Glob Change Biol 18:3026–3038

Welladsen HM, Southgate PC, Heimann K (2010) The effects of exposure to near-future levels of ocean acidification on shell characteristics of *Pinctada fucata* (Bivalvia: Pteriidae). Molluscan Res 30:125–130

➤ White MM, McCorkle DC, Mullineaux LS, Cohen AL (2013) Early exposure of bay scallops (*Argopecten irradians*) to high CO_2 causes a decrease in larval shell growth. PLoS ONE 8:e61065

➤ Wood HL, Spicer JI, Widdicombe S (2008) Ocean acidification may increase calcification rates, but at a cost. Proc R Soc Lond B Biol Sci 275:1767–1773

3

Effect of ectoparasite infestation density and life-history stages on the swimming performance of Atlantic salmon *Salmo salar*

S. Bui[1,*], T. Dempster[1,2], M. Remen[2], F. Oppedal[2]

[1]Sustainable Aquaculture Laboratory – Temperate and Tropical (SALTT), School of BioSciences, University of Melbourne, Victoria 3010, Australia
[2]Institute of Marine Research, Matredal 5984, Norway

ABSTRACT: To overcome sustainability obstacles and improve operations, the Atlantic salmon farming industry is testing novel approaches to production. Redistributing farm sites to offshore locations is one such solution; however, tolerance to high-current velocity sites must be considered, particularly if fish health status is compromised by parasites. We tested the effect of parasite density and life-history stage on the swimming performance of Atlantic salmon *Salmo salar* using a swim flume. Salmon with 3 different salmon lice *Lepeophtheirus salmonis* densities (0, 0.02 ± 0.01 and 0.11 ± 0.01 lice cm^{-2} [mean ± SE]) were tested across the 4 major life-history stages of lice (copepodid, chalimus, pre-adult and adult) for critical swimming performance (U_{crit}). Salmon U_{crit} declined slightly by a mean of 0.04 to 0.10 body lengths s^{-1} with high parasite densities compared to uninfested and low densities, across the lice stages, while progression through the parasite life-history stages had little effect on swimming performance. Our results suggest that increasing infestation density of salmon lice incurs negative fitness consequences for farmed Atlantic salmon held in high-current velocity sites, with little difference in costs associated with attachment by different life-history stages of the lice.

KEY WORDS: Salmonid · Salmon louse · Critical swimming speed · Swim flume · Copepod · Exposed aquaculture

INTRODUCTION

Numerous performance measures combine to determine individual fitness, and for aquatic animals, swimming is a key component of fitness in a challenging environment (Videler 1993). Swimming performance plays a vital role in predator avoidance, feeding behaviours, social interactions and migrations, and impairment of maximal swimming capacity may reduce an individual's ability to evade attack, hunt prey, keep up with a school or avoid suboptimal environmental conditions (Plaut 2001). Swimming performance can be reduced by parasites that force hosts to divert some of their energetic budget towards dealing with infestation, via immune or physi-

ological responses, and away from locomotion or metabolic activities (Barber et al. 2000).

The ectoparasitic sea louse *Lepeophtheirus salmonis* infests salmonids across the northern hemisphere, and represents a substantial problem in the aquaculture of Atlantic salmon *Salmo salar* (Torrissen et al. 2013, Murray et al. 2016). The parasite negatively affects the welfare status of farmed salmon, causes production and economic losses, and depresses wild salmonid populations in nearby environments due to parasite spill-back (Costello 2006, 2009, Krkošek et al. 2007, Thorstad et al. 2015). Sea lice have planktonic and attached stages that utilise a single salmonid host (Costello 2006). The free-swimming copepodid stage exhibits host-searching

*Corresponding author: samanthab@unimelb.edu.au

behaviours, and when a salmonid is encountered, they attach and progress through their life-history stages on the host (Johnson & Albright 1991). Infestation is associated with negative physiological and immune responses, with the different life-history stages inflicting varying levels of reaction (Grimnes & Jakobsen 1996, Wagner et al. 2008). Current knowledge predicts that most physiological harm occurs to the host immediately after the lice moult from the attached chalimus stage to the mobile pre-adult stage, largely due to the change in feeding behaviour (Jónsdóttir et al. 1992). Infestations are energetically costly, with infestations by adult lice reducing the critical swimming speed for Atlantic salmon (Wagner et al. 2003, Wagner et al. 2008).

Any impairment of fitness caused by sea lice infestation is expected to vary with the degree of their interaction with the host, through feeding characteristics, infestation density, or life-history stage (Wagner et al. 2008). The domesticated habitat of farmed salmon results in an altered host–parasite interaction; e.g. in Norway, regulations stipulate a threshold for parasite intensity, requiring that fish must be treated when parasite levels reach an average of 0.5 adult female sea lice per fish (Torrissen et al. 2013). Farmed salmon individuals therefore rarely carry high loads of adults, but will often harbor higher densities of earlier life-history stages. In contrast, wild salmonids can carry all life-history stages in high densities (Thorstad et al. 2015). The impact of parasite burden on individual fitness and survival from reduced swimming capabilities is likely to differ substantially between farmed and wild salmon. Farmed salmon are restricted to their caged micro-environment, where fluctuating current velocities require fish to alter swimming performance to maintain position in the school and avoid the net wall (Johansson et al. 2014). Wild salmon post-smolts must maintain their swimming performance to reach feeding areas in the ocean, and to migrate successfully up their natal stream and survive until spawning. In both cases, ectoparasitic infestation will decrease performance capacity (Wagner et al. 2003). However, we have little knowledge on the magnitude with which various sea lice stages diminish performance.

Measurement of critical swimming speed (U_{crit}) is a common method of assessing the swimming performance of fish (Brett 1964, Hammer 1995), particularly when testing the effects of a biotic or abiotic factor on physiological performance (Kolok 1999). U_{crit} is measured by placing a fish in a flume or flow tunnel, forcing them to swim against incrementally increasing

current velocities, and using the time and velocity at exhaustion to calculate critical swimming speed for that individual (Brett 1964). Although the ecological relevance of U_{crit} is debated, the general consensus is that it yields comparable data on the swimming ability of fish and accurately reflects physical status, from which fitness and survival can be inferred (Plaut 2001).

Here, we investigated the relationship between parasite load and swimming performance (as measured by U_{crit}) in post-smolt Atlantic salmon across the major life-history stages of salmon lice (copepodid, chalimus, pre-adult and adult stages). As the size of parasites increases and feeding characteristics intensify as life-history stages progress, particularly with the greater consumption rate and mobility of later stages, we predicted that their effect on swimming performance should scale similarly.

MATERIALS AND METHODS

Post-smolt Atlantic salmon *Salmo salar* were produced and held at the Norwegian Institute of Marine Research facilities. Experimental fish (n = 330, mean ± SE: 80 ± 1 g weight, 19.6 ± 0.1 cm fork length) were netted from their holding tank and transferred to an anaesthetic bath (Finquel MS-222, tricaine methanesulfonate, 10 g 100 l^{-1}). When fully sedated, individuals were implanted with passive integrated transponder (PIT; Glass tag unique, 3.85 × 23 mm, Trac-ID Systems AS) tags into their abdominal cavity, then transferred into one of 3 holding tanks (Ø = 3 m, 5.3 m^3) for recovery. A total of 110 fish were held in each tank at 15°C with a natural light regime, and allowed to recover for 7 d before infestation.

Sea lice production and infestation

Holding tanks were randomly assigned treatments of lice infestation pressures: 0, 30 or 90 *Lepeophtheirus salmonis* copepodids per fish. Egg strings were collected from gravid females held at 15°C and incubated at the same temperature. When hatched and >75 % of lice were at the copepodid stage (~6 d), the quantity of salmon louse copepodids was estimated using a microscope and counting chamber. Lice were divided into 2 containers for infestation ratios of 30 copepodids fish^{-1} and 90 copepodids fish^{-1} (low and high infestation pressures, respectively) to gain an expected parasite load of 10 and 30, based on a 30 % infestation success (Bjørn & Finstad 1998). During infestation, total water volume was reduced from 5.3 to 2.65 m^3 and flow stopped in all 3 holding tanks to prevent initial rising of water level and increase of infestation pressure. Copepodids were introduced to the 2 infestation tanks. After 20 min, flow was increased to a low rate (20 l min^{-1}) and after 3 h, was fully restored to 80 l min^{-1}. Oxygen was monitored throughout the infestation period to ensure levels were >80 % saturation at all times.

Swimming performance challenge

Performance challenges were conducted using a recirculating swim flume, constructed using polypropylene pipes forming an oval shape with 2 longer parallel sides, one of which contained the test chamber. The chamber was 248 cm long with an internal diameter of 36 cm, with grids at the front and back of the chamber, and openings on the top where exhausted fish could be taken out. The current velocity was generated by a motor-driven propeller (Flygt 4630, 11° propeller blade, Xylem Water Solutions Norge AS) mounted within another section of the flume, and propeller speed (in rounds per minute, rpm) was controlled by a panel located a distance from the flume, to prevent disturbance during the test. Calibration was made between rpm and cm s^{-1} in the flume, and challenge velocities were converted between cm s^{-1} and body lengths (BL) s^{-1} to be specific to the size range of fish we used. A resting tank, honeycomb filter and reduction cone filter were also within the flume to control and fluid movement through the chamber. New seawater was provided to the system in between the swim chamber and propeller, sourced from an adjacent tank, and water flowed out via an outlet located behind the swim chamber.

Swimming performance challenges involved a total of 30 fish in the chamber, 10 from each of the 3 infestation pressure groups. Swim flume runs occurred in line with the predicted growth rate of the sea lice based on water temperatures, to capture periods when lice had moulted into specific life-history stages. Four stages (and therefore 4 sets of runs) were tested: the copepodid stage (1 d post-infestation; dpi), chalimus stages (7 dpi), pre-adult stages (14 dpi) and adult stage (20 dpi). For each lice stage, 3 replicate performance challenges were conducted on consecutive days, except for the copepodid stage. The time frame before which sea lice moulted into the next stage was ~3 d, only allowing for 2 replicate challenges for the copepodid stage. Five runs contained <30 fish due to mortality or, in the

adult sea lice group, lack of parasitised fish for the high-range infestation pressure group. Runs never included <25 individuals in the swim flume.

When transferring fish into the swim flume, the water level was reduced in the holding tanks and the fish were lightly sedated (Finquel MS-222, tricaine methanesulfonate, 1 g 100 l^{-1}) until they were unresponsive to external stimuli. Random individuals were caught by hand and placed in a small container with full anaesthetic (Finquel, 10 g 100 l^{-1}). Upon loss of consciousness, parasite load was assessed, PIT-ID recorded, and the fish was put into the swim flume with no flow. MS-222 has little to no effect on crustaceans at doses applicable to fish (Schmit & Mezquita 2010) and therefore was not expected to affect sea lice attachment. Recovery of sedated fish was monitored inside the flume, and when all fish had regained normal swimming activity, the swim flume was set to a velocity of ~0.5 BL s^{-1}. Following standard protocols (see Plaut 2001 and references within), fish were left to acclimatize overnight, after which any effect of anaesthesia on swimming performance is negligible (Hayashida et al. 2013). Some mortalities occurred overnight due to fish escaping from the chamber.

The protocol for performance testing used increasing velocity increments of 0.5 BL s^{-1}, following established methodology (Kolok 1999). Velocities inside the swim chamber were confirmed using a hand-held velocimeter (Vane Wheel FA, with ZS25 insertion probe connected to handheld flowtherm NT, by Höntzsch) mounted in the centre of the chamber. We recorded velocities with increasing speed intervals, without fish and with a group of 25 fish (pilot). The relationship between propeller speed and predicted current velocity was accurate with and without fish (±1.5 cm s^{-1}), so the velocimeter was removed for performance challenges to prevent fish from using it as a hydrodynamic shelter.

After the acclimatisation period, the challenge began at 0.5 BL s^{-1}, and increased by increments of 0.5 BL s^{-1} at 20 min intervals (Kolok 1999). During the challenge, a camera was mounted behind the back grill facing the swimming chamber, so that fish could be observed without disturbance. Exhausted fish were detected using the camera and were rapidly removed from the chamber by hand. Fish were considered exhausted when they no longer swam against the current and lay on the back grill with no attempt to escape when touched. The velocity at which exhaustion occurred and the time elapsed since the beginning of that speed increment was recorded for each individual. Fish were euthanised immediately when taken out of the flume using an overdose of anaesthesia. Individuals were continuously removed from the flume upon exhaustion and the velocity increased incrementally until all fish were fatigued. Following the trial, all fish were weighed, measured, and had their PIT-ID recorded.

Critical swimming speed was calculated by: $U_{crit} = V + (t\Delta t^{-1})\Delta v$, where V = highest velocity maintained (cm s^{-1}), t = time elapsed at final velocity (min), Δt = time increment and Δv = velocity increment (Brett 1964). As critical swimming speed values have a strong relationship with fish length, we converted U_{crit} (cm s^{-1}) to relative U_{crit} (BL s^{-1}) by dividing the absolute swimming speed by the individual's fork length (cm).

Statistical analysis

As the settlement success of parasites increases with host size and the available surface area (Tucker et al. 2002), we standardised infestation intensity to fish body size. The total number of attached lice was converted to lice density using a model formula for fish surface area (S), where area (cm^2) was calculated using $S = 0.72L^{1.88}$ (L = total length; O'Shea et al. 2006). The relationship among parameters was assessed using multiple regression (Field et al. 2012) in R version 3.1.0 (R Core Team 2015). To investigate which factors influenced U_{crit} (cm s^{-1}), we included lice stage (categorical factor

converted to dummy variables for contrasts against the base category, copepodid stage), lice density, and fish length as predictor variables in the full model and tested using the linear model function. Diagnostic plots were assessed to check model assumptions. Models including only one predictor variable term were also run to examine the improvement of the full model with the inclusion of the variables in question. Pair-wise comparisons of the models using ANOVAs were conducted to determine the improvement of the full model with inclusions of predictor variables.

RESULTS

The infestation protocol was successful, with 100 % prevalence of sea lice at the copepodid stage. For the copepodid, chalimus and pre-adult stages, average lice density per fish increased with the initial infestation intensities of 0, 30 and 90 copepodids per fish (Table 1). However, pre-adult stages of sea lice did not survive well as abundance in the low and high infestation pressure tanks declined after sea lice had progressed to the adult stage (Table 1).

Mean U_{crit}, averaged across the lice stages and fish lengths, was (mean ± SE) 80.59 ± 0.52 cm s^{-1} (4.09 ± 0.03 BL s^{-1}) for unparasitised fish, 80.86 ± 0.52 cm s^{-1} (4.15 ± 0.03 BL s^{-1}) for fish with low parasite load, and 78.55 ± 0.46 cm s^{-1} (4.05 ± 0.03 BL s^{-1}) for heavily parasitised individuals (Table 1). Swimming performance

Table 1. Levels of parasite *Lepeophtheirus salmonis* load (lice density, corrected for body size of host Atlantic salmon *Salmo salar*), total abundance on the fish, and critical swimming speed (U_{crit}, mean ± SE) for each infestation intensity and sea lice stage (COP: copepodid, CH: chalimus, PA: pre-adult, A: adult). Infestation intensities are 0 (no lice), 30 (mid-range) and 90 (high-range) copepodids per fish. BL: body length

| Lice stage | n | Lice density (lice cm^{-2}) | | | Total abundance (lice fish^{-1}) | | U_{crit} (cm s^{-1}) | U_{crit} (BL s^{-1}) |
		Min.	Max.	Avg.	Min.	Max.		
No lice								
COP	19	0	0	0	0	0	74.88 ± 1.36	3.92 ± 0.08
CH	30	0	0	0	0	0	80.95 ± 0.70	4.22 ± 0.05
PA	30	0	0	0	0	0	81.59 ± 1.08	3.97 ± 0.06
A	32	0	0	0	0	0	82.75 ± 0.66	4.20 ± 0.04
Mid-range infestation								
COP	17	0.01	0.12	0.04	2	27	79.22 ± 1.10	4.13 ± 0.07
CH	30	0.01	0.16	0.03	1	26	79.40 ± 1.06	4.08 ± 0.06
PA	29	0.01	0.10	0.02	1	25	80.36 ± 0.77	4.12 ± 0.05
A	26	0.00	0.02	<0.01	0	4	84.19 ± 0.98	4.28 ± 0.05
High-range infestation								
COP	19	0.05	0.39	0.13	9	69	76.99 ± 1.35	3.90 ± 0.06
CH	30	0.06	0.60	0.19	14	102	78.28 ± 0.91	4.09 ± 0.07
PA	30	0.04	0.29	0.11	7	52	78.82 ± 0.60	4.03 ± 0.04
A	24	0.01	0.05	0.02	1	10	79.76 ± 0.95	4.13 ± 0.06

was negatively correlated to lice density and fish length; however, sea lice life-history stage had little influence on the mean U_{crit} across densities (Fig. 1, Table 2). When the 3 factors were incorporated into the multiple regression model, lice stages, lice density and fish length significantly contributed to U_{crit} (R^2 = 0.31, $F_{5,310}$ = 28.3, p < 0.001; Table 2). Lice density had a large effect on U_{crit} (β = −12.47) in the model, yet comparisons of single factor regressions showed that length had the strongest influence on performance compared to lice stage and lice density (R^2 = 0.18 compared to 0.11 and 0.07, respectively; Table 2). However, standardized β values revealed a lesser importance of length to the chalimus and pre-adult lice stage model factors (Table 2). When tested in individual regression models, all factors contributed significantly to swimming performance (Table 2), and the full model including all parameters was a signifi-

Table 2. Model parameter outputs for the effect of sea lice (*Lepeophtheirus salmonis*) stage, sea lice density and fork length of host Atlantic salmon *Salmo salar* on critical swimming speed (U_{crit}, cm s^{-1}). Results are shown for the full model (lice stage + density + length) and for single parameter models. COP: copepodid, CH: chalimus, PA: pre-adult, A: adult lice stages. Stand.: standardised, CI: confidence interval. Significant differences at p < 0.05

Factor	Single model R^2	p	β	Full model Stand. β	p	CI 2.5%	CI 97.5%
Lice stage	0.11	<0.001	–	–	–	–	–
CH vs. COP	–	–	2.96	0.65	<0.001	1.48	4.45
PA vs. COP			2.39	0.68	0.002	0.90	3.89
A vs. COP	–	–	4.45	0.07	<0.001	2.92	6.00
Lice density	0.07	<0.001	−12.47	-0.19	<0.001	−18.96	−5.99
Length	0.18	<0.001	1.44	0.32	<0.001	1.11	1.78

cant improvement on the individual models ($F_{2,310}$ = 45.89, p < 0.001). Plots of predicted values against standardised residuals (not shown) revealed even dispersion of data points around zero, satisfying assumptions of linearity, randomness and homoscedasticity. Q-Q plots indicate that observed residuals are within normal deviations from normality.

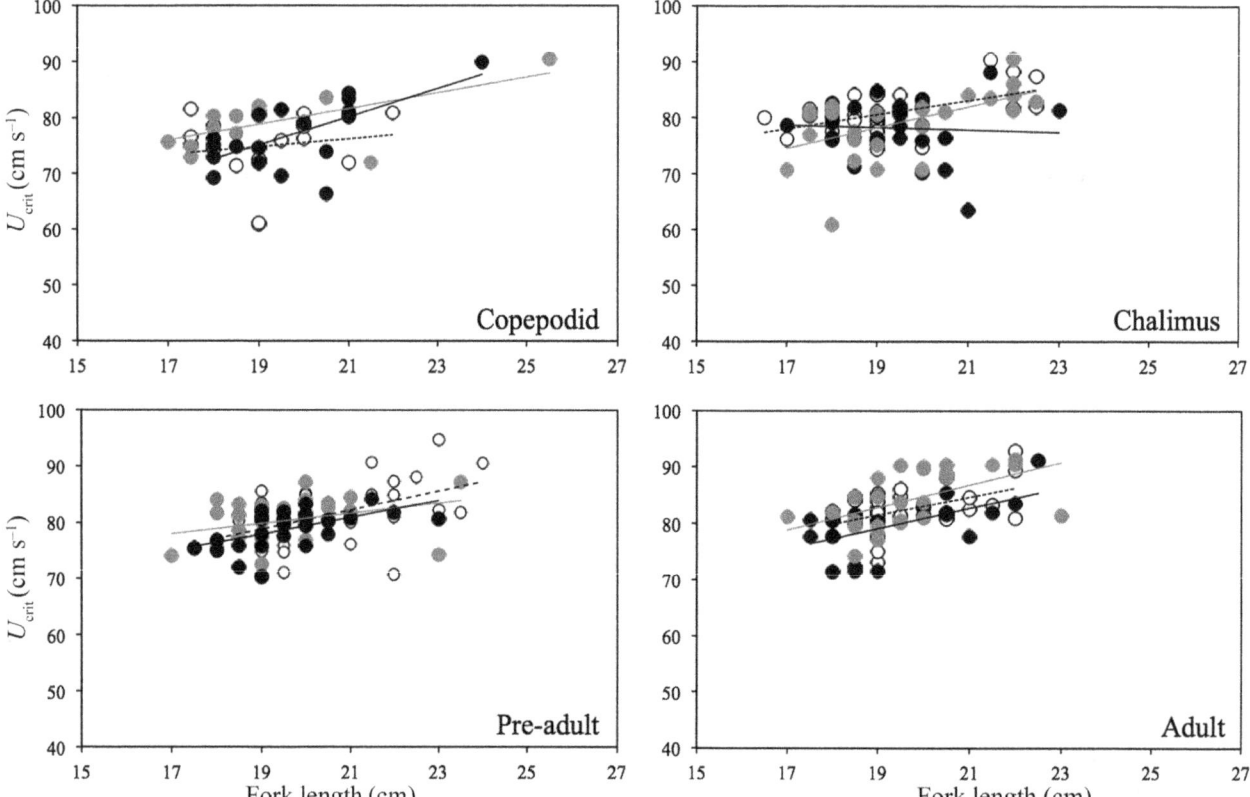

Fig. 1. Relationship between fish length and critical swimming speed (U_{crit}) of host Atlantic salmon *Salmo salar* as tested in the swim flume, at the 4 sea lice (*Lepeophtheirus salmonis*) life-history stages examined. Increasing infestation densities are categorically represented with no lice (grey circles, grey line), mid-range (30 copepodids fish^{-1}; open circles, dotted line), and high-range infestation density (90 copepodids fish^{-1}; black circles, black line)

DISCUSSION

Sea lice stages and swimming performance

Sea lice life-history stage, infestation intensity, and fish length had a varying yet cumulative effect on swimming performance. Our study is one of few that have tested the impact of juvenile lice stages on the swimming performance of Atlantic salmon. Nendick et al. (2011) investigated the performance of wild juvenile pink salmon *Oncorhynchus gorbuscha* infested with chalimus and pre-adult 1 stages and found that they had a large negative effect on performance (when fish were 0.34 g in average body mass), irrelevant of parasite density. However, the interaction between pink salmon and *Lepeophtheirus salmonis* differs from that observed with Atlantic salmon, where migrating juveniles exposed to infective copepids are much smaller and exhibit a more dramatic physiological and immunological response to infestation (Wagner et al. 2008, Brauner et al. 2012). The physiological impacts of copepodid and chalimus stages are not as pronounced as those of the mobile stages, yet spikes in immune and histopathological responses have been described in Atlantic salmon infested by lice at these attached stages (Jónsdóttir et al. 1992, Tadiso et al. 2011). We found that critical swimming performance was relatively similar between infested and uninfested salmon at the copepodid life stage; however, at the chalimus stage, high parasite loads resulted in a larger decline in U_{crit} with length. There is little difference between the copepodid and chalimus stages in terms of host physiological response (Bowers et al. 2000, Wagner et al. 2008), and our results suggest that infestation with copepodids imparts only a marginal cost on performance.

Pre-adult and adult *L. salmonis* are the sea lice stages expected to have the greatest physiological impact on their host (Jónsdóttir et al. 1992, Grimnes & Jakobsen 1996, Bowers et al. 2000) and their hosts' swimming performance (Wagner et al. 2003, Mages & Dill 2010). When sea lice moult into the mobile pre-adult stage, feeding behaviour changes and their consumption increases, which can result in host blood loss and osmoregulatory failure (Brandal et al. 1976, Grimnes & Jakobsen 1996, Sackville et al. 2011). Wagner & McKinley (2004) predicted that the feeding rates of *L. salmonis* could induce anaemia, reducing the critical swimming performance of Atlantic salmon. In our study, there was no scaling effect of sea lice life-history stage on critical swimming performance. Standardized β values indicate that chalimus and pre-adult stages have a more positive impact on U_{crit} compared to adult stages (Table 2) when parasite density and length are held constant. However, mean U_{crit} among the stages varied by 4.97 cm s^{-1} (0.20 BL s^{-1}) for mid-range densities and by 1.83 cm s^{-1} (0.23 BL s^{-1}) for high-range densities, indicating a small effect size due to lice stage. Although our results arose with higher infestation densities at the chalimus and pre-adult stages compared to adults, it is possible that the cost of infestation with high abundance of younger life-history stages is less than that of adult lice when present in low abundances.

Parasite density and swimming performance

Increasing sea lice density had a negative impact on U_{crit}, which was more pronounced when density was >0.7 lice cm^{-2} (i.e. high-range infestation density). Wagner et al. (2003) demonstrated a similar decline in U_{crit}, from 2.6 BL s^{-1} in unparasitised Atlantic salmon, to 2.4 and 2.1 BL s^{-1} in low and highly infested fish, respectively. In their trial, infestation intensity was low at 0.02 lice g^{-1}, and high at 0.13 lice g^{-1}, whereas our fish had a mean of 0.06 and 0.29 lice g^{-1} for our mid- and high-range infestation pressure groups (across all lice stages), respectively. However, we observed higher U_{crit} for infested fish, with means of 4.15 and 4.04 BL s^{-1} for mid- and high-range infestation levels. This relationship has also been observed in wild juvenile pink salmon *O. gorbuscha*, with high infestation intensities (1–5 adult females per fish; Mages & Dill 2010). Infestations of up to 0.04 adult sea lice g^{-1} are enough to induce physiological changes in the epithelial layer and at the molecular level (Nolan et al. 1999), which supports the evidence for reduced critical swimming speed at low infestation intensities. Future studies should compare the impact of sea lice infestation on performance of wild and farmed salmonids, to determine if selection pressures for infestation tolerance differ between the 2 as a result of their divergent ecological profiles (e.g. Gross 1998).

Implications for aquaculture management

Multiple environmental issues have led to increasing interest in culturing salmon at more exposed offshore sites (Holmer 2010). Locations considered suitable for farm sites are generally associated with high current velocities that improve production parameters such as fish welfare, the rapid dispersion of waste, oxygen and temperature conditions, and the avoidance of nutrient run-off from rivers (Benetti et al. 2010, Huntingford & Kadri 2013). Given lice infestation rates decline at high swimming velocities (1.4 BL sec^{-1}; Samsing et al. 2014), shifts towards faster current sites will become more prevalent. The implementation of off-shore salmon aquaculture requires a deep understanding of the welfare implications of being held at high water current velocities (Tudorache et al. 2013). To avoid negative physiological and welfare impacts, current velocities must not approach the limit for sustained swimming in Atlantic salmon (Jørgensen & Jobling 1993, Solstorm et al. 2015, 2016), a limit that shifts depending on biotic factors such as body size and parasite load. Although swimming speeds between 0.2 and 3.0 BL s^{-1} have been observed for salmon held in sea cages (Dempster et al. 2008, Korsøen et al. 2009), critical swimming speed has been experimentally determined at 1.9–8.3 BL s^{-1} for a range of size cohorts without infestation (Peake 2008). Here we have shown that critical swimming performance is reduced with increasing parasite load, and advise that considerations of acceptable current velocities for new farm sites take into account potential infestation pressure in combination with sustained swimming limits for the size group of salmon. Although present governmental legislation is strict for infestation prevalence and intensity in farms, particularly in Norway (Torrissen et al. 2013), observed sea lice loads are not always negligible (www.lusedata.no), and this may affect the ability of fish to hold their position in the water current if >0.02 lice cm^{-2} (Wagner et al. 2003, present study).

Implications for wild salmonids

The impact of increased infestation pressure arising from salmon farming on the health of wild salmonid populations remains a contentious issue (Krkošek et al. 2007, 2014, Jackson et al. 2013). Evidence for lethal thresholds are supported by results from monitoring programs of wild populations: surveys have rarely found post-smolts with sea lice loads >0.1 lice g^{-1}, while the threshold has recently been classified as 100% mortality for sea trout with >0.3 lice g^{-1} (Taranger et al. 2015). Even with low infestation intensity, our study shows that parasite loads of at least 0.01 lice g^{-1} (0.006 lice cm^{-2}), at any life-history stage, can reduce the critical swimming performance of post-smolts. This may contribute to their

declining survival with sea lice infestation, as the migration speed of post-smolts is vital for their survival through coastal areas where predation risk is high (Thorstad et al. 2012). Additionally, impaired swimming performance will reduce efficient predator avoidance and feeding behaviours in strong currents.

Increasing infestation level is likely to reduce the host's ability to sustain swimming performance in the pelagic phase, due to diminished aerobic capacity, osmoregulatory function and risk of secondary infestation (Tully & Nolan 2002, Wagner et al. 2003, Taranger et al. 2015), particularly after sea lice transition into the mobile pre-adult and adult stages. Hatchery-reared Atlantic salmon that were treated for sea lice before release had a higher rate of recapture compared to untreated salmon, suggesting that protection against infestation during out-migration facilitates survival (Krkošek et al. 2013, Skilbrei et al. 2013). Smolts treated against lice may grow more rapidly than untreated controls during their first year at sea (Skilbrei & Wennevik 2006), and untreated smolts stay longer at sea before they mature and return to the rivers (Vollset et al. 2014). Current evidence suggests that infestation with sea lice is the underlying cause of these observations, whereby our results on swimming performance could add to the multiple mechanisms of fitness decline with infestation.

Acknowledgements. We thank T. Vågseth and staff from the Institute of Marine Research at the Matre research station for their technical assistance and O. Skilbrei for comments on the manuscript. This research was funded by an Australian Research Council Future Fellowship grant (awarded to T. Dempster), and carried out partially within the Norwegian Research Council Centre for Research-Based Innovation on Exposed Aquaculture Operations (237790). This experiment was conducted according to the regulations stipulated by the Norwegian Regulation on Animal Experimentation (application ID: 7131).

LITERATURE CITED

Barber I, Hoare D, Krause J (2000) Effects of parasites on fish behaviour: a review and evolutionary perspective. Rev Fish Biol Fish 10:131–165

Benetti DD, Benetti GI, Rivera JA, Sardenberg B, O'Hanlon B (2010) Site selection criteria for open ocean aquaculture. Mar Technol Soc J 44:22–35

Bjørn PA, Finstad B (1998) The development of salmon lice (*Lepeophtheirus salmonis*) on artificially infected post smolts of sea trout (*Salmo trutta*). Can J Zool 76:970–977

Bowers JM, Mustafa A, Speare DJ, Conboy GA, Brimacombe M, Sims DE, Burka JF (2000) The physiological response of Atlantic salmon, *Salmo salar* L., to a single experimental challenge with sea lice, *Lepeophtheirus salmonis*. J Fish Dis 23:165–172

Brandal PO, Egidius E, Romslo I (1976) Host blood: a major food component for the parasitic copepod *Lepeophtheirus salmonis*, Krøyeri, 1838 (Crustacea: Caligidae). Norw J Zool 24:341–343

ä Brauner CJ, Sackville M, Gallagher Z, Tang S, Nendick L, Farrell AP (2012) Physiological consequences of the salmon louse (*Lepeophtheirus salmonis*) on juvenile pink salmon (*Oncorhynchus gorbuscha*): implications for wild salmon ecology and management, and for salmon aquaculture. Philos Trans R Soc Lond B 367:1770–1779

Brett JR (1964) The respiratory metabolism and swimming performance of young sockeye salmon. J Fish Res Board Can 21:1183–1226

ä Costello MJ (2006) Ecology of sea lice parasitic on farmed and wild fish. Trends Parasitol 22:475–483

ä Costello MJ (2009) How sea lice from salmon farms may cause wild salmonid declines in Europe and North America and be a threat to fishes elsewhere. Philos Trans R Lond Soc B 276:3385–3394

ä Dempster T, Juell JE, Fosseidengen JE, Fredheim A (2008) Behaviour and growth of Atlantic salmon (*Salmo salar* L.) subjected to short-term submergence in commercial scale sea-cages. Aquaculture 276:103–111

Field A, Miles J, Field Z (2012) Discovering statistics using R. Sage Publications, London

➤ Grimnes A, Jakobsen PJ (1996) The physiological effects of salmon lice (*Lepeophtheirus salmonis*) infection on post smolts of Atlantic salmon (*Salmo salar*). J Fish Biol 48: 1179–1194

ä Gross MR (1998) One species with two biologies: Atlantic salmon (*Salmo salar*) in the wild and in aquaculture. Can J Fish Aquat Sci 55(Suppl 1):131–144

ä Hammer C (1995) Fatigue and exercise tests with fish. Comp Biochem Physiol A Mol Integr Physiol 112:1–20

ä Hayashida K, Nii H, Tsuji T, Miyoshi K, Hamamoto S, Ueda H (2013) Effects of anesthesia and surgery on U_{crit} performance and MO_2 in chum salmon, *Oncorhynchus keta*. Fish Physiol Biochem 39:907–915

➤ Holmer M (2010) Environmental issues of fish farming in offshore waters: perspectives, concerns and research needs. Aquacult Environ Interact 1:57–70

Huntingford F, Kadri S (2013) Exercise, stress and welfare. In: Palstra AP, Planas JB (eds) Swimming physiology of fish: towards using exercise to farm a fit fish in sustainable aquaculture. Springer, London, p 161–174

ä Jackson D, Cotter D, Newell J, McEvoy S and others (2013) Impact of *Lepeophtheirus salmonis* infestations on migrating Atlantic salmon, *Salmo salar* L., smolts at eight locations in Ireland with an analysis of lice-induced marine mortality. J Fish Dis 36:273–281

ä Johansson D, Laursen F, Fernö A, Fosseidengen JE and others (2014) The interaction between water currents and salmon swimming behaviour in sea cages. PLoS ONE 9:e97635

ä Johnson SC, Albright LJ (1991) Development, growth, and survival of *Lepeophtheirus salmonis* (Copepoda: Caligidae) under laboratory conditions. J Mar Biol Assoc UK 71:425–436

ä Jónsdóttir H, Bron JE, Wootten R, Turnbull JF (1992) The histopathology associated with the pre-adult and adult stages of *Lepeophtheirus salmonis* on the Atlantic salmon, *Salmo salar* L. J Fish Dis 15:521–527

ä Jørgensen EH, Jobling M (1993) The effects of exercise on growth, food utilization and osmoregulatory capacity of juvenile Atlantic salmon, *Salmo salar*. Aquaculture 116: 233–246

ä Kolok AS (1999) Interindividual variation in the prolonged locomotor performance of ectothermic vertebrates: a comparison of fish and herpetofaunal methodologies and a brief review of the recent fish literature. Can J Fish Aquat Sci 56:700–710

ä Korsøen Ø, Dempster T, Fjelldal PG, Oppedal F, Kristiansen TS (2009) Long-term culture of Atlantic salmon (*Salmo salar* L.) in submerged cages during winter affects behaviour, growth and condition. Aquaculture 296:373–381

Krkošek M, Ford JS, Morton A, Lele S, Myers RA, Lewis MA (2007) Declining wild salmon populations in relation to parasites from farm salmon. Science 318:1772–1775

Krkošek M, Revie CW, Gargan PG, Skilbrei OT, Finstad B, Todd CD (2013) Impact of parasites on salmon recruitment in the Northeast Atlantic Ocean. Proc R Soc Lond B Biol Sci 280:20122359

Krkošek M, Revie CW, Finstad B, Todd CD (2014) Comment on Jackson et al. 'Impact of *Lepeophtheirus salmonis* infestations on migrating Atlantic salmon, *Salmo salar* L., smolts at eight locations in Ireland with an analysis of lice-induced marine mortality'. J Fish Dis 37:415–417

ä Mages PA, Dill LM (2010) The effect of sea lice (*Lepeophtheirus salmonis*) on juvenile pink salmon (*Oncorhynchus gorbuscha*) swimming endurance. Can J Fish Aquat Sci 67:2045–2051

ä Murray AG, Wardeh M, McIntyre KM (2016) Using the H-index to assess disease priorities for salmon aquaculture. Prev Vet Med 126:199–207

ä Nendick L, Sackville M, Tang S, Brauner CJ, Farrell AP (2011) Sea lice infection of juvenile pink salmon (*Oncorhynchus gorbuscha*): effects on swimming performance and post-exercise ion balance. Can J Fish Aquat Sci 68: 241–249

ä Nolan DT, Reilly P, Wendelaar Bonga SE (1999) Infection with low numbers of the sea louse *Lepeophtheirus salmonis* induces stress-related effects in postsmolt Atlantic salmon (*Salmo salar*). Can J Fish Aquat Sci 56: 947–959

ä O'Shea B, Mordue-Luntz AJ, Fryer RJ, Pert CC, Bricknell IR (2006) Determination of the surface area of a fish. J Fish Dis 29:437–440

Peake SJ (2008) Swimming performance and behaviour of fish species endemic to Newfoundland and Labrador: a literature review for the purpose of establishing design and water velocity criteria for fishways and culverts. Canadian Manuscript Report of Fisheries and Aquatic Sciences No. 2843, Oceans and Habitat Mangement Branch, Fisheries and Oceans Canada, Ottawa

➤ Plaut I (2001) Critical swimming speed: its ecological relevance. Comp Biochem Physiol Part A Mol Integr Physiol 131:41–50

R Core Team (2015) R: a language and environment for statistical computing. R Foundation for Statistical Computing, Vienna. www.R-project.org/

➤ Sackville M, Tang S, Nendick L, Farrell AP, Brauner CJ (2011) Pink salmon (*Oncorhynchus gorbuscha*) osmoregulatory development plays a key role in sea louse (*Lepeophtheirus salmonis*) tolerance. Can J Fish Aquat Sci 68:1087–1096

➤ Samsing F, Oppedal F, Johansson D, Bui S, Dempster T (2014) High host densities dilute sea lice *Lepeophtheirus salmonis* loads on individual Atlantic salmon, but do not reduce lice infection success. Aquacult Environ Interact 6:81–89

Schmit O, Mezquita F (2010) Experimental test on the use of

MS-222 for Ostracod anaesthesia: concentration, immersion period and recovery time. J Limnol 69:350–352

▶ Skilbrei OT, Wennevik V (2006) Survival and growth of sea-ranched Atlantic salmon, Salmo salar L., treated against sea lice prior to release. ICES J Mar Sci 63:1317–1325

▶ Skilbrei OT, Finstad B, Urdal K, Bakke G, Kroglund F, Strand R (2013) Impact of early salmon louse, Lepeophtheirus salmonis, infestation and differences in survival and marine growth of sea-ranched Atlantic salmon, Salmo salar L., smolts 1997–2009. J Fish Dis 36:249–260

▶ Solstorm F, Solstorm D, Oppedal F, Fernö A, Fraser TWK, Olsen RE (2015) Fast water currents reduce production performance of post-smolt Atlantic salmon Salmo salar. Aquacult Environ Interact 7:125–134

▶ Solstorm F, Solstorm D, Oppedal F, Olsen RE, Stien LH, Fernö A (2016) Not too slow and not too fast: water currents affect group structure, aggression and welfare in post-smolt Atlantic salmon Salmo salar. Aquacult Environ Interact 8:339–347

▶ Tadiso TM, Krasnov A, Skugor S, Afanasyev S, Hordvik I, Nilsen F (2011) Gene expression analyses of immune responses in Atlantic salmon during early stages of infection by salmon louse (Lepeophtheirus salmonis) revealed bi-phasic responses coinciding with the copepod-chalimus transition. BMC Genomics 12:141

▶ Taranger GL, Karlsen Ø, Bannister RJ, Glover KA and others (2015) Risk assessment of the environmental impact of Norwegian Atlantic salmon farming. ICES J Mar Sci 72:997–1021

▶ Thorstad EB, Whoriskey F, Uglem I, Moore A, Rikardsen AH, Finstad B (2012) A critical life stage of the Atlantic salmon Salmo salar: behaviour and survival during the smolt and initial post-smolt migration. J Fish Biol 81:500–542

▶ Thorstad EB, Todd CD, Uglem I, Bjørn PA and others (2015) Effects of salmon lice Lepeophtheirus salmonis on wild sea trout Salmo trutta—a literature review. Aquacult Environ Interact 7:91–113

▶ Torrissen O, Jones S, Asche F, Guttormsen A and others (2013) Salmon lice—impact on wild salmonids and salmon aquaculture. J Fish Dis 36:171–194

▶ Tucker CS, Sommerville C, Wootten R (2002) Does size really matter? Effects of fish surface area on the settlement and initial survival of Lepeophtheirus salmonis, an ectoparasite of Atlantic salmon Salmo salar. Dis Aquat Org 49:145–152

Tudorache C, de Boeck G, Claireaux G (2013) Forced and preferred swimming speeds of fish: a methodological approach. In: Palstra AP, Planas JB (eds) Swimming physiology of fish: towards using exercise to farm a fit fish in sustainable aquaculture. Springer, London, p 81–108

▶ Tully O, Nolan DT (2002) A review of the population biology and host-parasite interactions of the sea louse Lepeophtheirus salmonis (Copepoda: Caligidae). Parasitology 124:S165–S182

Videler JJ (1993) Fish swimming. Chapman & Hall, London

▶ Vollset KW, Barlaup BT, Skoglund H, Normann ES, Skilbrei OT (2014) Salmon lice increase age-at-maturity of Atlantic salmon (Salmo salar). Biol Lett 10:20130896

▶ Wagner GN, McKinley RS (2004) Anaemia and salmonid swimming performance: the potential effects of sub-lethal sea lice infection. J Fish Biol 64:1027–1038

▶ Wagner GN, McKinley RS, Bjørn PA, Finstad B (2003) Physiological impact of sea lice on swimming performance of Atlantic salmon. J Fish Biol 62:1000–1009

▶ Wagner GN, Fast MD, Johnson SC (2008) Physiology and immunology of Lepeophtheirus salmonis infections of salmonids. Trends Parasitol 24:176–183

4

Effect of water temperature on mortality of Pacific oysters *Crassostrea gigas* associated with microvariant ostreid herpesvirus 1 (OsHV-1 µVar)

Maximilian de Kantzow, Paul Hick*, Joy A. Becker, Richard J. Whittington

Faculty of Veterinary Science, The University of Sydney, 425 Werombi Road, Camden, NSW 2570, Australia

ABSTRACT: The ostreid herpesvirus 1 microvariant (OsHV-1 µVar) causes mass mortality of Pacific oysters *Crassostrea gigas*. Water temperature can directly influence the incidence of disease or correlate with seasonal changes in the environment and oyster physiology that modify the susceptibility of the oysters to disease. The effect of water temperature was evaluated in controlled laboratory conditions by intramuscular injection of OsHV-1 µVar after acclimation of 8 mo old spat and 17 mo old adult oysters at 4 different temperatures (14, 18, 22 and 26°C). Mortality was 84 and 77% at 26 and 22°C, respectively, compared to 23% at 18°C and nil at 14°C. There was a statistically significant interaction between the dose of OsHV-1 µVar and water temperature. At 18°C, mortality occurred exclusively at a dose of 10^6 OsHV-1 µVar genome copies per oyster whereas at the higher temperatures, oysters challenged with 10^3 copies per oyster also died. Mortality did not occur at 14°C and OsHV-1 µVar was detected in tissues of only 1% of the oysters after 14 d. When accounting for temperature and dose, spat (8 mo) were 2.7 times more likely to die than adults (17 mo). Our study confirms a direct effect of water temperature on infection and disease caused by OsHV-1 µVar. We identified a threshold water temperature of between 14 and 18°C below which productive infection does not occur and the requirement for a higher dose of OsHV-1 µVar to initiate infection at 18°C than at 22°C. These results have implications for predicting and managing disease outbreaks caused by OsHV-1 µVar.

KEY WORDS: Ostreid herpesvirus 1 · OsHV-1 µVar · Pacific oyster · *Crassostrea gigas* · Water temperature · Laboratory challenge · Dose-response · Disease susceptibility

INTRODUCTION

A microvariant genotype of the species *Ostreid herpesvirus 1* (OsHV-1) was first identified in 2008 in France (Segarra et al. 2010). This pathogen was shown to be associated with recurrent mass mortality events in the Pacific oyster *Crassostrea gigas*, with up to 100% mortality of juvenile oysters (Schikorski et al. 2011b, EFSA 2015). OsHV-1 is a member of the family *Malacoherpesviridae* within the order *Herpesvirales* (Davison et al. 2005, ICTV 2013). The World Organisation for Animal Health (OIE) defines micro-variant genotypes of OsHV-1 (OsHV-1 µVar) as those which are characterised by a deletion in the microsatellite locus upstream of open reading frame (ORF) 4, and several polymorphisms compared with the reference genotype (GenBank accession number AY509253) in ORF 4 (C region) and ORF 42/43 encoding an inhibitor of apoptosis. This definition incorporates several variations on the genotype first described as an OsHV-1 microvariant by Segarra et al. (2010).

Disease caused by OsHV-1 µVar has since been identified in several countries including Australia

*Corresponding author: paul.hick@sydney.edu.au

(Jenkins et al. 2013), New Zealand (Bingham et al. 2013), Ireland (Peeler et al. 2012, Clegg et al. 2014), The Netherlands (Gittenberger et al. 2016), Italy (Domeneghetti et al. 2014), and Spain (Aranguren et al. 2012, Roque et al. 2012), where it has caused serious disruption or complete cessation of Pacific oyster production (Paul-Pont et al. 2014, EFSA 2015). Additionally, OsHV-1 µVar has been detected in locations where the disease is absent (Dundon et al. 2011, Shimahara et al. 2012, Jee et al. 2013). The incidence of disease is strongly seasonal with occurrence limited from spring to autumn in waterways where OsHV-1 µVar is endemic (Oden et al. 2011, Peeler et al. 2012, Pernet et al. 2012, Paul-Pont et al. 2014, Renault et al. 2014).

Elevation of water temperature during the spring and summer was associated with mass mortality events in Australia, France and Ireland (Pernet et al. 2012, Clegg et al. 2014, Paul-Pont et al. 2014, Renault et al. 2014). A threshold temperature above which mortality occurred has been reported to be ~16°C in France and Ireland (Pernet et al. 2012, Clegg et al. 2014, Renault et al. 2014, Petton et al. 2015). Furthermore, in France a water temperature of 24°C, above which mortalities cease to occur, has been suggested as upper threshold temperature (Pernet et al. 2012). In New South Wales, Australia, disease occurred when the mean water temperature was between 19 and 24°C, although OsHV-1 µVar DNA was present when the water temperature was lower (Paul-Pont et al. 2013a, 2014). In some cases, a sudden change in water temperature by several degrees Celsius rather than reaching a mean threshold temperature has preceded mortality associated with OsHV-1 µVar (Clegg et al. 2014, Renault et al. 2014). However, disease has not always coincided with similar rapid changes in water temperature, suggesting that short- and long-term temperature changes and the presence of the virus are not the only triggers of mortality (Paul-Pont et al. 2013b, Clegg et al. 2014). The temperature of oysters present in the intertidal zone should not be assumed to be equal to the water temperature. As oysters may not be submerged at all times, temperature will also be influenced by the air temperature and solar radiation (Helmuth 1999).

Similar to expression of disease, Petton et al. (2013) found that transmission of the virus most efficiently occurred in the range of 16–22°C and did not occur at 13°C. Other studies showed that an OsVH-1 infected tissue preparation remained infectious for 54 h at 16°C compared to 33 h at 25°C (Martenot et al. 2015) and that OsHV-1 in seawater retained infectivity for

2 d at 20°C (Hick et al. 2016). Previously infected oysters showed high survival, and low quantities of OsHV-1 µVar DNA were detected when held at <14°C, but high mortality and increased quantity of OsHV-1 µVar DNA occurred when these oysters were subsequently exposed to 21°C (Pernet et al. 2015). This might indicate that a subclinical or latent infection with OsHV-1 µVar occurs at low temperature (Pernet et al. 2015, Petton et al. 2015).

While a seasonal increase in water temperature above 16–18°C is associated with recurrent disease outbreaks caused by OsHV-1 µVar, a range of concurrent seasonal changes in the host and environment may be more important in the pathobiology of the disease. Seasonal changes include the availability and type of food and the metabolic changes of the oyster (Soletchnik et al. 2006). The incidence of disease has varied with factors including: seasonal changes in reproductive effort and energy balance (Pernet et al. 2012, 2014); feeding (Evans et al. 2015); and management practices such as farming structures, immersion time and the density of oysters (Pernet et al. 2012, Paul-Pont et al. 2013a, Normand et al. 2014, Petton et al. 2015, Whittington et al. 2015a). Furthermore, the immune function of oysters is altered by water temperature (Green et al. 2014a,b).

Defining the effect of water temperature on disease caused by OsHV-1 µVar is not possible from field observations alone. The aim of this study was to assess the direct effect of water temperature on the outcome of OsHV-1 µVar infection under controlled laboratory conditions. Infection by intramuscular injection enabled administration of a measured dose of OsHV-1 µVar DNA to all oysters at the same time and eliminated variation due to factors that influence virus transmission.

MATERIALS AND METHODS

Oysters

Two batches of triploid Pacific oysters *Crassostrea gigas* were produced in a Tasmanian hatchery and grown under commercial conditions on fixed long lines in the Shoalhaven River, New South Wales, Australia. Each batch was certified to be OsHV-1 µVar free by a government laboratory, and this estuary is considered free of OsHV-1 µVar (Herbert 2011). The 2 batches were transferred to the laboratory in January 2015 at the ages of 8 mo (spat, Batch ID: SPL14B, length 61.6 ± 8.4 mm [mean ± SD], weight 8.5 ± 3.7 g) or 17 mo (adults, Batch ID:

SPL13B, 48.9 ± 6.4 mm, 11.6 ± 4.6 g). The batch of adult oysters had a greater mass but shorter shell length because of a deeper cup shape compared to younger oysters, reflecting previous growing conditions. Thus, rapidly growing young oysters and slower growing older oysters from commercial farming conditions were tested.

Husbandry and acclimation

The experiment was conducted in a temperature-controlled level 2 physical containment aquatic animal facility (University of Sydney, Camden Campus). The air temperature was maintained at 22°C with a 12 h photoperiod for the duration of the trial. The oysters were housed in 15 l plastic tanks, 6 of which were connected in parallel in a 250 l recirculation system; there were 4 separate systems. Each system was attached to an independent bio-filter (Fluval 406) and a UV unit, both of which were removed for 2 h d^{-1} for feeding. The systems contained artificial seawater (ASW) with a salinity of 30 ppt prepared from dechlorinated water and artificial sea salt (Red Sea Salt). Water temperature was maintained for each system using thermostatically controlled aquarium heaters (AquaOne) and in-line chiller units (DBA-110, Daeil). Temperature data loggers (Thermocron) recording the temperature every 30 min were placed in the sump and in tanks 1 and 4 of each system. Salinity, total ammonia nitrogen and pH were measured every second day (API salt water master test kit, Aquarium Pharmaceuticals). Negative control oysters that were injected with an inoculum that was free of OsHV-1 μVar were maintained separately in 5 l of aerated ASW in vessels floated in each sump of each system so as to provide the same water temperature profile for acclimation and during the trial. Exposure to light, concentration of feed and stocking density were similar to that of the challenged oysters without the potential for waterborne exposure to OsHV-1 μVar. Water quality for the negative controls was maintained by exchanging 50% of the ASW each day.

Oysters were fed commercial algal mix *Isochrysis*, *Pavlova*, *Thalassiosira weissflogii* and *Tetraselmis* (Shellfish diet 1800, Reed Mariculture). A maintenance ration was calculated based on the oysters' wet meat weight according to the manufacturer's instructions and corresponded to 12.5 ml d^{-1} for each recirculation system and 1 ml d^{-1} for the controls (Reed Mariculture).

Oysters were purged without feed in ASW overnight and 10 adult (SPL13B) and 10 spat (SPL14B) oysters were randomly assigned to each of the 6 tanks in each recirculation system. A polypropylene mesh divider separated the age groups within each tank. Six oysters of each age were randomly assigned as negative controls at each temperature. The water temperature in each recirculation system was initially 20°C for 2 d, reflecting water temperature in the field, before being adjusted by 1°C d^{-1} until reaching the target temperatures: 14, 18, 22 and 26°C. Oysters were held at their target temperature for at least 48 h prior to inoculation.

OsHV-1 μVar inoculum

A fresh stock of OsHV-1 μVar was prepared immediately prior to the trial by amplification of an OsHV-1 μVar stock in donor oysters (SPL13B, n = 8). The source of the stock virus was oysters from the Georges River (NSW, Australia) that were naturally infected with OsHV-1 μVar in 2011. The the identity of the μVar virus was confirmed by sequencing. A 0.2 μm filtered tissue homogenate was prepared as described by Paul-Pont et al. (2015). The virus stock was archived at −80°C with 10% v/v glycerol and 10% v/v foetal bovine serum (Sigma). The donor oysters were relaxed in 50 g MgCl$_2$ (Sigma) per liter of dechlorinated tap water prior to injection of 100 μl of a 1/100 dilution of the virus stock in sterile ASW into the adductor muscle. The donor oysters were maintained in ASW at 20°C for 4 d without feed.

The entire mantle and gill was excised from the donor oysters and homogenised by stomaching at maximum speed for 2 min (MiniMix, Crown Scientific) with 10 ml ASW. The homogenate was made up to 1/10 w/v with additional sterile ASW and centrifuged at 1000 × g for 10 min at 4°C. The supernatant was filtered to 0.2 μm using syringe filters (Minisart, Sartorius). The OsHV-1 μVar genome was quantified by qPCR, and the clarified and filtered tissue homogenate was stored at 4°C for 1 d prior to administration. Immediately prior to use the homogenate was diluted to 1 in 50 and 1 in 5000 v/v with 0.2 μm-filtered, sterile ASW to produce inocula with a dose of 10^6 and 10^3 OsHV-1 μVar genome equivalent copies per 100 μl. A negative control inoculum was prepared by the same procedure from a cryopreserved tissue homogenate prepared from oysters that tested negative for OsHV-1 μVar DNA.

Inoculation of oysters with OsHV-1

Oysters were relaxed in a 50 g $MgCl_2$ (Sigma) per liter of dechlorinated water maintained at the temperature of each treatment group. The oysters were challenged by injection of 10^6 or 10^3 OsHV-1 µVar genome equivalent copies per oyster in a total volume of 100 µl into the adductor muscle using a 25 gauge needle and 1 ml syringe. Half of the tanks in each system were assigned the high dose of 10^6 OsHV-1 µVar genome copies and half the low dose of 10^3 OsHV-1 µVar genome copies. The same inoculum was administered to all oysters regardless of size. The negative control homogenate was diluted 1 in 50 w/v with 0.2 µm filtered-sterile ASW prior to injection of 100 µl into the adductor muscle of the control oysters.

Experimental design

For each of the 4 target water temperatures, 1 recirculation system consisting of 6 tanks was maintained. Each tank contained 10 oysters from the adult batch that were physically separated with a mesh divider from 10 oysters from the spat batch. There were 120 oysters at each temperature, half of which were challenged with the higher dose and the other with the lower dose of OsHV-1 µVar. There were 12 negative control oysters for each water temperature, half from each age batch.

Observation and sample collection

Every oyster was inspected every 12 h, and dead oysters were removed and stored at −80°C. Oysters were considered to be dead when they were open, non-responsive to disturbance of the tank and did not retract the mantle following stimulation with a 22 gauge needle. Feeding was monitored by visual observation of the rate at which algae were cleared from the water. A random sample of 20 live oysters was taken at 7 d post challenge to establish the prevalence of the virus in the 14 and 18°C treatment groups. All surviving oysters were sampled at the end of the trial, 14 d post challenge.

Detection and quantification of OsHV-1 µVar

Tissues were processed according to previously described methods (Paul-Pont et al. 2013a, Whittington

et al. 2015b). Briefly, a sample of gill and mantle was excised (pooled weight 0.08–0.12 g) and placed in a 2 ml screw cap tube with 1 ml of ultrapure water (Ultrapur) and 0.4 g of 0.1 mm silica zirconia beads (Daintree Scientific). The samples were homogenised using a TissueLyser II (Qiagen) for 120 s; the tubes were inverted for a second 120 s homogenisation cycle and then centrifuged at 900 × g for 10 min. The supernatant was stored at −80°C. Nucleic acids were purified from 50 µl of supernatant with the Mag-MAX-96 Viral Isolation Kit (Ambicon, Life Technologies) and a magnetic particle processor (MagMAX Express 96 Applied Biosystems, Life Technologies) according to manufacturer's directions with the AM1836 deep-well standard program (Ambicon, Life Technologies). Purified nucleic acids were eluted in 75 µl elution buffer and stored at −20°C.

A real-time quantitative PCR (qPCR) assay targeting ORF 99 of the OsHV-1 µVar genome (B region) that encodes an inhibitor of apoptosis was modified from the method described by Martenot et al. (2010). Duplicate 25 µl reactions were prepared with the Ag Path ID One-step RT-PCR kit (Life Technologies), BF and B4 primers (900 nm), OsHV-1B probe (250 nm) and 5 µl of undiluted nucleic acid template from the sample of each individual oyster.

The thermocycling program conducted with a real time PCR system (Mx3000P, Stratagene) was 95°C for 10 min followed by 40 cycles of 95°C for 15 s and 60°C for 45 s. Each plate of 96 reactions included a 10-fold dilution series of the plasmid pOSHV1-Breg for quantification (Paul-Pont et al. 2013b); duplicate OsHV-1 µVar positive and negative tissue homogenates (extraction controls); and nil template controls. The ROX normalised FAM fluorescence signal was analysed with a propriety algorithm (Stratagene) to identify positive results and estimate the quantity of OsHV-1 µVar DNA in samples in which both replicates had a logarithmic increase above the threshold value. Samples which satisfied the criteria for detection, but produced a threshold cycle (Ct) value greater than the range of the standard curve, were indicated as below limit of quantification (BLOQ) and were not included in quantitative analyses.

Statistical analysis

Survival analyses were conducted according to the method described in Dohoo et al. (2003). The survival time was defined as the number of hours from OsHV-1 µVar exposure to mortality for oysters that tested positive for OsHV-1 µVar DNA by qPCR. Kaplan-

Table 1. Prevalence and quantity of OsHV-1 µVar DNA (range of OsHV-1 µVar genome equivalents mg^{-1} tissue) in apparently healthy Pacific oysters *Crassostrea gigas* at 7 and 14 d post challenge and total mortality after 14 d, with viral concentration detected in oyster tissues at the time of death. Prevalence calculation: at 7 d post challenge, from samples randomly selected from all the apparently healthy oysters in each group; at 14 d post challenge, considering all surviving oysters. BLOQ: below the limit of quantification. For dead oysters, the number of oysters with greater than 10^4 OsHV-1 µVar genome copies mg^{-1} is given in parentheses; a threshold associated with mortality Oden et al. (2011). n/s: not sampled

| Water tempe- rature (°C) | Apparently healthy oysters | | | | | | Dead oysters | | |
| | 7 d post challenge | | | 14 d post challenge | | | | | |
	n	Prevalence (%)	OsHV-1 µVar quantity	n	Prevalence (%)	OsHV-1 µVar quantity	n	Mortality (%)	OsHV-1 µVar quantity
14	20	30.0	BLOQ	109	1.0	BLOQ	1	1.0	0
18	20	65.0	BLOQ – 1.20×10^3	90	38.6	$0 – 3.39 \times 10^2$	23	22.8	$1.12 \times 10^2 – 7.08 \times 10^4$ (7)
22	0	n/s	–	34	2.7	1.26×10^3	98	76.9	$6.17 \times 10^1 – 6.76 \times 10^5$ (79)
26	0	n/s	–	20	10.0	BLOQ	112	83.8	$5.37 \times 10^2 – 1.58 \times 10^7$ (106)

Meier survival curves were plotted for each set of replicate treatment groups and for each experiment factor. Curves were compared using the Mantel-Cox log-rank test. The replicate treatment groups were not significantly different (all p > 0.05), so they were combined for further statistical analysis. Oysters that were alive at the end of the trial were considered to be censored at the time of sampling, regardless of OsHV-1 DNA test status. A Cox proportional hazards (PH) model with shared frailty to account for clustering was fitted for treatment groups at 18, 22 and 26°C using the Breslow method for tied events (STATA statistical software Version 13). A significant interaction between water temperature and dose was identified and these factors could not be separated. The assumptions of the PH model were assessed graphically using log-cumulative hazard plots for each parameter in the model and statistically using the scaled Schoenfeld residuals. The goodness of fit was confirmed by the Nelson-Aalen cumulative hazard and Cox-Snell residue plots. The model was assessed under conditions of complete positive and negative correlation.

The quantities of OsHV-1 µVar DNA were log$_{10}$ transformed to satisfy the assumption of normality and compared across different treatments using a restricted maximum likelihood, linear mixed model (REML) (GenStat, 16th edition, 2015, VSN International). Water temperature, dose of virus and age of the oyster were considered as fixed effects, with tank as a random effect. Pairwise comparisons of estimated means used the least significant differences, with significance accepted at p < 0.05.

RESULTS

The water temperature remained within 1°C of the target temperature for each system after the acclima-tion period, with total ammonia nitrogen < 0.5 mg l^{-1} and pH 8.2. Mortality did not occur, nor was OsHV-1 µVar DNA detected in the control oysters at any temperature (n = 48). The control oysters were considered to be healthy based on their opening and closing responses throughout the trial.

There was no mortality associated with the OsHV-1 µVar challenge in oysters maintained at 14°C (Table 1), although 1 oyster died at 72 h post challenge and tested negative for OsHV-1 µVar DNA. The total mortality at 18°C was 23% (Table 1), but none of the oysters injected with the lower dose of OsHV-1 µVar died at this temperature. The mortality at 22 and 26°C was 77 and 84%, respectively.

All oysters that died at 18, 22 and 26°C tested positive for OsHV-1 µVar DNA with between 6.17×10^1 and 1.58×10^7 OsHV-1 µVar genome copies per mg of tissue. The quantity of OsHV-1 µVar DNA detected at the time of death was higher than in survivors and was not affected by the initial dose (p = 0.51). The quantity of viral DNA at the time of death for oysters at 18°C was approx. 5- and 6-fold lower than at 22 and 26°C, respectively (Fig. 1a). Similarly, accounting for temperature, there was 2.5 times more OsHV-1 µVar DNA detected in the spat batch of oysters compared to the adult batch (Fig. 1b).

There were a low prevalence and low quantities of OsHV-1 µVar DNA in oysters surviving after 14 d (Table 1). A single surviving oyster at 14°C, one at 22°C and 2 survivors at 26°C tested positive whereas 38.6% of 90 surviving oysters at 18°C were positive for OsHV-1 µVar DNA (not shown). The dose strongly influenced the OsHV-1 µVar status of survivors at 18°C; 82% of those receiving the higher dose remained positive compared to 8.2% of those receiving the lower dose. The prevalence of OsHV-1 µVar DNA 7 d after challenge in live oysters at 14°C was 30%, and the quantity of virus was very low. At this

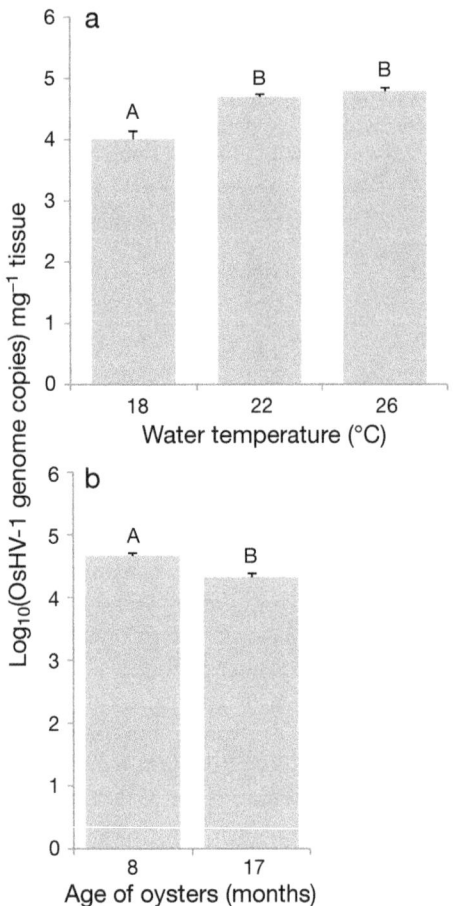

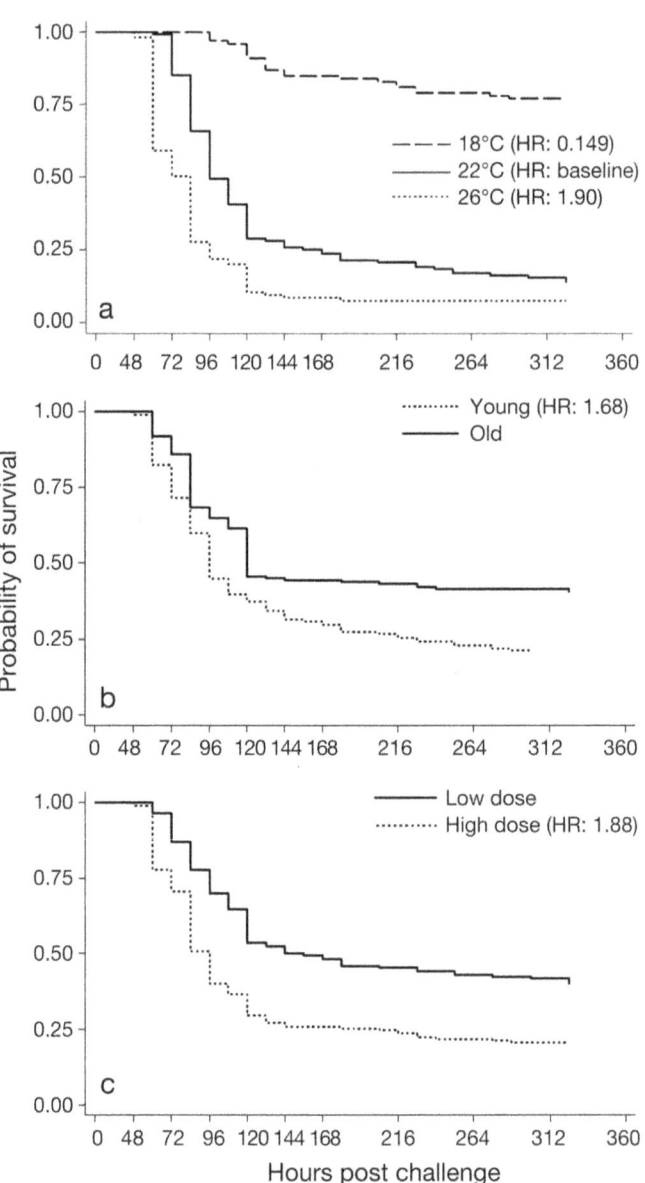

Fig. 1. Quantity (mean + SE) of OsVH-1 µVar DNA (log-transformed no. of OsHV-1 µVar genome copies mg^{-1} tissue) in Pacific oysters *Crassostrea gigas* that died from OsHV-1 µVar challenge at (a) different water temperatures, accounting for age, and (b) different ages, accounting for water temperature. The dose of OsHV-1 µVar did not affect the quantity of viral DNA in the oysters that died (p = 0.51). Groups not sharing uppercase letters are significantly different (p < 0.05, restricted maximum likelihood, linear mixed model)

Fig. 2. Kaplan-Meier survival curves and hazard ratio (HR) for Pacific oysters *Crassostrea gigas* challenged with OsHV-1 µVar for each tested experimental factor: (a) water temperature (oysters at 14°C were negative for OsHV-1 µVar and there was a single mortality), (b) age (8 mo old spat vs. 17 mo old adults), and (c) OsHV-1 µVar dose level (low and high dose: 10^3 and 10^6 OsHV-1 µVar genome copies mg^{-1} tissue, respectively)

time, OsHV-1 µVar DNA was detected in 100% of live oysters at 18°C that were administered the higher dose, but in only 33% of the oysters challenged with the low dose (n = 10).

Mortality occurred in a single episode with no secondary peak as might have occurred if there was secondary infection caused by cohabitation with diseased oysters (Fig. 2). Initially, water temperature, age and dose of OsHV-1 µVar were examined in isolation of the other factors, and there was no evidence of any violations of the assumptions of proportional hazards (Fig. 2). On any given day, all oysters held at 26°C were 1.9 times more likely to die with OsHV-1 µVar infection compared to all oysters held at 22°C (i.e. the hazard ratio [HR]; 95% confidence interval

[CI]: 1.45–2.50). The median survival time was 84 and 96 h for oysters held at 26 and 22°C, respectively (Fig. 2A). There was a protective effect for oysters at 18°C, with a hazard ratio <1 when compared to oysters held at 22°C (Fig. 2A). Over the experimental period, all oysters from the spat batch were 1.68 (1.29–2.18, 95% CI) times more likely to die with high quantities of OsHV-1 µVar DNA compared to

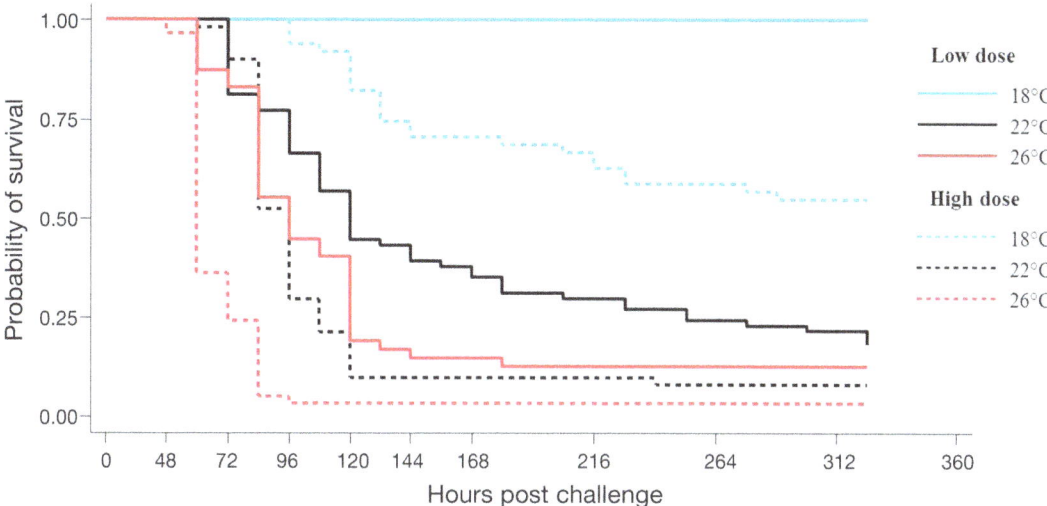

Fig. 3. Kaplan-Meier survival curves for Pacific oysters *Crassostrea gigas* challenged with OsHV-1 μVar showing the interaction between water temperature (18, 22, and 26°C) and dose (10^3 and 10^6 genome copies mg^{-1} tissue)

the older oysters (Fig. 2B). There was nearly twice the risk (HR 1.88; 1.45–2.44 95% CI) of dying with OsHV-1 μVar infection in all oysters given the high dose of OsHV-1 μVar compared to the low dose exposure (Fig. 2C). The median survival time for all oysters given the high dose of OsHV-1 μVar was 96 h compared to 144 h for the low dose.

The Cox PH model enabled a multivariable analysis. There was a significant interaction between water temperature and dose, indicating that the effect of water temperature was different depending on the challenge dose. The factors could not be separated and were considered as 6 separate treatment groups (Fig. 3). The oysters at 14°C were excluded from the PH model because mortality related to OsHV-1 μVar infection did not occur. The final model included the treatment groups shown in Table 2 and was significant (p < 0.001). On any given day, oysters held at 26°C were 2.1 and 3.6 times more likely to die with OsHV-1 μVar infection compared to oysters at 22°C when challenged with low and high doses, respectively (Table 2). On any given day, oysters given a high dose of OsHV-1 were 2.1 and 3.8 times more likely to die with OsHV-1 μVar infection compared to oysters challenged with the lower dose when held at 22 and 26°C, respectively (Table 2). There was a protective effect for oysters at 18°C compared to the higher temperatures (Fig. 3, Table 2). The assumption of proportional hazards was not violated for age, as it did not interact with either water temperature or dose of OsHV-1 μVar (all p > 0.1). Accounting for water temperature and dose, the young oysters were 2.7 (2.1–3.6 95% CI) times likely to succumb to OsHV-1 μVar infection com-

Table 2. Hazard ratios for Pacific oysters *Crassostrea gigas* challenged with OsHV-1 μVar for each tested experimental factor (temperatures 18, 22, 26°C; doses 10^3 [low] and 10^6 [high] OsHV-1 μVar copies per oyster; age [8 mo old spat]). There was a significant interaction between water temperature and dose of OsHV-1 μVar. –: no mortality observed for the low dose at 18°C

Factor	Level	Hazard ratio	
		Point estimate	95% CI
Low dose	18 vs. 22°C	–	–
	26 vs. 22°C	2.1	1.5–2.9
High dose	18 vs. 22°C	0.2	0.1–0.2
	26 vs. 22°C	3.6	2.7–4.9
22°C	High vs. low dose	2.1	1.7–2.7
26°C	High vs. low dose	3.8	2.5–5.8
Age	Spat (8 mo)	2.7	2.1–3.6

pared to older oysters (Table 2). Residual analysis determined the PH model fit to the data and did not detect a significant effect when considered under conditions of complete positive and complete negative censoring.

DISCUSSION

This study demonstrates the important direct effect of water temperature on the expression of disease caused by OsHV-1 μVar when administered by injection into the adductor muscle under controlled laboratory conditions. The intramuscular injection challenge method was first described by Schikorski et al. (2011b) in France and developed further in Australia by Paul-Pont et al. (2015). This allowed pathogenesis to be examined separately from factors that affect transmission and bypasses the external defences of the oyster. The quantity of OsHV-1 µVar DNA in the inoculum administered to each oyster was measured, and the infection was not dependent on the transmission of variable quantities of virus present in the environment or shed by donor animals, as is the case with cohabitation models (Schikorski et al. 2011a, Petton et al. 2013, Evans et al. 2015). Unlike field studies, the laboratory environment enabled water temperature to be evaluated *per se*, with control of other factors that are also seasonally variable such as photoperiod and feed availability. The infection model did not introduce variability in factors other than controlled OsHV-1 µVar exposure that might have influenced pathogenesis, such as co-infection with *Vibrio* spp. (Petton et al. 2013, 2015). Our experiment used 2 batches of oysters to demonstrate a marked and direct effect of water temperature on the outcome of challenge with OsHV-1 µVar. The results are consistent with effects of water temperatures observed during disease in field conditions (Paul-Pont et al. 2014, Whittington et al. 2015a). Further work is required to replicate the experiment, test temperatures between 14 and 18°C in more detail and evaluate oysters with different genetic and life histories.

This study supports the observations of others who have shown a strong association between the onset of mortality and water temperature that is associated with warmer seasons (Petton et al. 2013, Clegg et al. 2014, Paul-Pont et al. 2014, Renault et al. 2014, Pernet et al. 2015). In the present study, mortality related to challenge with OsHV-1 µVar did not occur in oysters at 14°C. OsHV-1 µVar infection may not have been able to establish at this temperature because the virus could not replicate or was cleared by an innate immune response. Alternatively, a subclinical infection may have been established that was not detected by qPCR (Pernet et al. 2015). The risk of disease in these oysters if water temperature subsequently increased was not determined. However, at 18°C, OsHV-1 µVar DNA persisted for at least 14 d in oysters that had been given the low dose, suggesting that viral replication may have occurred without inducing mortality. The incubation period for OsHV-1 µVar-associated mortality in naïve oysters exposed to infected oysters from the field at 17.5°C was 6 d (Petton et al. 2013). A procedure for demonstrating latent infection that involves warming oysters to 21°C has been reported (Pernet et al. 2015, Petton et al. 2015).

In this trial, mortality associated with OsHV-1 µVar replication, as indicated by increased quantities of viral DNA at the time of death, occurred at 18°C but not at 14°C in conditions that were otherwise equal. A strong, direct effect of water temperature on the host–pathogen interaction was indicated, with increased mortality at higher water temperatures. The quantity of viral DNA that was detected in oysters at the time of death was higher with each increase in temperature from 18 to 26°C. The disease outcome might also indicate a deleterious increase in the immune response to the virus that is promoted by the higher temperature (Green et al. 2014a). The effect of both water temperature and the rate of temperature change on the immune responses of the oyster, and the transcriptional changes in the virus genome present opportunities for further research.

In the present study, the threshold dose of OsHV-1 µVar for disease expression was temperature dependent, and for a given dose, total mortality was higher at higher water temperatures. A clear implication of these results is that the dose of virus is likely to be a very important risk factor for an outbreak of disease caused by OsHV-1 µVar. There is an apparent difference between the threshold temperatures at which OsHV-1 µVar-associated mortalities occur in France (16°C) compared to Australia (21°C) (Pernet et al. 2012, Jenkins et al. 2013, Paul-Pont et al. 2014, Renault et al. 2014). It is possible that this difference is due to virus strain differences or differences in oyster genetics between Europe and Australia. Australian and French OsHV-1 µVar isolates are closely related on available genetic sequence; however, comparative analysis has focused on phylogenetic similarities and not virulence traits (Jenkins et al. 2013). Alternatively, the differences might reflect different methods for measuring and reporting water temperature. There can be considerable diel variation in water temperature in the shallow near-shore waters where oysters are grown (Kaplan et al. 2003, Lucas et al. 2006). Thus, an average water temperature obtained in a main channel might not be completely informative.

The risk for OsHV-1 µVar epidemics developing in aquaculture will increase with temperatures greater than 18°C due to a lower lethal dose of OsHV-1 µVar and a higher replication rate of the virus. Thus, both the water temperature and the amount of OsHV-1 µVar present will influence the time of the first outbreak of the season. The dose effect also offers a possible explanation for the sometimes high variation in mortality of oysters of the same age in disease outbreaks in the field (Paul-Pont et al. 2013a, 2014, Clegg et al. 2014). While the disease risk will be influenced by many factors, including those that affect transmission, the importance of dose and temperature was highlighted by this laboratory investigation, which controlled for extrinsic factors that complicate field observations. Disease mitigation strategies have been proposed based on field observations of factors that influence disease (Petton et al. 2015, Whittington et al. 2015a). Improved disease management and prediction of epidemics can be achieved based on better knowledge of the direct influence of water temperature and initial quantity of OsHV-1 μVar on pathogenesis and total mortality.

The environments where oysters are cultivated, including the hydrodynamics of the bay or estuary, are suspected to play a major role in the transmission of the virus through water connectivity (Pernet et al. 2012, Paul-Pont et al. 2013b). Hydrodynamics also influence the local water temperature in the areas where oysters are grown. Generally, these are relatively shallow, likely to be warmed by the sun and subjected to greater temperature fluctuation than the deeper areas that are well mixed by currents (Pernet et al. 2012). Oysters grown on long-lines or in trays in the intertidal zone experience time out of water at low tide and may therefore be relatively heated or cooled compared to the ambient water temperature during this time (Paul-Pont et al. 2013a). The observations of oysters immersed at a static temperature may not provide a complete indication of the disease outcomes expected in cultivated oysters subject to the fluctuations of an inter-tidal environment.

CONCLUSION

This study confirms that water temperature directly influences mortality caused by OsHV-1 μVar infection of oysters, consistent with the pattern of recurrent seasonal disease outbreaks. Higher water temperatures, in the range of 18–26°C probably increase the risk of epidemics associated with OsHV-1 μVar due to a lower lethal dose and a higher viral replication rate. Mortality of oysters injected with OsHV-1 μVar in laboratory conditions depended on the temperature of the water when other factors were controlled. Mortality did not occur and there was no evidence of viral replication at 14°C. Yet the same challenge resulted in mortality exceeding 75% with a high rate of viral replication at 22 and 26°C. The influence of the dose of OsHV-1 was evident at the intermediate temperature of 18°C, where the lower dose did not result in mortality and the higher dose resulted in less severe disease compared to that which occurred at higher temperatures.

Acknowledgements. This project was funded by the Fisheries Research and Development Corporation and The University of Sydney. Oysters were donated by Shellfish Culture, Tasmania and were grown by Leon and Angela Riepsamen, Goodnight Oysters, Greenwell Point. Thanks to Alison Tweedie, Olivia Evans, Navneet Dhand, Craig Kristo, Stuart Glover and Slavicka Patten for technical assistance.

LITERATURE CITED

Aranguren R, Costa MM, Novoa B, Figueras A (2012) Detection of herpesvirus variant (OsHV-1μvar) in Pacific oysters (*Crassostrea gigas*) in Spain and development of a rapid method for its differential diagnosis. Bull Eur Assoc Fish Pathol 32:24–29

Bingham P, Brangenberg N, Williams R, van Andel M (2013) Marine and freshwater investigation into the first diagnosis of ostreid herpesvirus type 1 in Pacific oysters. Surveillance (Wellingt) 40:20–24

ä Clegg TA, Morrissey T, Geoghegan F, Martin SW, Lyons K, Ashe S, More SJ (2014) Risk factors associated with increased mortality of farmed Pacific oysters in Ireland during 2011. Prev Vet Med 113:257–267

ä Davison AJ, Trus BL, Cheng NQ, Steven AC and others (2005) A novel class of herpesvirus with bivalve hosts. J Gen Virol 86:41–53

Dohoo IR, Martin W, Stryhn H (2003) Veterinary epidemiologic research, 2nd edn. VER Inc, Charlottetown

ä Domeneghetti S, Varotto L, Civettini M, Rosani U and others (2014) Mortality occurrence and pathogen detection in *Crassostrea gigas* and *Mytilus galloprovincialis* close-growing in shallow waters (Goro lagoon, Italy). Fish Shellfish Immunol 41:37–44

ä Dundon WG, Arzul I, Omnes E, Robert M and others (2011) Detection of type 1 Ostreid herpes variant (OsHV-1 μvar) with no associated mortality in French-origin Pacific cupped oyster *Crassostrea gigas* farmed in Italy. Aquaculture 314:49–52

EFSA (European Food Safety Authority) (2015) Oyster mortality. EFSA J 13:4122–4159

ä Evans O, Hick P, Dhand N, Whittington RJ (2015) Transmission of Ostreid herpesvirus-1 in *Crassostrea gigas* by cohabitation: effects of food and number of infected donor oysters. Aquacult Environ Interact 7:281–295

ä Gittenberger A, Voorbergen-Laarman MA, Engelsma MY (2016) Ostreid herpesvirus OsHV-1 μVar in Pacific oysters *Crassostrea gigas* (Thunberg 1793) of the Wadden Sea, a UNESCO world heritage site. J Fish Dis 39:105–109

ä Green TJ, Montagnani C, Benkendorff K, Robinson N, Speck P (2014a) Ontogeny and water temperature influences the antiviral response of the Pacific oyster, *Crassostrea gigas*. Fish Shellfish Immunol 36:151–157

ä Green TJ, Robinson N, Chataway T, Benkendorff K, O'Connor W, Speck P (2014b) Evidence that the major hemolymph protein of the Pacific oyster, *Crassostrea gigas*, has antiviral activity against herpesviruses. Antiviral Res 110:168–174

ä Helmuth B (1999) Thermal biology of rocky intertidal mussels: quantifying body temperatures using climatological data. Ecology 80:15–34

Herbert B (2011) Aquatic animal health. Anim Health Surveill Q Rep 16:8–9

ä Hick P, Evans O, Looi R, English C, Whittington RJ (2016)

Stability of Ostreid herpesvirus-1 (OsHV-1) and assessment of disinfection of seawater and oyster tissues using a bioassay. Aquaculture 450:412–421

ICTV (International Committee on Taxonomy of Viruses) (2013) Virus taxonomy 2013 release. http://ictvonline.org/virusTaxonomy.asp

► Jee BY, Lee SJ, Cho MY, Lee SJ and others (2013) Detection of Ostreid Herpesvirus 1 from adult Pacific oysters *Crassostrea gigas* cultured in Korea. Fish Aquat Sci 16: 131–135

► Jenkins C, Hick P, Gabor M, Spiers Z and others (2013) Identification and characterisation of an ostreid herpesvirus-1 microvariant (OsHV-1 μ-var) in *Crassostrea gigas* (Pacific oysters) in Australia. Dis Aquat Org 105:109–126

► Kaplan DM, Largier JL, Navarrete S, Guiñez R, Castilla JC (2003) Large diurnal temperature fluctuations in the nearshore water column. Estuar Coast Shelf Sci 57:385–398

► Lucas LV, Sereno DM, Burau JR, Schraga TS and others (2006) Intradaily variability of water quality in a shallow tidal lagoon: mechanisms and implications. Estuar Coasts 29:711–730

► Martenot C, Oden E, Travaille E, Malas JP, Houssin M (2010) Comparison of two real-time PCR methods for detection of ostreid herpesvirus 1 in the Pacific oyster *Crassostrea gigas*. J Virol Methods 170:86–89

► Martenot C, Denechere L, Hubert P, Metayer L and others (2015) Virulence of Ostreid herpesvirus 1 μVar in sea water at 16 °C and 25 °C. Aquaculture 439:1–6

► Normand J, Blin JL, Jouaux A (2014) Rearing practices identified as risk factors for ostreid herpesvirus 1 (OsHV-1) infection in Pacific oyster *Crassostrea gigas* spat. Dis Aquat Org 110:201–211

► Oden E, Martenot C, Berthaux M, Travaille E, Malas JP, Houssin M (2011) Quantification of ostreid herpesvirus 1 (OsHV-1) in *Crassostrea gigas* by real-time PCR: determination of a viral load threshold to prevent summer mortalities. Aquaculture 317:27–31

► Paul-Pont I, Dhand NK, Whittington RJ (2013a) Influence of husbandry practices on OsHV-1 associated mortality of Pacific oysters *Crassostrea gigas*. Aquaculture 412–413:202–214

► Paul-Pont I, Dhand NK, Whittington RJ (2013b) Spatial distribution of mortality in Pacific oysters *Crassostrea gigas*: reflection on mechanisms of OsHV-1 transmission. Dis Aquat Org 105:127–138

Paul-Pont I, Evans O, Dhand NK, Rubio A, Coad P, Whittington RJ (2014) Descriptive epidemiology of mass mortality due to Ostreid herpesvirus-1 (OsHV-1) in commercially farmed Pacific oysters (*Crassostrea gigas*) in the Hawkesbury River estuary. Aquacult 422–423:146–159

► Paul-Pont I, Evans O, Dhand NK, Whittington RJ (2015) Experimental infections of Pacific oyster *Crassostrea gigas* using the Australian ostreid herpesvirus-1 (OsHV-1) μVar strain. Dis Aquat Org 113:137–147

► Peeler EJ, Reese AR, Cheslett DL, Geoghegan F, Power A, Thrush MA (2012) Investigation of mortality in Pacific oysters associated with Ostreid herpesvirus-1 μVar in the Republic of Ireland in 2009. Prev Vet Med 105:136–143

► Pernet F, Barret J, Le Gall P, Corporeau C and others (2012) Mass mortalities of Pacific oysters *Crassostrea gigas* reflect infectious diseases and vary with farming prac-

tices in the Mediterranean Thau lagoon, France. Aquacult Environ Interact 2:215–237

► Pernet F, Lagarde F, Le Gall P, D'Orbcastel ER (2014) Associations between farming practices and disease mortality of Pacific oyster *Crassostrea gigas* in a Mediterranean lagoon. Aquacult Environ Interact 5:99–106

► Pernet F, Tamayo D, Petton B (2015) Influence of low temperatures on the survival of the Pacific oyster (*Crassostrea gigas*) infected with ostreid herpes virus type 1. Aquaculture 445:57–62

► Petton B, Pernet F, Robert R, Boudry P (2013) Temperature influence on pathogen transmission and subsequent mortalities in juvenile Pacific oysters *Crassostrea gigas*. Aquacult Environ Interact 3:257–273

► Petton B, Boudry P, Alunno-Bruscia M, Pernet F (2015) Factors influencing disease-induced mortality of Pacific oysters *Crassostrea gigas*. Aquacult Environ Interact 6: 205–222

► Renault T, Bouquet AL, Maurice JT, Lupo C, Blachier P (2014) Ostreid herpesvirus 1 infection among Pacific oyster (*Crassostrea gigas*) spat: relevance of water temperature to virus replication and circulation prior to the onset of mortality. Appl Environ Microbiol 80: 5419–5426

► Roque A, Carrasco N, Andree KB, Lacuesta B and others (2012) First report of OsHV-1 microvar in Pacific oyster (*Crassostrea gigas*) cultured in Spain. Aquaculture 324–325:303–306

► Schikorski D, Faury N, Pepin JF, Saulnier D, Tourbiez D, Renault T (2011a) Experimental ostreid herpesvirus 1 infection of the Pacific oyster *Crassostrea gigas*: kinetics of virus DNA detection by q-PCR in seawater and in oyster samples. Virus Res 155:28–34

► Schikorski D, Renault T, Saulnier D, Faury N, Moreau P, Pepin JF (2011b) Experimental infection of Pacific oyster *Crassostrea gigas* spat by ostreid herpesvirus 1: demonstration of oyster spat susceptibility. Vet Res 42:27

► Segarra A, Pepin JF, Arzul I, Morga B, Faury N, Renault T (2010) Detection and description of a particular Ostreid herpesvirus 1 genotype associated with massive mortality outbreaks of Pacific oysters, *Crassostrea gigas*, in France in 2008. Virus Res 153:92–99

► Shimahara Y, Kurita J, Kiryu I, Nishioka T and others (2012) Surveillance of type 1 Ostreid herpesvirus (OsHV-1) variants in Japan. Fish Pathol 47:129–136

► Soletchnik P, Faury N, Goulletquer P (2006) Seasonal changes in carbohydrate metabolism and its relationship with summer mortality of Pacific oyster *Crassostrea gigas* (Thunberg) in Marennes-Oléron bay (France). Aquaculture 252:328–338

► Whittington RJ, Dhand NK, Evans O, Paul-Pont I (2015a) Further observations on the influence of husbandry practices on OsHV-1 μVar mortality in Pacific oysters *Crassostrea gigas*: age, cultivation structures and growing height. Aquaculture 438:82–97

► Whittington RJ, Hick P, Evans O, Rubio A, Alford B, Dhand N, Paul-Pont I (2015b) Protection of Pacific oyster (*Crassostrea gigas*) spat from mortality due to ostreid herpesvirus 1 (OsHV-1 μVar) using simple treatments of incoming seawater in land-based upwellers. Aquaculture 437:10–20

5

Effects of water spinach *Ipomoea aquatica* cultivation on water quality and performance of Chinese soft-shelled turtle *Pelodiscus sinensis* pond culture

Wei Li[1,2], Huaiyu Ding[3], Fengyin Zhang[4], Tanglin Zhang[1], Jiashou Liu[1], Zhongjie Li[1,*]

[1]State Key Laboratory of Freshwater Ecology and Biotechnology, Institute of Hydrobiology, Chinese Academy of Sciences, Wuhan 430072, China
[2]School of Aquatic and Fishery Sciences, University of Washington, Box 355020, Seattle, Washington 98195-5020, USA
[3]Jiangsu Engineering Laboratory for Breeding of Special Aquatic Organisms, Huaiyin Normal University, Huaian 223300, China
[4]College of Life Sciences, Jianghan University, Wuhan 430056, China

ABSTRACT: The Chinese soft-shelled turtle *Pelodiscus sinensis* is a highly valued freshwater species cultured in China. A 122 d experiment was conducted to assess the effects of water spinach *Ipomoea aquatica* cultivation in floating beds on water quality, and growth performance and economic return of *P. sinensis* cultured in ponds. Two treatments, each in triplicate, with and without *I. aquatica* cultivation were designed. Results showed that the levels of total ammonia nitrogen (TAN), total nitrogen (TN), total phosphorus (TP), chlorophyll *a* (chl *a*) and turbidity in treatments with *I. aquatica* cultivation (IAC) were significantly ($p < 0.05$) lower than those in treatments without *I. aquatica* (control). Mean TN and TP concentrations in the IAC treatment were 27.9 and 42.5%, respectively, lower than in the control treatment at the end of the experiment. The presence of *I. aquatica* also has a positive effect on the performance of *P. sinensis*. Although no significant difference was found in specific growth rate (SGR) between the 2 treatments, mean survival rates, production and net income were significantly higher in the IAC treatment compared to the control ($p < 0.05$). These results suggest that *I. aquatica* cultivation in the pond system of turtles has a synergistic effect on overall economic return and is effective at improving turtle growth performance and water quality.

KEY WORDS: *Pelodiscus sinensis* · *Ipomoea aquatica* · Aquaculture effluent · Growth performance · Economic return

INTRODUCTION

The Chinese soft-shelled turtle *Pelodiscus sinensis* is the most common turtle species cultured in China (Li et al. 2013). Soft-shelled turtles have long been a part of traditional Chinese cuisine and are much sought after as a high-priced food item in restaurants. In the past, the great bulk of soft-shelled turtles consumed were wild-caught, leading to an over exploitation of wild turtles that resulted in a 'turtle cri-

sis' in the country (Li et al. 2013). However, aquaculture developments over the past 4 decades have enabled the increasing demand for turtles to be met (Shi et al. 2008). The soft-shelled turtle is a highly valued aquaculture species in China with the 2014 yield reaching 341 000 t (Fisheries Administration of the People's Republic of China 2015). However, low survival rates and poor flesh quality of farmed turtles often compromise production and economic benefits (Ding 2000). Generally, the culture environment is an important factor influencing the survival, growth and quality of cultured species (Neal et al. 2010, Brito et al. 2014, Price et al. 2015). Also, the impacts of aquaculture effluents on water quality and diversity of adjacent natural water bodies have been addressed by many studies (Xie et al. 2004, De Silva 2012, Herbeck et al. 2014). Hence, improving culture environment is not only beneficial for better performance and flesh quality of the species, but also eases the negative impacts on the environment.

There are many approaches to improving a culture environment, and phytoremediation, especially using aquatic plants, is currently attracting much attention. It is considered promising because of its low cost, non-intrusiveness and safety (Li et al. 2009, Mook et al. 2012). Aquatic plants can effectively reduce total nitrogen, total phosphorus and chemical oxygen demand (Sooknah & Wilkie 2004). Water spinach *Ipomoea aquatica* Forsk. is an important crop in oriental cuisine and is also used in animal feeds. Water spinach grows well in moist soil or wetland systems and is commonly found creeping along muddy stream banks or floating in freshwater marshes and ponds (Li & Li 2009, Jampeetong et al. 2012). Floating bed technology makes it possible to plant this species on the water surface of fishponds, which not only improves the culture environment, but also provides additional economic benefits (Li & Li 2009).

Previous studies have demonstrated that planting *I. aquatica* in fishponds can efficiently remove nutrients and improve water quality (Li & Li 2009, Dai et al. 2012). Accordingly, we hypothesized that *I. aquatica* co-cultured in the pond systems of turtles could maintain water quality and provide better economic return. In this study, we compared selected water quality parameters, nutrient contents, and growth, survival and production of *P. sinensis* in ponds with and without *I. aquatica* cultivation. The aim of this study was to examine the effects of *I. aquatica* cultivation on water quality, growth performance and economic returns of *P. sinensis* cultured in earthen ponds. Results from this study will be useful to bring about efficient management and to maximize profitability of *P. sinensis* pond culture.

MATERIALS AND METHODS

Study location and experimental procedure

The experiment was conducted at a turtle breeding base in Honghu, Hubei Province, Central China, from June to October 2013. Juvenile *Pelodiscus sinensis*

(mean [$\pm$SD] body weight 27.7 $\pm$ 2.1 g) were obtained from a turtle farm in Honghu and were acclimated for 6 wk in three 1 ha ponds that were filled to a depth of 100 cm prior to the commencement of the trials. During this period, turtles were maintained under an ambient day:night light cycle, in pond water temperature that ranged from 20.8 to 29.6°C and were fed daily with a commercial feed (46% crude protein).

Two experimental treatments, with and without *Ipomoea aquatica* cultivation, each in triplicate, were used in the study. The experiment was conducted in 6 randomly selected ponds each of 3000 m^2 (100 × 30 m) at the turtle breeding base. After acclimation for 6 wk, 3000 healthy turtles (mean body weight 55.6 $\pm$ 3.3 g) were selected (without considering the sex) and stocked randomly in each pond with a water depth of 1.2 m. There were no significant differences (*t*-test, p > 0.05) in body weight between the treatments. After stocking, 150 kg of silver carp *Hypophthalmichthys molitrix* (mean body weight 153.7 $\pm$ 9.9 g) and 100 kg of bighead carp *Aristichthys nobilis* (mean body weight 205.3 $\pm$ 15.2 g) were stocked into each pond to regulate water quality, as this is a common practice that is performed in Chinese turtle farming. The culture procedures complied with the animal welfare laws of the Government of China and the ethical rules of the Institutional Animal Care and Use Committee of the Institute of Hydrobiology (Approval ID: Keshuizhuan 08529).

Prior to stocking, water samples (500 ml) were collected from all experimental ponds. Turbidity, total ammonia nitrogen (TAN), total nitrogen (TN), total phosphorus (TP) and chemical oxygen demand (COD$_{Mn}$) were determined according to standard methods (APHA 1992). In the *I. aquatica* cultivation (IAC) treatment, each pond was planted with *I. aquatica* (cut stems 20.4 $\pm$ 3.1 cm) on floating beds of bamboo frames (80 × 1.25 m) with 20 cm plant spacing and 25 cm row pitch. A polyethylene net with 3 cm mesh was fixed on the frame. Three floating beds were set up in each pond, and the total area covered approximated 10% of each pond.

P. sinensis juveniles were reared for 4 mo and fed twice daily at 08:00 to 09:00 h and 16:00 to 17:00 h with a commercial dry pellet feed of 46% crude protein content (Hangzhou Haihuang Feed Development) and minced fillet of silver carp *H. molitrix* (approximately 15% crude protein content). The daily ration consisted of approximately 2% of the body weight of commercial feed and 6% body weight of minced fillet. This ration was chosen based on observations made during the acclimation period, and on the amount of feed that the juvenile turtles would consume in 1 h. The body weight of the turtles was measured every 2 wk, and uneaten food was estimated. The daily feed allowance was adjusted every 2 wk based on these observations.

During the experimental period, natural water from a nearby lake was supplemented when the water level of the ponds decreased below 10.0 cm. From July to October, *I. aquatica* was harvested every month, and weight and revenue recorded. At the end of the trial, the number and individual weights of turtles were recorded. Growth performance of the turtles was evaluated using survival, relative weight gain (WG), specific growth rate (SGR), absolute growth rate (AGR) and protein efficiency ratio (PER) (Nuwansi et al. 2016).

Water was sampled monthly at around 10:00 h for each pond and monitored for turbidity (HACH 2100Q) *in situ*. TAN, TN, TP and COD_{Mn} were analyzed following the standard methods for the examination of water and wastewater (APHA 1992). Chlorophyll *a* (chl *a*) concentration was determined using a fluorometer with methanol extraction of the filtrate (Holm-Hansen & Riemann 1978).

Economic analysis

The economic analysis followed that of Gomes et al. (2006). The total revenue (in US dollars; $1.00 US = $6.20 RMB) included that of harvested *P. sinensis*, silver and bighead carp and *I. aquatica*, sold at $14.52 kg^{-1}, $0.48 kg^{-1}, $1.13 kg^{-1} and $0.32 kg^{-1} in the local market, respectively. The total cost consisted of the cost of *P. sinensis* juveniles, fish juveniles, *I. aquatica* stems, commercial feed, minced fillet, floating bed, labor (4 mo), labor for picking *I. aquatica* and pond rent. *P. sinensis* juveniles, *I. aquatica* stems, commercial feed and minced fillet costs were $1.61 ind.$^{-1}$, $0.48 kg^{-1}, $1.29 kg^{-1} and $0.39 kg^{-1}, respectively. The costs of silver and bighead carp juveniles were $0.77 kg^{-1} and $1.29 kg^{-1}, respectively. Floating bed costs included the material, assembly and fixing in ponds. Labor costs included 4 farmers that handled routine work and their benefits in compliance with the laws of the country. Pond rental was $483.87 ha^{-1}.

Statistical analysis

All data are expressed as mean ± SD. A *t*-test was used to determine differences between treatments. Differences were considered significant at p < 0.05.

All analyses were performed using SPS 16.0 statistical package (SPSS).

RESULTS

Water quality

Trends in TAN, TN, TP, COD_{Mn}, chl *a* concentrations and turbidity between the 2 treatments did not differ significantly (p < 0.05) at the commencement of the experiment. A high degree of fluctuation in TAN and TN levels was observed during the culture period. TAN concentrations varied significantly between treatments and sampling dates, with lower mean values in IAC treatments compared to the control treatments in October (p < 0.05; Fig. 1A). TN concentrations in the IAC treatment were significantly lower than those in the control treatment in the last 4 mo (p < 0.05; Fig. 1B). The mean TN concentration in the IAC treatment was 27.9% lower compared to the control at the end of the experiment. TP concentrations were significantly affected by the presence of *Ipomoea aquatica* (p < 0.05; Fig. 1C); the mean TP concentration in the IAC treatment was 42.5% lower compared to the control treatment at the end of the experiment. The COD_{Mn} levels did not significantly differ between the 2 treatments during the whole experimental period (p > 0.05; Fig. 1D), however chl *a* concentration and turbidity were significantly influenced by *I. aquatica*, with lower mean values in IAC ponds compared to the controls from August to October, respectively (p < 0.05; Fig. 1E,F).

Growth performance

For *Pelodiscus sinensis*, mean survival rate in the IAC treatment was significantly higher than that in the control treatment (p < 0.05). No significant differences between the 2 treatments were observed in initial weight, final weight, WG, SGR or AGR, indicating that growth of *P. sinensis* was not significantly influenced by *I. aquatica* cultivation. Significant differences were observed in mean production in the IAC treatment, which was significantly higher (2496.7 ± 58.4 kg ha^{-1}) than that in the control (2180.0 ± 49.7 kg ha^{-1}) treatment (p < 0.05). There was no significant difference in PER between the 2 treatments (p > 0.05) (Table 1).

The survival rate of silver carp was not significantly different between treatments (p > 0.05), but final weight, WG, SGR and AGR in the IAC treatment was

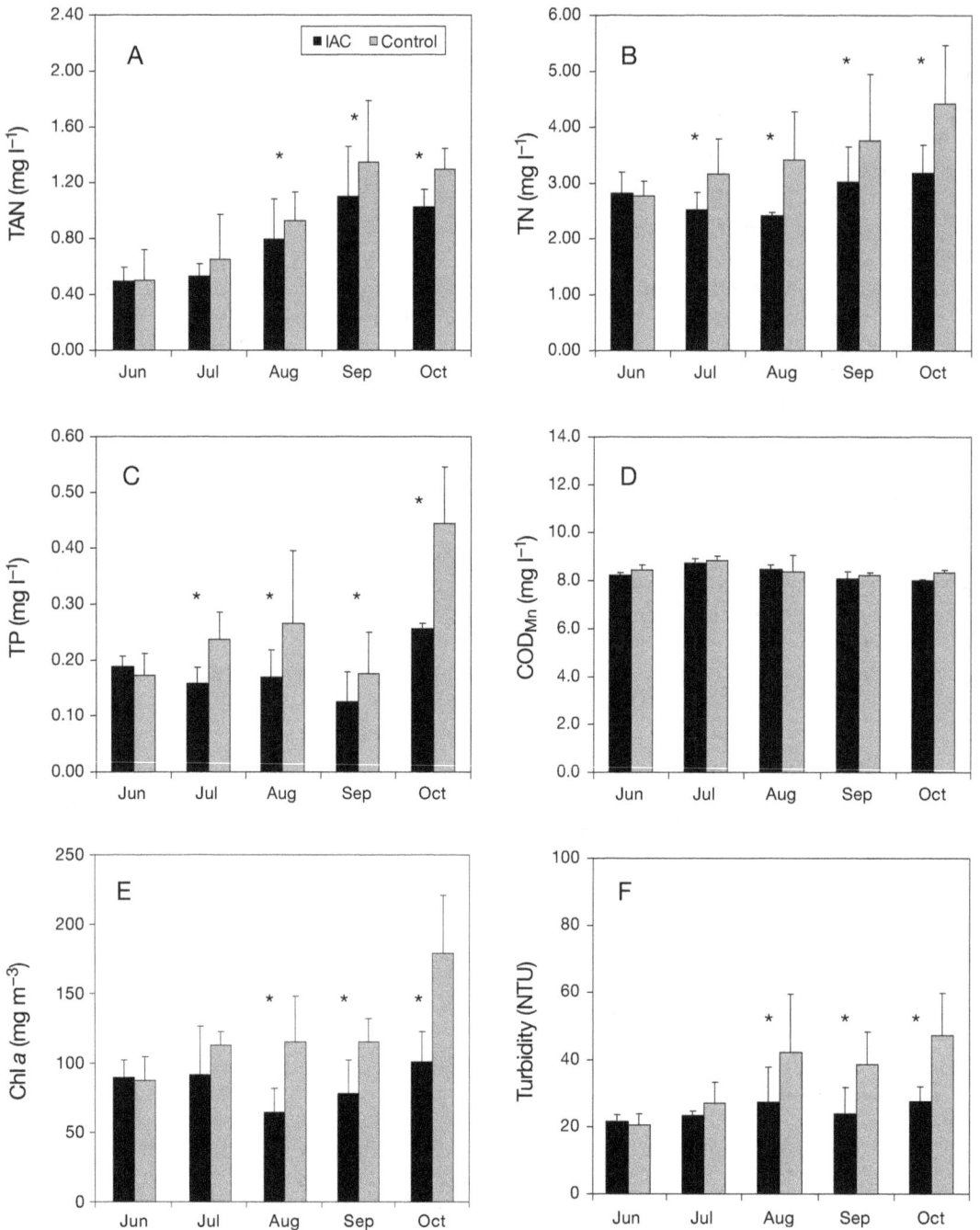

Fig. 1. Fluctuations in (A) total ammonia nitrogen (TAN), (B) total nitrogen (TN), (C) total phosphorus (TP), (D) chemical oxygen demand (COD_{Mn}), (E) chlorophyll *a* (chl *a*) concentrations and (F) turbidity in the Chinese soft-shelled turtle *Pelodiscus sinensis* ponds during the 4 mo experimental period. Values are means (±SD) of 3 replicate ponds per sampling time in each treatment. IAC: *Ipomea aquatica* treatment. (*) indicates significant difference between treatments (p < 0.05)

significantly lower than those in the control treatment (p < 0.05). For bighead carp, mean survival rate in the IAC treatment was higher than that in the control treatment, while final weight, WG, SGR and AGR in the IAC treatment was lower than those in the control treatment, but these differences were not significant (p > 0.05) (Table 1).

Economic analysis

The cost–benefit analysis of *P. sinensis* juveniles cultured in ponds for 4 mo under 2 treatments is shown in Table 2. Total revenue and total costs were both higher in the IAC treatment than in the control and net income was higher in the IAC treatment. The

Table 1. Growth performance of Chinese soft-shelled turtle *Pelodiscus sinensis* and 2 carp species (silver carp *Hypophthalmichthys molitrix* and bighead carp *Aristichthys nobilis*) cultured for 4 mo under 2 different treatments in aquaculture ponds. Data are means ± SD of 3 replicate ponds. For each row, means with different superscript letters are significantly different from each other (p < 0.05). IAC: *Ipomoea aquatica* treatment; WG: weight gain; SGR: specific growth rate; AGR: absolute growth rate; PER: protein efficiency ratio

Parameter	Turtle		Silver carp		Bighead carp	
	IAC	Control	IAC	Control	IAC	Control
Survival (%)	81.3 ± 2.7[a]	72.4 ± 1.6[b]	93.3 ± 3.3[a]	90.1 ± 2.8[a]	91.7 ± 2.9[a]	89.3 ± 2.2[a]
Initial weight (g)	55.3 ± 2.1[a]	57.1 ± 2.6[a]	153.7 ± 9.9[a]	153.7 ± 9.9[a]	205.3 ± 15.2[a]	205.3 ± 15.2[a]
Final weight (g)	307.2 ± 5.5[a]	301.1 ± 3.7[a]	680.6 ± 23.6[a]	777.0 ± 20.9[b]	785.9 ± 25.3[a]	869.0 ± 30.7[a]
WG (g)	251.9 ± 5.1[a]	244.0 ± 3.3[a]	526.9 ± 15.4[a]	623.3 ± 14.9[b]	580.6 ± 19.8[a]	663.7 ± 23.9[a]
SGR (% d^{-1})	1.40 ± 0.03[a]	1.36 ± 0.02[a]	1.22 ± 0.04[a]	1.33 ± 0.02[b]	1.10 ± 0.03[a]	1.18 ± 0.03[a]
AGR (g d^{-1})	2.06 ± 0.11[a]	2.0 ± 0.09[a]	4.32 ± 0.16[a]	5.11 ± 0.21[b]	4.76 ± 0.19[a]	5.44 ± 0.23[a]
Production (kg ha^{-1})	2496.7 ± 58.4[a]	2180.0 ± 49.7[b]	2066.7 ± 63.7[a]	2276.7 ± 77.4[b]	1168.3 ± 82.1[a]	1260.0 ± 79.2[a]
PER	0.81 ± 0.04[a]	0.74 ± 0.03[a]				

Table 2. Cost and return analysis of Chinese soft-shelled turtle *Pelodiscus sinensis* cultured for 4 mo under 2 different treatments in ponds. IAC: *Ipomoea aquatica* treatment. Values are in US dollars ($1.00 US = $6.20 RMB)

Parameter	Treatment	
	IAC	Control
Total revenue	**12 871.38**	**10 251.06**
P. sinensis harvest (kg)	749.00	654.00
Revenue kg^{-1} of *P. sinensis*	14.52	14.52
Revenue of *P. sinensis*	10875.48	9496.08
Fish (bighead carp and	620.00 +	683.00 +
silver carp) harvest (kg)	350.50	378.00
Revenue kg^{-1} of fish	0.48/1.13	0.48/1.13
Revenue of fish	693.66	754.98
I. aquatica harvest (kg)	4069.50	0
Revenue kg^{-1} of *I. aquatica*	0.32	0.32
Revenue of *I. aquatica*	1302.24	0
Total cost	**9067.14**	**8140.77**
P. sinensis juveniles	4838.71	4838.71
Fish juveniles	232.25	232.25
I. aquatica stems	217.74	0
Commercial feed	1347.26	1229.59
Minced fillet	1212.47	1106.67
Floating bed investment	129.03	0
Labor (4 mo)	387.10	387.10
Cost for picking *I. aquatica*	387.10	0
Pond rent	145.16	145.16
Other costs	170.32	201.29
Net income	**3804.24**	**2110.29**
Net income ha^{-1}	**12 680.80**	**7034.30**

revenue obtained from *P. sinensis* was the dominant component of the total revenue in both treatments. In addition, the revenue from *I. aquatica* in the IAC treatment was $1302.24, accounting for 10.12% of total revenue. The major components of total cost were *P. sinensis* juveniles and feeds, with turtles representing 53.37 and 59.44% of the total cost for the

IAC and control treatments, respectively, and feeds representing 28.23 and 28.70%.

DISCUSSION

Water quality management is an important component of the production and quality of aquatic animals in aquaculture, and aquaculture effluent is recognized as a serious global problem because of its influence on surrounding watersheds (Othman et al. 2013, Zhang et al. 2014). Previous studies have reported that floating bed technology using aquatic vegetables in culture ponds is an effective method of removing nutrients and improving water quality (Li et al. 2007, Li & Li 2009, Song et al. 2009). In the present study, the concentrations of TAN, TN, TP, chl *a* and turbidity were significantly affected (lowered) by *Ipomoea aquatica* cultivation. The mean TN and TP concentrations in the IAC treatment were 27.9 and 42.5% lower compared to the control, respectively, suggesting that this aquatic vegetable can efficiently remove nitrogen and phosphorus nutrients from aquaculture wastewater in turtle culture ponds. Firstly, *I. aquatica* can accumulate nutrients in its leaves, stems and other tissues. In addition, periphyton that grow on the flourishing roots of *I. aquatica* may also contribute to absorbing nitrogen and phosphorus (Wu et al. 2014, Basílico et al. 2016). The low concentrations of chl *a* and turbidity in the IAC treatment were attributed to inhibition of algal growth due to the low concentration of TN and TP in the water and the reduced illumination induced by *I. aquatica* cultivation. Therefore, we suggest that using floating beds of aquatic vegetables in turtle culture ponds has the potential to improve the aquaculture environment and decrease wastewater discharge.

Silver carp and bighead carp, 2 planktivorous filter-feeding fishes, are usually co-cultured in turtle rearing ponds, and this is a common practice in Chinese turtle farming. These fish do not compete with the turtles for food, and also play a role in improving the water quality by feeding on phyto- and zooplankton. Accordingly, in our study, silver carp and bighead carp of the same biomass were stocked in both treatments. The study did not permit us to ascertain the interrelationship (if any) that existed between *I. aquatica* cultivation and growth of the 2 carp species. However, the efficiency of phosphorus and nitrogen absorption by *I. aquatica* could be explained further through comparing the biomass of the carps. The weight gain of 2 carp species was higher in the control ponds, suggesting that more nitrogen and phosphorus was accumulated by the carps in that treatment. In addition, concentrations of nitrogen and phosphorus in the control treatment were higher than those in the IAC treatment at the end of the experiment, indicating a reduction in the rate of nitrogen and phosphorus absorption by *I. aquatica* actually higher than 27.9 and 42.5%, respectively.

In the present study, there were no significant differences in WG, SGR, AGR and PER between the 2 treatments, but there was a difference in survival, suggesting that *I. aquatica* cultivation significantly affected survival rather than the growth of *Pelodiscus sinensis* in ponds. Although this is the first study to report the effects of *I. aquatica* cultivation on *P. sinensis* growth and survival in pond culture, similar observations on cultured fish species, such as crucian carp *Carassius auratus* (Chen et al. 2010), have been made. However, different results were found in fish–seaweed cohabiting systems (Lombardi et al. 2006, Portillo-Clark et al. 2013). Portillo-Clark et al. (2013) reported that cohabitation with green feather alga *Caulerpa sertularioides* had a significant positive effect on biomass and growth rather than survival rates of juvenile yellow leg shrimp *Farfantepenaeus californiensis*. The inconsistency was probably dependent on the difference of the interaction between co-cultured plants and animals. Similarly, shrimp and turtle cultivation generates residues that are very rich in nutrients, which provides favorable conditions for the co-cultured plants and the associated microbial community (Duan et al. 1995, Portillo-Clark et al. 2013). Green feather algae offer an abundant living surface for microbial communities because of their highly 'feathered' morphology, which may promote the development of microbial biofilms that can enrich the diet of shrimp and lead to increased growth (Portillo-Clark et al. 2013). In our study, *I. aquatica* was cultivated on floating beds, which can strip nutrients from water through the roots rather than provide food for turtles. This is likely the main reason why significant differences did not exist in the growth of turtles in our study. However, the floating beds were able to provide a platform for turtles to bask in the light, which is advantageous for killing bacteria and reducing the occurrence of disease (Huang & Ben 2001). This, in addition to a better aquaculture environment because the N and P nutrients were absorbed by *I. aquatica*, could explain the higher survival in the IAC treatment.

Several factors affect the economic return of aquaculture systems such as yield, sale prices, feed costs, fingerlings or juveniles costs, system investments and operating costs (Muangkeow et al. 2007, de Oliveira et al. 2012). In the present study, total revenues and total costs were highest in the IAC treatment, as was the net income. These findings suggest that *I. aquatica* cultivation in the turtle ponds was effective at improving profitability. Economic analysis in this study mainly emphasized yield, turtle juveniles and feed costs. The feed costs accounted for 28.23 and 28.70% of the total costs in the IAC and control teatments, respectively, which is lower than that for cage production of various fish species, which range from 30 to 60% of total costs (Huguenin 1997, Silva et al. 2007). Compared to the control treatment, revenue of *I. aquatica* in the IAC treatment is one of the most important components (10.12%) of the total revenue. Thus, *I. aquatica*, as a co-product in turtle culture ponds, can generate extra economic benefits. The present findings suggest that farmers can use floating beds to cultivate *I. aquatica* in turtle culture ponds for maximizing profitability.

In summary, this study has demonstrated that co-culture of *I. aquatica* in *P. sinensis* culture ponds has positive effects on nutrient stripping and economic returns, which allows for optimal utilization of the available pond space, contributing substantially to improving the economic output and water quality. Further studies incorporating higher covered areas and in combination with stocking densities might maximize economic outputs and nutrient removal further.

Acknowledgements. Authors acknowledge the inputs of Prof. Sena S. De Silva, Deakin University, Australia and Casey Clark, University of Washington, USA in improving the manuscript. We are grateful to the anonymous reviews and give special thanks to the editor for valuable critique of the manuscript. The present research was supported by a grant from the National Scientific and Technological Supporting Program of China (No. 2012BAD27B02-6), the opening foundation of the Jiangsu Engineering Laboratory for Characteristic Aquatic Species Breeding (No. CASB1306), and the Science and Technology Service Network Initiative of the Chinese Academy of Sciences (KFJ-SW-STS-145).

LITERATURE CITED

APHA (American Public Health Association) (1992) Standard methods for the examination of water and wastewater, 18th edn. American Public Health Association, Washington, DC

ä Basílico G, de Cabo L, Magdaleno A, Faggi A (2016) Poultry effluent bio-treatment with Spirodela intermedia and periphyton in mesocosms with water recirculation. Water Air Soil Pollut 227:1–11

ä Brito LO, Arantes R, Magnotti C, Derner R, Pchara F, Olivera A, Vinatea L (2014) Water quality and growth of Pacific white shrimp Litopenaeus vannamei (Boone) in coculture with green seaweed Ulva lactuca (Linaeus) in intensive system. Aquacult Int 22:497–508

Chen J, Meng S, Hu G, Qu J, Fan L (2010) Effects of Ipomoea aquatica cultivation on artificial floating rafts on water quality of intensive aquaculture ponds. J Ecol Rur Environ 26:155–159 (in Chinese with English Abstract)

Dai X, Guo Y, Qian H, Hu W, Chen W (2012) The purification effect of three vegetables and different cultivation on aquaculture water from shrimp pond. J Shanghai Ocean Univ 21:777–783 (in Chinese with English Abstract)

ä de Oliveira EG, Pinheiro AB, de Oliveira VQ, da Silva ARM Jr and others (2012) Effects of stocking density on the performance of juvenile pirarucu (Arapaima gigas) in cages. Aquaculture 370-371:96–101

ä De Silva SS (2012) Aquaculture: a newly emergent food production sector—and perspectives of its impacts on biodiversity and conservation. Biodivers Conserv 21:3187–3220

Ding XM (2000) The situation and measures on cultivation of Trionyx sinensis. Chin Fish 1:12 (in Chinese with English Abstract)

ä Duan D, Xu L, Fei X, Xu H (1995) Marine organisms attached to seaweed surfaces in Jiaozhou Bay, China. World J Microb Biotechnol 11:351–352

Fisheries Administration of the People's Republic of China (2015) China fishery statistical yearbook. China Agriculture Press, Beijing

ä Gomes LC, Chagas EC, Martins-Junior H, Roubach R, Ono EA, Lourenço JNP (2006) Cage culture of tambaqui (Colossoma macropomum) in a central Amazon floodplain lake. Aquaculture 253:374–384

ä Herbeck LS, Sollich M, Unger D, Holmer M, Jennerjahn TC (2014) Impact of pond aquaculture effluents on seagrass performance in NE Hainan, tropical China. Mar Pollut Bull 85:190–203

ä Holm-Hansen O, Riemann B (1978) Chlorophyll a determination: improvements in methodology. Oikos 30:438–447

Huang LP, Ben XL (2001) Effects of environment on Chinese soft-shell turtle. Aquaculture 5:34–35 (in Chinese with English Abstract)

ä Huguenin J (1997) The design, operations and economics of cage culture systems. Aquacult Eng 16:167–203

ä Jampeetong A, Brix H, Kantawanichkul S (2012) Effects of inorganic nitrogen forms on growth, morphology, nitrogen uptake capacity and nutrient allocation of four trop-

ical aquatic macrophytes (Salvinia cucullata, Ipomoea aquatica, Cyperus involucratus and Vetiveria zizanioides). Aquat Bot 97:10–16

ä Li W, Li Z (2009) In situ nutrient removal from aquaculture wastewater by aquatic vegetable Ipomoea aquatica on floating beds. Water Sci Technol 59:1937–1943

ä Li M, Wu Y, Yu Z, Sheng G, Yu H (2007) Nitrogen removal from eutrophic water by floating-bed-grown water spinach (Ipomoea aquatica Forsk.) with ion implantation. Water Res 41:3152–3158

ä Li M, Wu Y, Yu Z, Sheng G, Yu H (2009) Enhanced nitrogen and phosphorus removal from eutrophic lake water by Ipomoea aquatica with low-energy ion implantation. Water Res 43:1247–1256

ä Li H, Zhou Z, Wu T, Wu Y, Ji X (2013) Do fluctuations in incubation temperature affect hatchling quality in the Chinese soft-shelled turtle Pelodiscus sinensis? Aquaculture 406-407:91–96

ä Lombardi JV, de Almeida MHL, Lima PRT, Salee BOJ, de Paula EJ (2006) Cage polyculture of the Pacific white shrimp Litopenaeus vannamei and the Philippines seaweed Kappaphycus alvarezii. Aquaculture 258:412–415

ä Mook WT, Chakrabarti MH, Aroua MK, Khan GMA, Ali BS, Islam MS, Abu Hassan MA (2012) Removal of total ammonia nitrogen (TAN), nitrate and total organic carbon (TOC) from aquaculture wastewater using electrochemical technology: a review. Desalination 285:1–13

ä Muangkeow B, Ikejima K, Powtongsook S, Yi Y (2007) Effects of white shrimp, Litopenaeus vannamei (Boone), and Nile tilapia, Oreochromis niloticus L., stocking density on growth, nutrient conversion rate and economic return in integrated closed recirculation system. Aquaculture 269:363–376

ä Neal RS, Coyle SD, Tidwell JH (2010) Evaluation of stocking density and light level on the growth and survival of the Pacific white shrimp, Litopenaeus vannamei, reared in zero-exchange systems. J World Aquacult Soc 41: 533–544

ä Nuwansi KKT, Verma AK, Prakash C, Tiwari VK, Chandrakant MH, Shete AP, Prabhath GPWA (2016) Effect of water flow rate on polyculture of koi carp (Cyprinus carpio var. koi) and goldfish (Carassius auratus) with water spinach (Ipomoea aquatica) in recirculating aquaponic system. Aquacult Int 24:385–393

ä Othman I, Anuar AN, Ujang Z, Rosman NH, Harun H, Chelliapan S (2013) Livestock wastewater treatment using aerobic granular sludge. Bioresour Technol 133: 630–634

ä Portillo-Clark G, Hernandez RC, Servin-Villegas R, Magallon-Barajas FJ (2013) Growth and survival of the juvenile yellowleg shrimp Farfantepenaeus californiensis cohabiting with the green feather alga Caulerpa sertularioides at different temperatures. Aquacult Res 44:22–30

ä Price C, Black KD, Hargrave BT, Morris JA Jr (2015) Marine cage culture and the environment: effects on water quality and primary production. Aquacult Environ Interact 6: 151–174

Shi H, Parham JF, Fan Z, Hong M, Yin F (2008) Evidence for the massive scale of turtle farming in China. Oryx 42: 147–150

ä Silva CR, Gomes LC, Brandão FR (2007) Effect of feeding rate and frequency on tambaqui (Colossoma macropomum) growth, production and feeding costs during the first growth phase in cages. Aquaculture 264:135–139

ä Song HL, Li XN, Lu XW, Inamori Y (2009) Investigation of

microcystin removal from eutrophic surface water by aquatic vegetable bed. Ecol Eng 35:1589–1598

▶ Sooknah RD, Wilkie AC (2004) Nutrient removal by floating aquatic macrophytes cultured in anaerobically digested flushed dairy manure wastewater. Ecol Eng 22:27–42

▶ Wu Y, Xia L, Yu Z, Shabbir S, Kerr PG (2014) In situ bioremediation of surface waters by periphytons. Bioresour Technol 151:367–372

▶ Xie B, Ding Z, Wang X (2004) Impact of intensive shrimp farming on the water quality of the adjacent coastal creeks from eastern China. Mar Pollut Bull 48:543–553

▶ Zhang Q, Achal V, Xu Y, Xiang WN (2014) Aquaculture wastewater quality improvement by water spinach (*Ipomoea aquatica* Forsskal) floating bed and ecological benefit assessment in ecological agriculture district. Aquacult Eng 60:48–55

6

Impact of environmental conditions on biomass yield, quality, and bio-mitigation capacity of *Saccharina latissima*

Annette Bruhn[1,*], Ditte Bruunshøj Tørring[2], Marianne Thomsen[3],
Paula Canal-Vergés[2], Mette Møller Nielsen[1,2], Michael Bo Rasmussen[1],
Karin Loft Eybye[5], Martin Mørk Larsen[4], Thorsten Johannes Skovbjerg Balsby[1],
Jens Kjerulf Petersen[2]

[1]Department of Bioscience, Aarhus University, Vejlsøvej 25, 8600 Silkeborg, Denmark

[2]Danish Shellfish Centre, Institute of Aquatic Resources, Technical University of Denmark, DTU-Aqua, Øroddevej 80, 7900 Nykøbing Mors, Denmark

[3]Department of Environmental Sciences, Aarhus University, Frederiksborgvej 399, 4000 Roskilde, Denmark

[4]Department of Bioscience, Aarhus University, Frederiksborgvej 399, 4000 Roskilde, Denmark

[5]Division of Life Science & Food Technology, Danish Technological Institute, Kongsvang Allé 29, 8000 Aarhus C, Denmark

ABSTRACT: Seaweeds are attractive as a sustainable aquaculture crop for food, feed, bioenergy and biomolecules. Further, the non-value ecosystem services of seaweed cultivation (i.e. nutrient recapture) are gaining interest as an instrument towards sustainable aquaculture and for fulfilling the aims of the EU Marine Strategy Framework Directive. Environmental factors determine the yield and quality of the cultivated seaweed biomass and, in return, the seaweed aquaculture affects the marine environment by nutrient assimilation. Consequently, site selection is critical for obtaining optimal biomass yield and quality and for successful bio-mitigation. In this study, 5 sites for cultivation of *Saccharina latissima* were selected within a eutrophic water body to guide site selection for future kelp cultivation activities. Results were coupled to marine monitoring data to explore the relationship between environmental conditions and cultivation success. The biomass yields fluctuated 10-fold between sites due to local variations in light and nutrient availability. Yields were generally low, i.e. up to 510 g fresh weight (FW) per meter seeded line; however, the dry matter contents of protein and high-value pigments were high (up to 17% protein and 0.1% fucoxanthin). Growth performance, biomass quality and bio-mitigation potential was restricted by low availability of light and bioavailable phosphorus, and biofouling through juvenile suspension feeders was a critical factor at all cultivation sites. At specific sites, the tissue metal contents (Pb and Hg) exceeded the limit values for feed or food. Our results emphasize the importance of careful site selection before establishing large-scale cultivation, and stress the challenges and benefits of kelp cultivation in eutrophic waters.

KEY WORDS: Eutrophication · Limfjorden · Seaweed farming · Metals · Nitrogen · Phosphorus · Site quality · Ecosystem service

INTRODUCTION

Cultivation of macroalgae is a rapidly growing industry in a global perspective (FAO 2016). The main driver is the establishment of a production of marine-based biomass for food, energy, protein and biomolecules (Bruton et al. 2009, Kraan 2013, Wei et al. 2013), but also exploitation of the bio-mitigation capacity of the algae is in focus (Troell et al. 1999, Castine et al. 2013, Marinho et al. 2015a). The non-use value eco-

*Corresponding author: anbr@bios.au.dk

system service (Daly 1998) provided by cultivated algae in terms of recapturing nutrients in coastal areas is of commercial and societal interest — as compensation for increased aquaculture activities (Sanderson et al. 2012, Handå et al. 2013, Smale et al. 2013, Holdt & Edwards 2014, Marinho et al. 2015a) or as a potential instrument for circular nutrient management, improving the ecological status of eutrophic marine areas (Seghetta et al. 2016) in line with the EU Marine Strategy Framework Directive (EU 2008a, 2014).

In Europe, the effort concerning cultivation of large brown algae (Laminariales), in particular, is increasing. The most commonly cultivated brown algae species in Europe, *Saccharina latissima* ((Linnaeus) C.E. Lane, C. Mayes, Druehl & G.W. Saunders), has been cultivated on a smaller or larger scale in Ireland, Scotland, Germany, Holland, Spain, Norway, Faroe Islands and Denmark (i.e. Buck & Buchholz 2004, 2005, Buck et al. 2008, Werner et al. 2009, Wegeberg 2010, Edwards & Watson 2011, Forbord et al. 2012, Handå et al. 2013, Peteiro & Freire 2013b, Wegeberg et al. 2013, Marinho et al. 2015a). The achieved biomass yields and the biochemical composition of the biomass vary considerably seasonally and spatially, primarily because of different environmental conditions (Edwards & Watson 2011, Handå et al. 2013, Peteiro & Freire 2013a,b, Marinho et al. 2015a). In return, the algae production also exerts an impact on the environmental conditions through the removal of nutrients (Troell et al. 1999, Stephens et al. 2014, Marinho et al. 2015a). Thus, algae cultivation sites should be carefully selected for optimizing biomass production, biomass quality as well as the non-value ecosystem services (Kerrison et al. 2015). Further, the seasonal timing of the deployment and harvest of the algae needs to be optimized according to local environmental conditions. The focus of the optimization, i.e. high protein yield or high carbohydrate yield, will depend on the final application of the biomass. Sporophytes of *Laminaria* species store nutrients for length growth during periods when environmental nutrients concentrations are high (Bartsch et al. 2008). Thus, high environmental nutrient concentrations favour high tissue nitrogen (N) concentrations: in wild *S. latissima* up to 3.5% N of dry matter (DM) (Gevaert et al. 2001, Nielsen et al. 2014), up to 5.0% N of DM when cultivated in close proximity to fish aquaculture (Handå et al. 2013, Marinho et al. 2015a) and even up to 6.7% N of DM when cultivated under highly eutrophic conditions (Nielsen 2015). The molar N:phosphorus (P) ratio is commonly in the range of 9–25:1 (Atkinson & Smith 1983), and P con-

centrations of up to 0.8% of DM are reported in nutrient-rich waters (Marinho et al. 2015a). High tissue N concentrations reflect a correspondingly high content of proteins (Manns et al. 2014, Marinho et al. 2015b, Angell et al. 2016). Consequently, both the biomass quality and the bio-mitigation capacity of the produced algae increase in nutrient-rich waters, increasing the value of the biomass, as well as improving the environmental condition of the water body through harvest, and thus removal of nutrients.

Of the 21 Danish water bodies, Limfjorden receives the highest annual net supply of nutrients (8.2 t N and 0.30 t P km^{-2} y^{-1}; Seghetta et al. 2016). These high nutrient loadings have caused a regime shift in the fjord from benthic to pelagic primary production (Krause-Jensen et al. 2012). The high pelagic primary production supports a substantial stock of benthic suspension feeders, including blue mussels *Mytilus edulis* L., supporting a local mussel fishery (Maar et al. 2010, Timmermann et al. 2014). Mussel farming has been successfully tested as an instrument to recapture nutrients and improve the ecological status of Limfjorden (Petersen et al. 2014), and farming of long-line blue mussels is an emerging business in Limfjorden. Along with the development of the mussel farming industry, interest in macroalgae cultivation is increasing, partly because the 2 crops may be cultivated using the same structures (Nielsen 2015). Due to the high environmental nutrient concentrations, cultivation of large brown algae in a water body like Limfjorden would theoretically hold a potential for the production of a *Saccharina* biomass with high protein content, representing a higher value for the food or feed market. At the same time the potential of seaweed cultivation as an instrument for circular nutrient management would be maximized. Despite the relatively small size of Limfjorden (1500 km^2), local environmental conditions differ considerably between the different basins (Maar et al. 2010). Cultivation of *S. latissima* has to date been documented only once at 1 site in Limfjorden, indicating a potential for cultivation of *S. latissima*. This study, however, also demonstrates the need for investigating optimal timing of cultivation and harvest in order to maximize biomass yield and avoid biofouling (Wegeberg 2010).

Testing and evaluating the interactions between local environmental conditions and biomass yield, quality and potential for bio-mitigation through nutrient recapture of cultivated kelps in coastal waters is needed before implementing cultivation on a larger scale. This applies not only to Limfjorden, but to any water body where macroalgae cultivation is intended.

The aim of this study was to compare the biomass yield, bio-mitigation capacity and nutritional quality for food and feed of *S. latissima* cultivated at 5 sites in Limfjorden as well as to explore the influence of local environmental conditions on these parameters with the purpose of guiding site selection and timing of harvest. The 5 selected cultivation sites each represented their basin in Limfjorden, with the basins characterized by different environmental conditions regarding salinity, turbidity, nutrient availability and sediment metal concentrations.

MATERIALS AND METHODS

Study area and cultivation sites

Limfjorden is a shallow, semi-enclosed estuary located between the North Sea and the Kattegat (Fig. 1). The total surface area of the fjord is ~1500 km² and the average depth is 4.6 m. The total catchment area is 7587 km² and is predominantly agricultural land. Despite a small tidal amplitude, tidal forces and wind are the drivers of the annual net flow of 6.8 km³ of water from the North Sea via the Thyborøn channel in the west through Limfjorden to the Kattegat. Limfjorden consists of several relatively shallow water basins connected by narrow and deep sounds. The big broads have water depths of 5–8 m, whereas the sounds have depths of 18–22 m, the deepest point being Oddesund (28 m). The average salinity varies from 32–34 in the western part to 19–25 in the central and eastern part (Lyngby et al. 1999, Markager et al. 2006, Krause-Jensen et al. 2012, Timmermann et al. 2014).

Five existing mussel farms were selected as experimental cultivation sites (Fig. 1, Table 1) for the

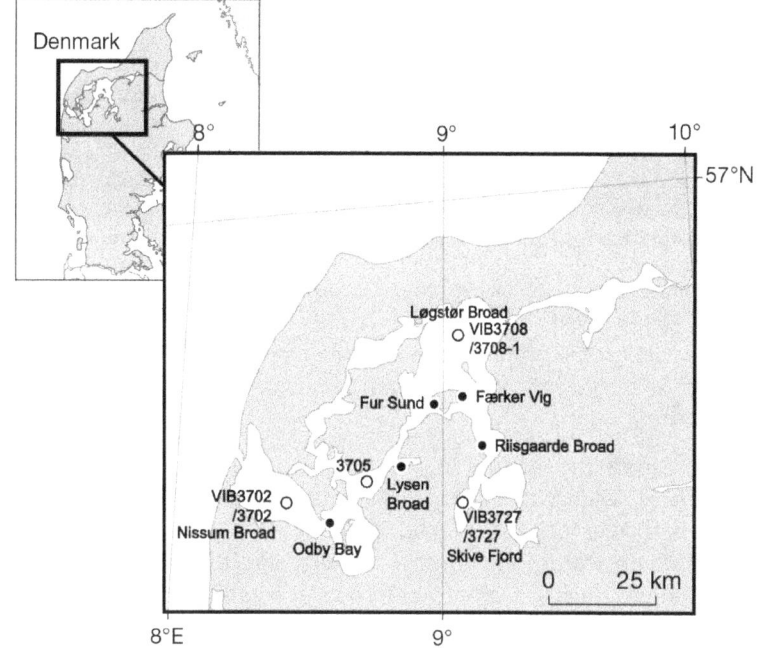

Fig. 1. Location of Limfjorden in Denmark and the 5 *Saccharina latissima* cultivation sites (filled circles) and 4 environmental monitoring stations (open circles). Stns VIB3702, VIB3708 and VIB 3727 are pelagic stations for monitoring water quality. Stns 3702, 3705, 3708-1 and 3727 are stations for monitoring benthic metal concentrations

following reasons: (1) they were each located in a distinct basin of Limfjorden; (2) aquaculture licenses were already active; (3) the mussel cultivation structures could be used for the seaweed cultivation; and (4) the 5 basins were covered by the Danish National Monitoring and Assessment Program for the Aquatic and Terrestrial Environment (NOVANA).

Environmental data

For each of the 5 cultivation sites, the data on biomass yield and quality were coupled to environmental data from an environmental monitoring station located centrally within each basin (Fig. 1, Table 2).

Table 1. Location, size and characteristics of the 5 *Saccharina latissima* cultivation sites in Limfjorden, Denmark. N: north, E: east, S: south, W: west

Site	Position		Basin	Size (m × m)	Depth (m)	Sea bed classification	Degree of exposure
	Latitude (°N)	Longitude (°E)					
Odby Bay	56.577	8.570	Nissum/Kaas Broad	300 × 300	4	Soft mud	Exposed to winds from E and S
Lysen Broad	56.692	8.841	Sallingsund	250 × 500	2–5	Fine sand with clay	Protected
Fur Sund	56.816	8.968	Fur Sund	250 × 750	5	Rocky/sandy	Exposed to strong currents
Færker Vig	56.834	9.073	Løgstør Broad	300 × 300	4	Hard sand/stone	Protected
Riisgaarde Broad	56.736	9.151	Riisgaarde Broad/ Skive Fjord	250 × 500	10	Soft mud	Exposed to winds from N, E, S

Table 2. Deployment and sampling dates at the 5 *Saccharina latissima* cultivation sites in Limfjorden, batches of seeded lines, as well as identification numbers of and distance to the environmental monitoring stations (pelagic and sediment stations, see Fig. 1) from which data were used for analyses

Site	Deployment date	Sampling 1 date	Sampling 2 date	Batch	Pelagic station	Sediment station	Distance (km)
Odby Bay	Dec 6, 2011	Apr 11, 2012	Jun 12, 2012	1	VIB3702	3702	14
Lysen Broad	Dec 6, 2011	Apr 11, 2012	Jun 12, 2012	1	–[a]	3705	8.5
Fur Sund	Dec 6, 2011	Apr 11, 2012	Jun 12, 2012	1	VIB3708	3708-1	20
Færker Vig	Oct 28, 2011	Apr 11, 2012	May 25, 2012	2	VIB3708	3708-1	14
Riisgarde Broad	Dec 6, 2011	Apr 11, 2012	Jun 12, 2012	1	VIB3727	3727	14

[a]No pelagic monitoring station was within proximity to the cultivation site at Lysen Broad

Data from the monitoring stations were retrieved from NOVANA through the National Database for Marine Data (ODAM) (Fig. 1, Table 1).

Data on water temperature, salinity, turbidity and concentrations of oxygen, chlorophyll *a* (chl *a*), inorganic nutrients (dissolved inorganic N [DIN = NO_2^--N, NO_3^--N, NH_4^+-N], dissolved inorganic bioavailable P [ortho-P]) and sediment metals were collected and analysed using standard methods according to the current national Technical Instructions for Marine Monitoring (Markager 2004, Pedersen et al. 2004, Larsen 2013, Markager & Fossing 2013, Vang 2013, Vang & Hansen 2013). Sampling was performed on average every 2–3 wk. Sampling of sediment was performed every 1–5 yr. By trapezoidal integration, all pelagic environmental data were calculated into weighted averages over 2 periods up to the time point of each biomass sampling — early spring: the period of detectable growth from 1 February 2012 to Sampling 1, 11 April 2012; and late spring: the last part of the grow-out period from Sampling 1 (11 April 2012) to Sampling 2 (25 May or 12 June 2012) (see next section and Table 2).

Data regarding temperature, salinity and turbidity were differentiated according to the actual cultivation depths (1.5 and 2.5 m, respectively). Data regarding nutrients, oxygen and chl *a* were only available from 1 m of depth, but no significant stratification prevailed during the cultivation period. Sediment metal concentration data were averaged for each station over a period covering the preceding 10 yr (2003–2012). Data on local incoming light was supplied from the Danish Meteorological Institute.

Ideally, cultivation sites and monitoring stations could have been geographically closer. However, the data from the environmental monitoring stations was considered as being representative for the cultivation sites despite the distances of 8–20 km between monitoring station and cultivation site for a number of reasons: (1) other studies correlating monitoring data and macrovegetation performance in Limfjorden generally achieve good correlations (e.g. Krause-Jensen et al. 2012), (2) the experimental period from winter to early summer is a period of maximal wind-driven circulation (Wiles et al. 2006), and absence of vertical stratification (Christiansen et al. 2006), (3) mixing was confirmed as no stratification was observed during the experimental period, and (4) sites and stations were located in the more open parts of the basins in proximity to point sources of run-off from land. Coupling of biomass yield and quality to environmental data for the cultivation site at Lysen Broad was not possible, as only data on sediment chemistry was available from the environmental monitoring station in this basin.

Cultivation and sampling of *Saccharina latissima*

Two batches of *S. latissima* seeded lines were used in the cultivation experiment (Table 2). Batch 1 consisted of 500 m of ready-made seeded line (diameter: 6 mm) produced by direct sporulation (Wegeberg 2010) at Blue Food A/S, Denmark. This batch was delivered to the Danish Shellfish Centre on 5 December 2011, kept in running seawater overnight and deployed the following day at 4 sites: Odby Bay, Lysen Broad, Fur Sund and Riisgaarde Broad. Batch 2 was deployed at Færker Vig and consisted of 125 m of seeded line (diameter: 6 mm), also produced through direct sporulation, but at the Danish Shellfish Centre during August 2011. Both batches were produced from fertile material from a *S. latissima* population in the Danish Belt Sea, and visual inspection of the lines upon deployment did not reveal any difference between the 2 batches in quality, density or size of the juvenile sporophytes. Length of the seedlings at deployment was ~1 mm. All lines were deployed as vertical droppers, each 2.5 m long, attached to a horizontal long-line with a 50 cm tethering line. The droppers were interspaced by 40 cm along the horizontal long-line. During the grow-out period, the horizontal long-lines were kept 50 cm below the water surface to avoid disturbance by floating ice and heavy storms. Consequently, the seeded lines were positioned between 1 and 3.5 m depth.

Sampling of biomass was performed twice by random selection of 3 droppers from each site: Sampling 1 on 11 April 2012 and Sampling 2 on either 25 May or 12 June 2012 (Table 2). The lines were brought to the laboratory where the upper 2 m of each line was divided into 2 sections: the upper section represented the seeded line hanging in 1–2 m depth (average 1.5 m), and the lower section representing the seeded line hanging at a depth of 2–3 m (average 2.5 m). The remaining 50 cm of each seeded line with the attached bottom weight was discarded due to lack of biomass. The following parameters were recorded for both sections of the lines: total weight of sample (seeded line + algae + epiphytes), weight of seeded line, weight of algae, weight and taxonomy of dominating biofouling epiphytic organisms, and finally, average sporophyte frond length, based on 15 randomly selected sporophytes. After sampling, tissue samples were stored at −20°C until biochemical analyses were performed. Due to increasingly heavy biofouling by epiphytic organisms over time, algae material harvested from late May and onwards (Sampling 2) was fully covered with epiphytic organisms such as ascidians and juvenile mussels, and thus considered unsuitable for food or feed applications. Therefore only algae material sampled in April (Sampling 1) was used in the biochemical analyses. Due to very limited biomass harvested from Fur Sund at 2.5 m in April, only pigment analyses were performed on this biomass.

Calculations of growth rates and biomass yields

Specific growth rates (SGRs) were calculated from measurements of the fresh weight (FW) per running meter of seeded line as:

$$SGR(\%) = 100 \times \frac{\ln\left(\frac{FW_t}{FW_0}\right)}{t} \qquad (1)$$

where FW_0 and FW_t corresponded to the fresh weight of S. latissima per m of seeded line at time 0 and after t days of cultivation, respectively. Biomass yields were reported as g FW per m of seeded line (g

FW m^{-1}). The average frond length of the S. latissima sporophytes was calculated as an average length ± SE of the 15 randomly selected sporophytes from each sample.

Saccharina latissima tissue biochemistry

DM, ash, carbon (C), N and P. Algae samples were freeze-dried at −40°C and homogenized by dry milling. DM content was calculated as percentage of FW. A known amount of dry algae was combusted at 550°C for 2 h, and the ash fraction was calculated as percentage of DM. Concentrations of C and N in the freeze-dried algae tissue were analysed by Pregl-Dumas ignition in pure oxygen atmosphere followed by chromatographic separation of C and N with detection of the individual elements by thermal conductivity (Culmo 2010). Total P content of the algae biomass was as analysed spectrophotometrically according to standard methods (Grasshoff et al. 1983). Prior to analysis, the dried and homogenized samples were heated at 550°C for 2 h, autoclaved with 2 M hydrogen chloride (HCl) (20 mg DM for 7 ml acid), and finally filtered through GFF filters (Whatman).

Metals. Metal concentrations (As, Cd, Hg, Pb) were determined by inductively coupled plasma-mass Spectrometry (ICP-MS). In short, a 0.2 g dry subsample was digested in a closed vessel microwave oven using 5 ml of nitric acid (7 M) and 1 ml of hydrogen peroxide, then diluted to 50 ml with milliQ water, followed by ICP-MS determination using internal standards of Rh, Ir and Ge to correct for drift (see Nielsen et al. 2012). Certified reference material of macroalgae from IAEA-140 (Coquery et al. 2000) was used for quality assurance.

Pigments. Pigment concentrations (chl a, fucoxanthin, violaxanthin and β-carotene) were determined using acetone extraction and quantification by HPLC as described in Boderskov et al. (2016). Pigment standards were obtained from DHI Laboratory Products.

Crude protein and amino acids (AAs). Crude protein and AA composition were analysed only for samples from Færker Vig. Total organic bound crude protein was determined by the Kjeldahl principle according to Nordic Committee on Food Analysis (2005). Protein content was calculated by multiplying the amount of N by a factor of 5 and expressed as percent of DM (Angell et al. 2016). The determination of AAs was done by HPLC according to EU 152/2009 (A) and ISO 13903:2005. AA contents were expressed as percentage of DM.

Data analysis

For comparing growth performance and biomass quality between sites and depths, 2-way ANOVA (using Tukey's post hoc analysis) and linear regression analyses were performed using JMP 10.0 (SAS Institute). Explorative data analysis was performed to identify significant correlation patterns between macroalgae growth and environmental parameters. Data were log transformed in order to obtain normal distribution and homogeneity of variance for the residuals of the models. Multivariate data analysis (MVDA) was performed to guide model selection of variables to be tested using general linear models (GLM). Partial least square regression (PLS-R) was used as explorative technique for pattern recognition using the Unscrambler v.10.2 (CAMO Software). Biomass yield and biofouling in early and late spring as well as bio-mitigation capacity, i.e. N and P content in the harvested seaweed biomass, were selected as Y-variables in the PLS-R models and modelled using environmental parameters, characterising the marine growth environment surrounding the individual cultivation sites, as original explanatory variables (data not shown). GLMs were used to assess the effect of light, salinity, availability of ortho-P, temperature and environmental N:P ratio (NP_E) on growth performance, biofouling, and biomass quality. The environmental parameters were selected as independent variables based on the indicative impact on the dependent variable (biomass growth parameters and quality), as observed from MVDA (data not shown). As several of the independent variables showed strong correlations (Pearson, Table S1 in the Supplement at www.int-res.com/articles/suppl/q008 p619_supp.pdf), the independent variables were split into 2 models to avoid issues with collinearity — Model 1: light, salinity, and ortho-P; Model 2: temperature, salinity and NP_E. These analyses were performed in SAS 9.3 (SAS Institute) using the Proc mixed function with cultivation site as a random factor. The level of significance applied was 0.05, unless mentioned otherwise.

RESULTS

Environmental conditions

The environmental conditions differed among the basins of Limfjorden (Fig. 2, Table 1). Differences were most pronounced with regard to salinity, light, and concentrations of inorganic nutrients and chl *a*.

Salinity

The salinity in the different basins decreased with increasing distance from the North Sea: Nissum Broad, 29.0–31.9; Løgstør Broad, 26.0–28.9; and Skive Fjord, 23.5–26.5. In Nissum Broad, the salinity increased slightly over the grow-out period, whereas in Løgstør Broad and Skive Fjord the salinity decreased over the period, reflecting a stronger influence of run-off from land (Fig. 2A). No pronounced stratification of the water column was observed from the monitoring data during the grow-out period at any of the stations (data not shown).

Temperature

Generally, the differences in temperature among stations were minor (<1°C), and even less between the 2 cultivation depths at any station. The temperatures experienced during the full grow-out period ranged from minimum temperatures in all basins measured on 1 February (between –0.2 and 1.5°C) to maximum temperatures in June (13.8–14.1°C) (Fig. 2B).

Inorganic nutrients

The average concentrations of DIN from deployment to April ranged between 20 and 40 µM; however, with concentrations up to 58 µM in Skive Fjord in winter and early spring (Fig. 2C). In late spring, between April and June, the DIN concentrations decreased <2 µM in Nissum Broad, but remained high between 10 and 20 µM in the other basins. In all periods, the highest DIN concentrations were measured near Skive Fjord and the lowest in Nissum Broad. Concentrations of ortho-P were high during the winter period (0.4–0.9 µM), but decreased below 0.1 µM during the spring bloom from February to April (Fig. 2D).

Pelagic chl *a*

In early spring, February and March, the phytoplankton concentrations peaked with 12 and 16 µg chl *a* l^{-1} in Løgstør and Nissum Broad, respectively (Fig. 2E). In Skive Fjord, the highest chl *a* concentrations were measured in early June (14 µg chl *a* l^{-1}).

Light

The photon flux density generally decreased by ~50% from 1.5 to 2.5 m, emphasizing the high turbidity of Limfjorden (Fig. 2F). The algae at 1.5 m experienced an average of 400–700 µmol photons m^{-2} s^{-1} in late

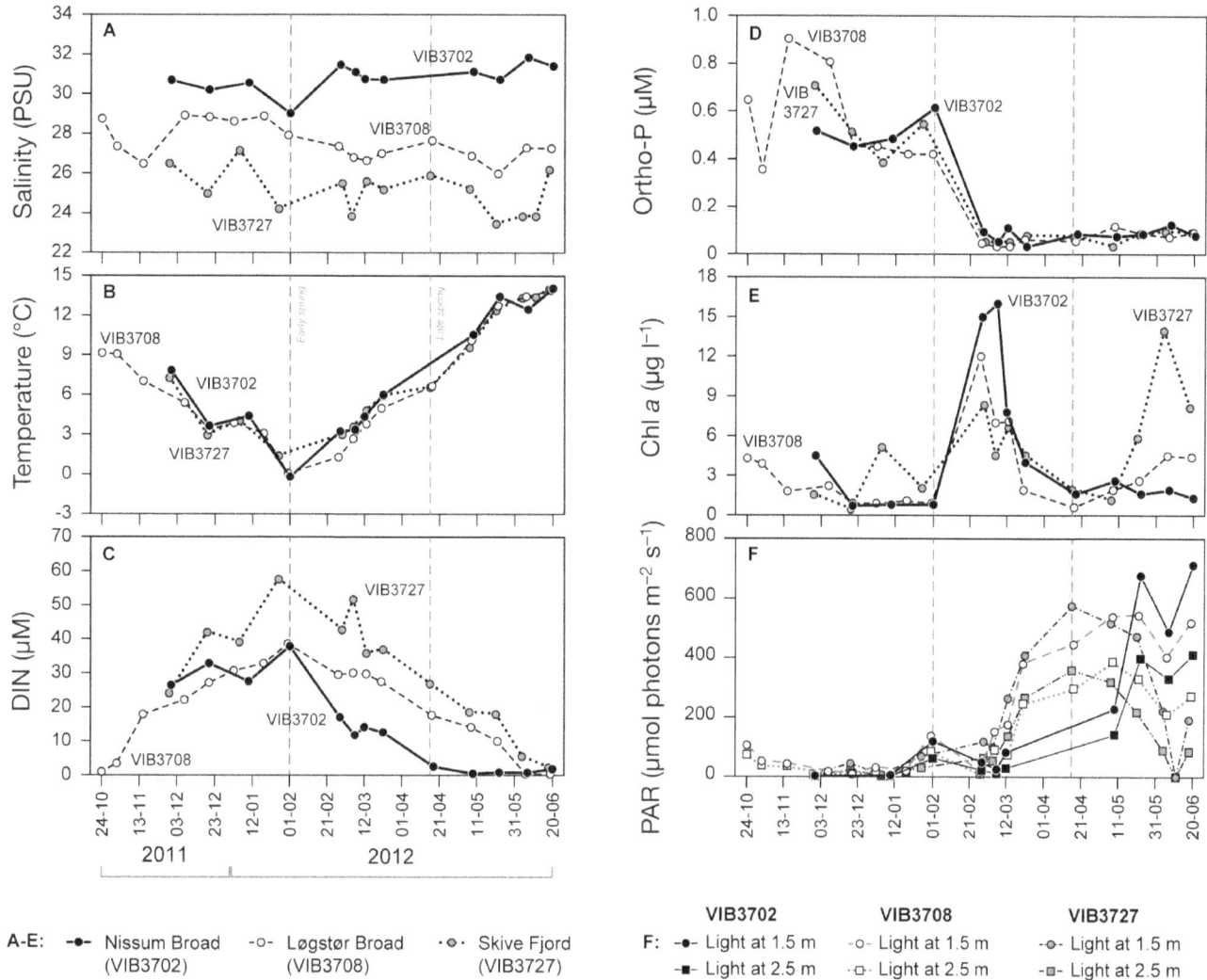

Fig. 2. Seasonal pattern of selected environmental parameters at the 3 pelagic stations (see Table 2) during the grow-out period of *Saccharina latissima* (October–December 2011 to June 2012): (A) salinity, (B) temperature, (C) dissolved inorganic nitrogen (DIN), (D) ortho-phosphate, (E) chlorophyll *a* and (F) photosynthetically active radiation (PAR) estimated at the cultivation depths (1.5 and 2.5 m). Data represent the actual measured values

spring, whereas the algae at 2.5 m only experienced up to 400 μmol photons m^{-2} s^{-1} in the same period.

Overall, a high degree of inter-correlation between the key environmental parameters was observed (Table S1 in the Supplement). In early spring (February to April) the concentration of phytoplankton biomass (chl *a*) correlated strongly to the concentrations of dissolved inorganic nutrients. During early spring, the concentrations of pelagic chl *a* correlated positively to ortho-P, and negatively to DIN concentrations, whereas light availability correlated negatively to ortho-P concentrations. The DIN concentrations were negatively correlated to salinity. In late spring, the pelagic phytoplankton biomass was negatively correlated to salinity and positively to temperature.

Saccharina latissima growth performance

Biomass yield, frond length, SGR and biofouling

At all cultivation sites, the biomass yields and frond lengths were higher at 1.5 m than at 2.5 m depth (Fig. 3A,B, Table 3). The highest biomass yield in April (mean ± SE: 510 ± 66 g FW m^{-1}) as well as the longest fronds in April and June (40.9 ± 3.7 cm in April and 33.7 ± 9.0 cm in June) were achieved at Færker Vig at 1.5 m (Fig. 3A,B, Table 4). In June, the biomass yield in Odby Bay and Færker Vig at 1.5 m was significantly higher than at the remaining 3 sites (Table 4). At 2.5 m, the highest biomass yield in June was obtained in Odby

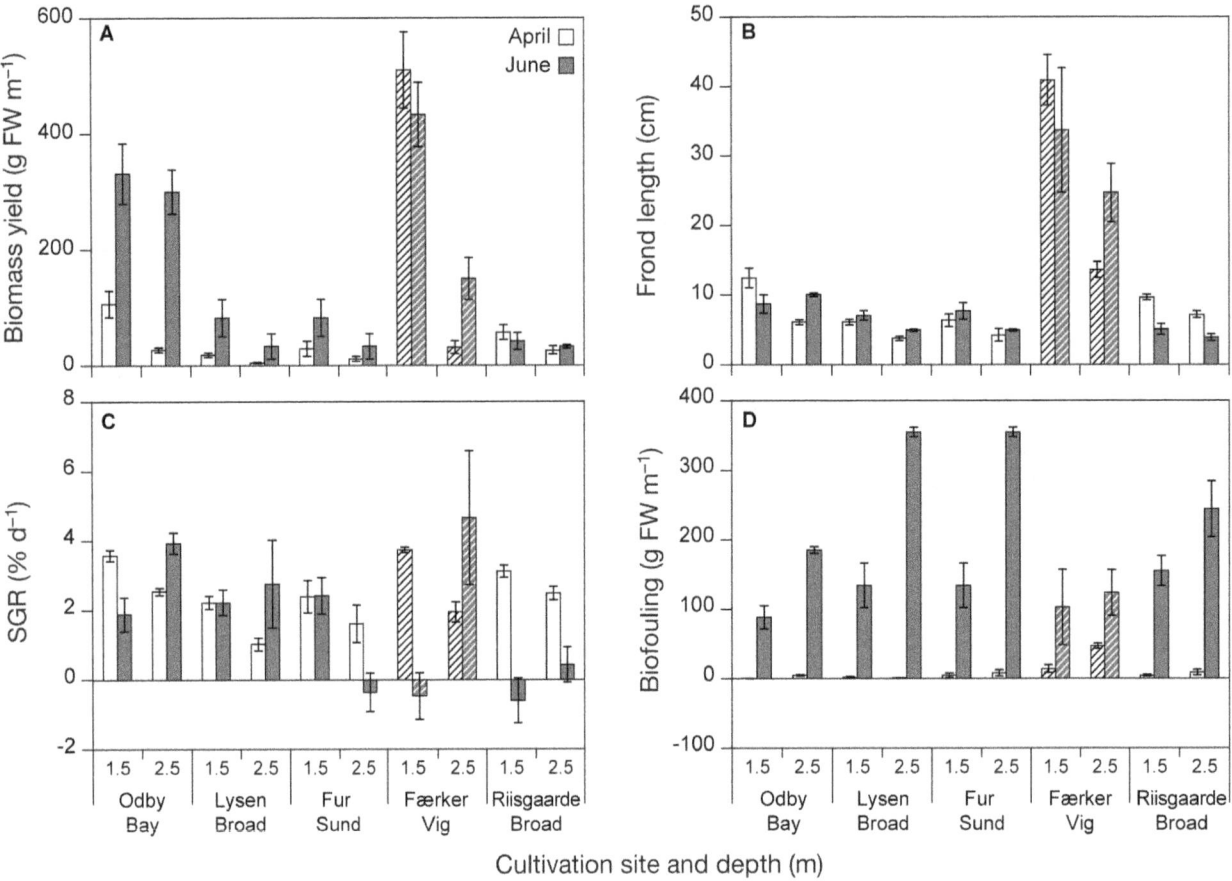

Fig. 3. Growth performance and biofouling of *Saccharina latissima* at the 5 cultivation sites (see Fig. 1): (A) biomass yields, (B) average frond lengths (n = 15), (C) specific growth rates (SGR), and (D) biofouling of *S. latissima* at 1.5 and 2.5 m depths at the 5 cultivation sites sampled in April (white bars) and June (grey bars), respectively. Solid bars represent batch 1 of seeded lines, and crossed bars represent batch 2. Data represent means ± SE, n = 3

Bay, whereas the longest fronds were found in Færker Vig (Fig. 3A,B, Table 4).

The SGR (in the period from deployment to April) reflected the same pattern as the biomass yield at 1.5 m depth: Færker Vig (3.8% d^{-1}) > Odby Bay (3.7% d^{-1}) > Riisgaarde Broad (3.1% d^{-1}) > Fur Sund (2.4% d^{-1}) > Lysen Broad (2.2% d^{-1}), and all with significantly higher SGRs at 1.5 m as compared to 2.5 m (Fig. 3C, Table 3). However, from April to June, the SGR decreased for the algae nearest to the surface (1.5 m) at Odby Bay, Færker Vig and Riisgaarde Broad and at the 2 latter sites to negative values. At Lysen Broad and Fur Sund, the SGR of the algae near the surface was constant throughout the full grow-out period. Regarding the algae growing at 2.5 m from April to June, diverging trends were observed: at 3 cultivation sites (Odby Bay, Lysen Broad and Færker Vig) the SGRs exceeded the SGRs at 1.5 m in the early growth period, whereas at the other 2 sites (Fur Sund and Riisgaarde Broad), the SGRs decreased to around or below zero.

The degree of biofouling increased dramatically at all sites from April to June, and was in June significantly higher at 2.5 m than at 1.5 m depth, with the one exception of Færker Vig (Fig. 3D). In June, the biomass yield of biofouling organisms (predominantly hydroids, juvenile *M. edulis* and ascidians) exceeded the biomass yields of *S. latissima* at 3 sites (Lysen Broad, Fur Sund and Riisgaarde Broad) at both depths (Fig. 3A,D).

In early spring, the growth performance (biomass increase [Fig. 4A], length growth and SGR) was positively correlated to the light availability, with also salinity and ortho-P availability being positively correlated to length growth and SGR (Fig. 4B, statistics are provided in Table S2 in the Supplement), respectively. The total biomass yield in June was negatively correlated to the degree of biofouling in late spring (linear regression, p = 0.003, R^2 = 0.28). (Fig. 4C). The biofouling in late spring was positively correlated to the sea temperature at the cultivation depth between April and June (Fig. 4D, Table S2).

Saccharina latissima biomass quality

DM, tissue N and P

The DM content of the algae varied between 6.3 and 16.8% of fresh weight (Fig. 5A). The C content generally ranged between 26.8 and 33.4% of DM, except at Fur Sund, where the C content was significantly lower (15.3–20.5% of DM). At Odby Bay and Fur Sund, the tissue C concentrations were significantly higher in the biomass closest to the surface. This was not the case at the other sites (Fig. 5B, Table 3).

The tissue N concentration in April was significantly higher in biomass from Odby Bay (4.5% of DM) than from any of the other cultivation sites (3.5–4.0% of DM) (Fig. 5C, Table 4). Only at Riisgaarde Broad was there a significantly higher N concentration in the algae cultivated at 2.5 m than at 1.5 m depth. The tissue P content was significantly higher in the algae cultivated at 2.5 m than at 1.5 m in Odby Bay and Riisgaarde Broad, where the P content in the algae from 2.5 m was up to 0.28% of DM compared to 0.11% of DM at 1.5 m (Fig. 5D).

The bio-mitigation capacity of N and P varied between sites and depths from (mean ± SE) 0.02 ± 0.01 to 1.84 ± 0.24 g N m^{-1} and 0.001 ± 0.0004 to 0.05 ± 0.01 g P m^{-1}, respectively (Fig. 5E,F), reflecting predominantly the large fluctuations in biomass yields (Fig. 3A).

The environmental concentration of ortho-P was positively related to the tissue DM and N contents (Fig. 4B), while not related to the tissue P content

Table 3. Dependency of *Saccharina latissima* growth performance and biomass quality on cultivation site and depth, as well as the interaction between the two. p-values from 2-way ANOVA, data were log transformed prior to analysis. Statistical significance (p > 0.05) is indicated in **bold**. A: April, J: June. (−) designates a negative correlation, otherwise the correlation is positive

Parameter	Cultivation site × Depth	Cultivation site	Depth
Growth performance			
Yield (g m^{-1}) (A)	0.064	**<0.001**	**<0.001(−)**
Yield (g m^{-1}) (J)	0.336	**<0.001**	**0.004(−)**
Length (cm) (A)	**0.014**	**<0.001**	**<0.001(−)**
Length (cm) (J)	0.398	**<0.001**	**0.036(−)**
SGR (% d^{-1}) (A)	0.535	**0.003**	**<0.001**
SGR (% d^{-1}) (J)	0.057	**0.025**	0.118
Biofouling (g m^{-1}) (A)	0.417	**<0.001**	0.097
Biofouling (g m^{-1}) (J)	0.937	0.058	**0.003**
Biomass quality (A)			
DM (% FW)	**0.021**	**<0.001**	0.643
C (% DM)	**0.030**	**<0.001**	**<0.001(−)**
N (% DM)	**0.001**	**<0.001**	0.959
P (% DM)	**0.006**	**0.008**	**<0.001**
Chl *a* (mg g DM^{-1})	**0.029**	**0.005**	**<0.001**
Fucoxanthin (mg g DM^{-1})	0.433	**0.006**	**<0.001**
Violaxanthin (mg g DM^{-1})	0.302	**0.003**	**0.004**
Beta-carotene (mg g DM^{-1})	0.316	**0.026**	**0.005**
As (mg kg DM^{-1})	0.249	**<0.001**	**0.003**
Hg (mg kg DM^{-1})	0.057	0.050	0.761
Pb (mg kg DM^{-1})	**0.003**	**<0.001**	**<0.001**
Cd (mg kg DM^{-1})	0.926	**<0.001**	0.831

(Table S2). Temperature was positively correlated to the tissue DM, N and P contents (Table S2). Light availability correlated positively to the tissue C content, but negatively to P content (Table S2).

Table 4. Tukey post-hoc pairwise comparisons of *Saccharina latissima* growth performance in April and June, and tissue biochemistry in April among cultivation sites at 1.5 and 2.5 m depth. Different letters are assigned to significantly different results. Fuco: fucoxanthin, Viola: violaxanthin, β-car: β-carotene

Site	Yield (Apr)	Yield (Jun)	Frond length (Apr)	Frond length (Jun)	Biofouling (Apr)	Biofouling (Jun)	DM	N	P	C	Chl *a*	Fuco	Viola	β-car
1.5 m depth														
FærkerVig	A	A	A	A	A	A	B	B	A	B	AB	B	AB	A
Odby Bay	B	A	B	B	B	A	AB	A	A	A	A	A	A	A
Riisgaarde Broad	B	B	B	B	AB	A	A	B	A	A	A	B	B	A
Fur Sund	B	B	B	B	AB	A	C	B	A	C	B	B	–	A
Lysen Broad	B	B	B	B	AB	A	A	B	A	A	A	AB	AB	A
2.5 m depth														
Færker Vig	A	B	A	A	A	C	B	B	B	B	BC	ABC	AB	A
Odby Bay	A	A	BC	B	B	BC	A	A	A	AB	A	A	A	A
Riisgaarde Broad	A	B	B	B	B	B	A	A	AB	A	AB	AB	AB	A
Fur Sund	A	B	BC	B	B	A	B	C	B	C	C	C		A
Lysen Broad	A	B	C	B	B	A	–	–	–	–	BC	BC	B	A

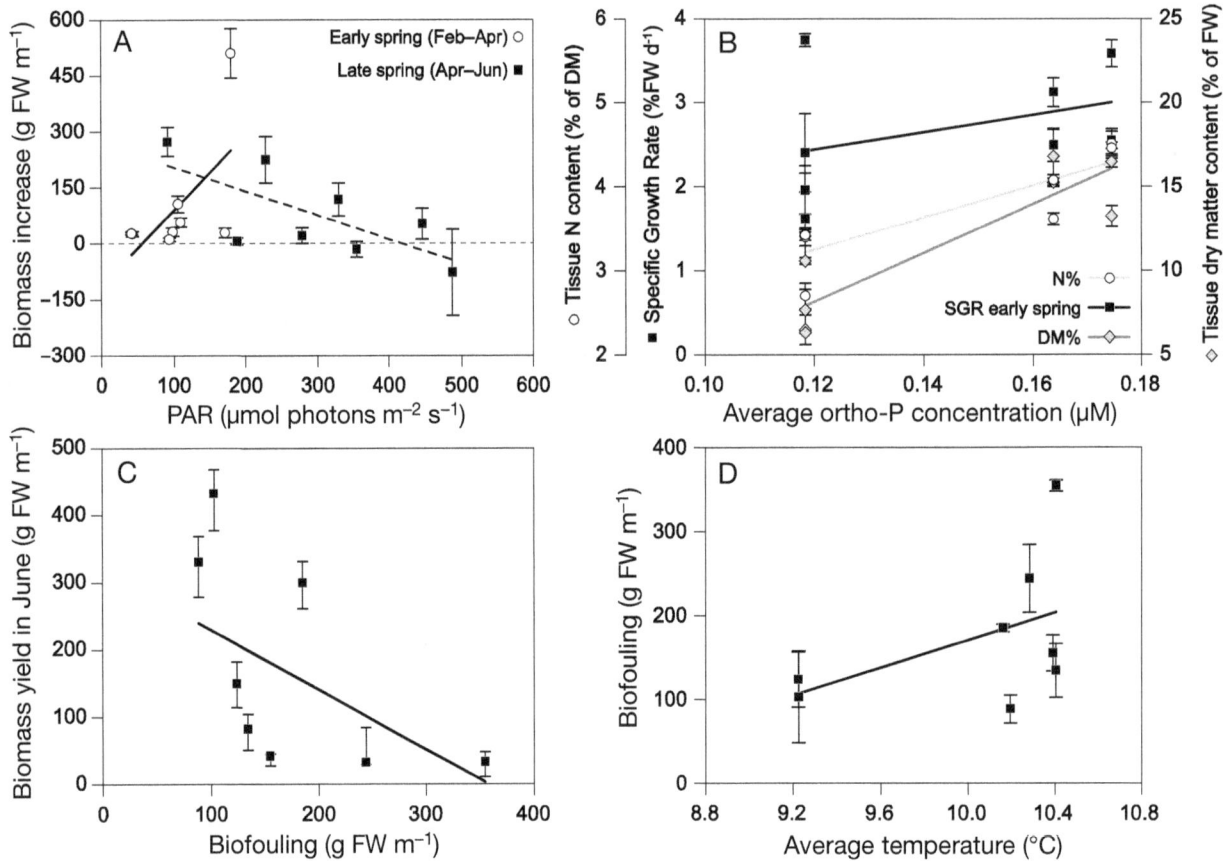

Fig. 4. Correlations between environmental parameters and *Saccharina latissima* biomass yields and quality. (A) Biomass increase or decrease in each of the 2 periods (early and late spring) as a function of light availability; (B) specific growth rate, tissue N and dry matter (DM) as a function of P availability in early spring; (C) biomass yield in June as a function of the biofouling in June (linear regression, p = 0.003, R^2 = 0.28); (D) biofouling in June as a function of average water temperature. Data
represent means ± SE, n = 3. Datapoints represent values for 4 stations (2 depths each). Significant correlations are indicated
by solid lines, non-significant relations by dotted lines. Statistics for correlations in (A,B,D) are given in Table S2 in the
Supplement.

The average molar tissue ratios of C:N and N:P
were (mean ± SE) 9.0 ± 0.3 and 69.7 ± 4.8, respectively, indicating strong P-limitation already in early
spring (data not shown).

Protein and AAs

The content of crude protein in the *S. latissima* biomass from Færker Vig was 17.0 ± 0.2% and 16.0 ±
0.1% of DM in the biomass at 1.5 m and 2.5 m,
respectively. The essential AAs (EAAs) constituted
23.8 ± 0.2 and 27 ± 1.9% of the total AAs (TAAs), at
1.5 m and 2.5 m depth, respectively. The specific
EAA, methionine, constituted 1.25 ± 0.04% (1.5 m)
and 1.37 ± 0.10% (2.5 m) of the TAAs, whereas
another EAA, lysine, constituted 3.25 ± 0.07% (1.5 m)
and 4.06 ± 0.51% of TAAs (2.5 m).

Pigments

The tissue pigment contents ranged from 1.19–
2.49 mg chl *a* g DM^{-1}, 0.62–1.09 mg fucoxanthin g
DM^{-1}, 0.01–0.04 mg violaxanthin g DM^{-1} and
0.01–0.03 mg β-carotene g DM^{-1} (Fig. 6). Higher
contents of fucoxanthin, violaxanthin and β-
carotene were found in algae cultivated at 2.5 m
depth, than at 1.5 m. Also, there was a significant
difference in the content of the 3 pigments among
sites (Table 3), with algae cultivated at Odby Bay
yielding the highest concentrations. Regarding
chl *a*, there was a significant interaction effect
between site and depth, with higher concentrations
of chl *a* at 2.5 m depth as compared to 1.5 m
(Table 3), except at Lysen Broad, where no significant difference in the chl *a* content between cultivation depths was observed (Fig. 6A, Table 3). The

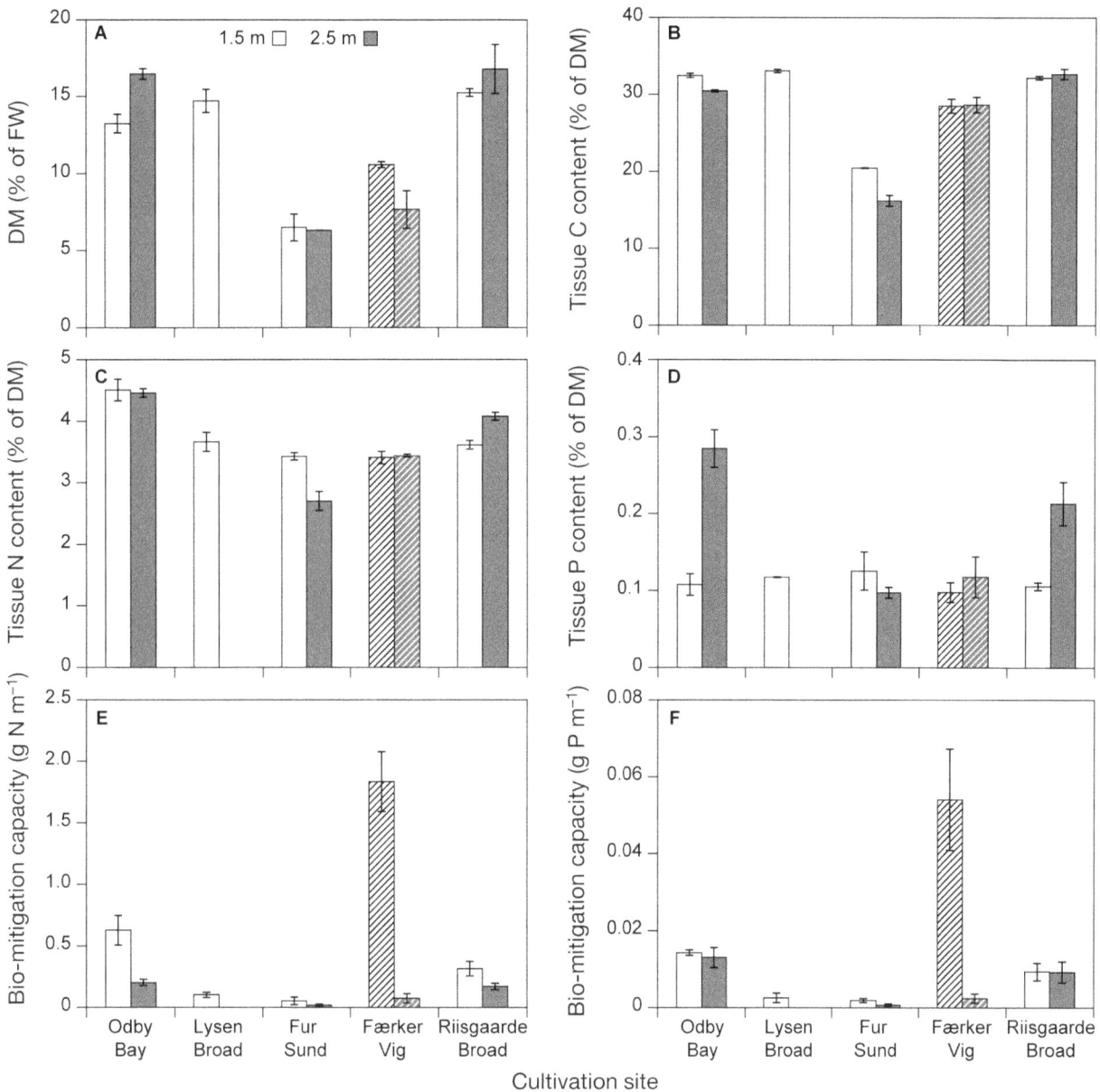

Fig. 5. Tissue concentrations of (A) dry matter (DM; % of fresh weight), (B) carbon (C), (C) nitrogen (N), (D) phosphorus (P), as well as bio-mitigation capacity of (E) N and (F) P of cultivation of *Saccharina latissima* harvested in April at the 5 cultivation sites at 1.5 (white bars) and 2.5 m depth (grey bars). Solid bars represent batch 1 of seeded lines, and crossed bars represent batch 2. Data represent means ± SE, n = 3

tissue concentrations of all pigments were negatively related to light availability (Table S2).

Harmful metals

The tissue concentrations of the harmful metals As Pb and Cd showed significant differences between sites and/or cultivation depths (Tables 3 & 5), with higher concentration of Pb and lower concentrations of As at Fur Sund as compared to the other sites. No

significant differences were observed in tissue Hg concentrations between sites and cultivation depths (Tables 3 & 5).

Metal concentrations ranged between (mean ± SE) 9.90 ± 0.93 and 31.67 ± 1.07 mg As kg DM^{-1}, 0.91 ± 0.13 and 1.72 ± 0.08 mg Cd kg DM^{-1}, 1.11 ± 0.20 and 17.60 ± 3.33 mg Pb kg DM^{-1}, and between 0.18 ± 0.01 and 1.03 ± 0.40 mg Hg kg DM^{-1} (Table 5). The tissue concentrations of As were positively correlated to SGR (linear regression, p = 0.004, R^2 = 0.285, F = 10.556, df = 23, slope = 6.207), whereas the tissue Cd

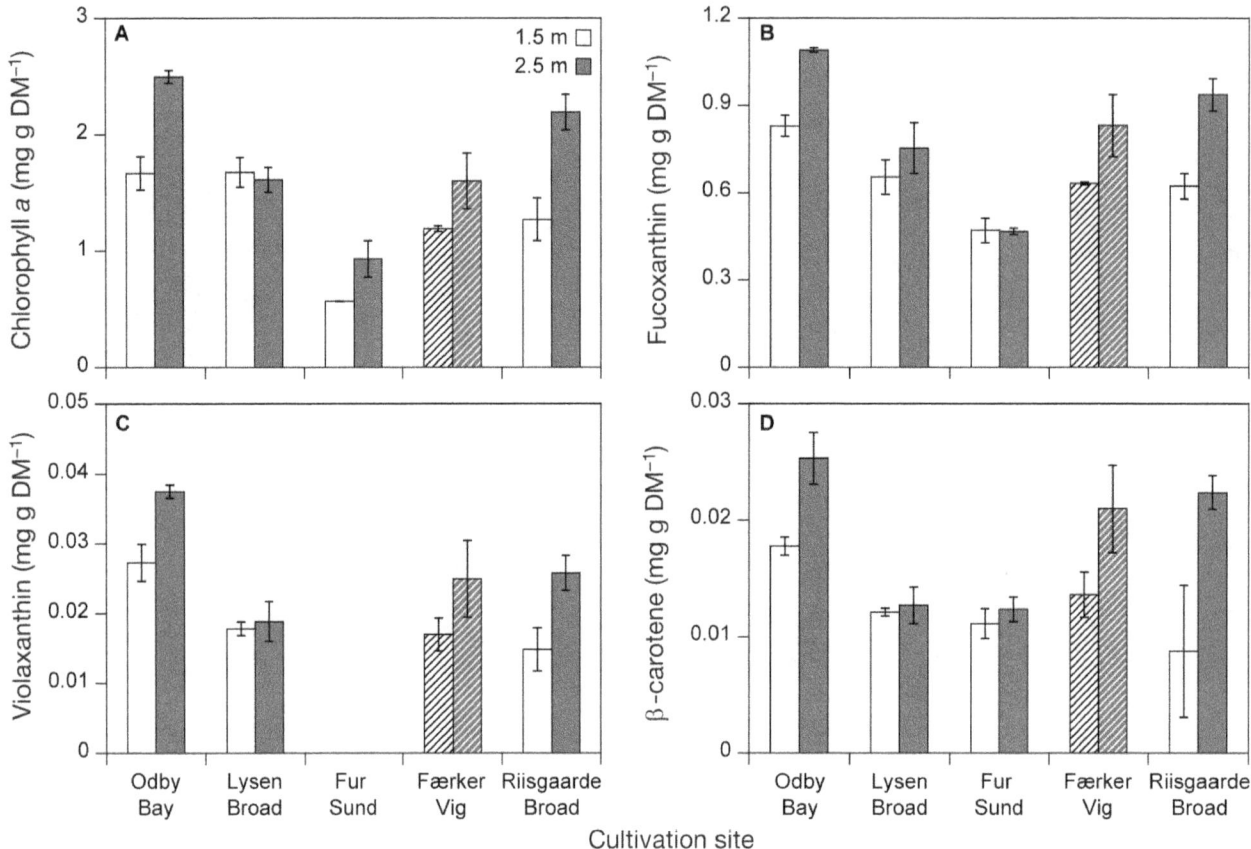

Fig. 6. Tissue concentrations of the pigments (A) chlorophyll *a*, (B) fucoxanthin, (C) violaxanthin and (D) β-carotene in *Saccharina latissima* harvested in April, at 1.5 (white bars) and 2.5 m depths (grey bars). Solid bars represent batch 1 of seeded lines, and crossed bars represent batch 2. Data represent means ± SE, n = 3

concentrations were positively correlated to sediment concentrations of Cd (linear regression, p < 0.0001, R^2 = 0.679, F = 51.878, df = 23, slope = 2.327).

DISCUSSION

Pelagic environment

The lack of correlation between the availability of DIN and ortho-P indicated different origin of the 2 nutrients. The availability of DIN was negatively correlated to salinity, indicating input with freshwater run-off from the surrounding agricultural areas. The effect of freshwater run-off was also observed in Løgstør Broad and Skive Fjord as a decrease in salinity over the cultivation period.

The positive correlation between ortho-P and chl *a* indicated that the availability of P was controlling the primary production in Limfjorden in early spring, with DIN concentrations being too high to be limiting. P-limitation has previously been observed in

eutrophic coastal regions, including parts of Limfjorden (Lyngby 1990, Holmboe et al. 1999, Lyngby et al. 1999, Pedersen et al. 2010), as a consequence of a more efficient sewage treatment reducing the emissions of P as compared to N to the marine environment (Conley et al. 2000, Kronvang et al. 2005). In late spring the lack of correlation between nutrients and pelagic phytoplankton biomass indicated that other factors came into play controlling phytoplankton biomass, potentially grazing (Maar et al. 2010), as also indicated by the increasing density of biofouling organisms (filter-feeders).

The general inverse reflection of the pelagic phytoplankton biomass (chl *a*) by the photon flux density at cultivation depth indicated a close coupling between pelagic phytoplankton density and turbidity, as is common for Limfjorden (Krause-Jensen et al. 2012). However, impaired light conditions were also observed in winter in particular in Nissum Broad and Skive Fjord, most likely as a consequence of high wind speeds causing resuspension (Nissum) and/or soft sediment that is easily resuspended (Skive).

Table 5. Tissue metal concentrations of the biomass harvested in April, expressed as ppm of fresh biomass (fresh weight) (mg kg FW^{-1}) and as ppm of dry matter (mg kg DM^{-1}). The concentrations are compared to the limit values of fresh biomass for food and food supplement according to the EU food legislation (EU 2008b), to the limit values of dry biomass for use in feed according to the EU feed legislation (EU 2013), as well as to the limit values according to the Danish regulations on sludge used as fertilizer (Danish Ministry of Environment 2006). Data are mean ± SE, n = 3. nd: no data. Underscored numbers indicate tissue concentrations exceeding limit values for food supplement. Numbers in **bold** indicate tissue concentrations exceeding limit values for feed, and numbers in *italics* indicate tissue concentrations exceeding limit values for use as fertilizer

Cultivation site	Depth (m)	As	Cd	Pb	Hg
Concentration in fresh biomass (mg kg FW^{-1})					
Odby Bay	1.5	4.02 ± 0.18	0.16 ± 0.02	0.15 ± 0.03	0.02 ± 0.00
	2.5	3.31 ± 0.21	0.19 ± 0.00	1.09 ± 0.03	0.08 ± 0.01
Lysen Broad	1.5	2.34 ± 1.17	0.16 ± 0.08	0.31 ± 0.18	0.04 ± 0.02
	2.5	nd	nd	nd	nd
Fur Sund	1.5	0.67 ± 0.02	0.06 ± 0.00	1.11 ± 0.15	nd
	2.5	0.42 ± 0.21	0.04 ± 0.02	0.68 ± 0.34	nd
Færker Vig	1.5	3.35 ± 0.07	0.11 ± 0.01	0.27 ± 0.08	0.04 ± 0.01
	2.5	2.08 ± 0.60	0.08 ± 0.02	0.26 ± 0.04	0.02 ± 0.01
Riisgaarde Broad	1.5	3.38 ± 0.18	0.26 ± 0.01	0.20 ± 0.02	<u>0.16 ± 0.06</u>
	2.5	3.18 ± 0.51	0.27 ± 0.03	0.43 ± 0.02	0.08 ± 0.01
Limit values, fresh biomass					
Mussels	–		1	1.5	0.5
Fish meat	–		0.05–0.3	0.3	0.05–1
Food supplement	–		1–3[a]	3	0.1
Concentration in dry biomass (mg kg DM^{-1})					
Odby Bay	1.5	30.37 ± 0.38	*1.19 ± 0.06*	1.11 ± 0.20	0.18 ± 0.01
	2.5	20.06 ± 0.87	*1.18 ± 0.02*	**6.65 ± 0.28**	0.51 ± 0.06
Lysen Broad	1.5	24.65 ± 1.09	*1.42 ± 0.06*	3.60 ± 0.70	0.32 ± 0.08
	2.5	nd	nd	nd	nd
Fur Sund	1.5	10.61 ± 1.38	*0.91 ± 0.13*	**17.60 ± 3.33**	nd
	2.5	9.90 ± 0.93	*0.94 ± 0.01*	**16.20 ± 0.16**	nd
Færker Vig	1.5	31.67 ± 1.07	*1.08 ± 0.11*	2.55 ± 0.83	0.39 ± 0.10
	2.5	26.11 ± 3.40	*1.04 ± 0.12*	3.66 ± 0.91	0.35 ± 0.15
Riisgaarde Broad	1.5	22.15 ± 0.83	*1.72 ± 0.08*	1.34 ± 0.10	*1.03 ± 0.40*
	2.5	18.65 ± 1.42	*1.63 ± 0.03*	2.63 ± 0.31	0.51 ± 0.07
Limit values, dry biomass					
Feed		40 (10[b])	1–2[b]	5[b]–10	
Sludge		25[c]	0.8	120	0.8

[a]Food supplement derived from seaweed; [b]complete feed (As, Cd for pet animals based on seaweed); [c]limit value only for use as fertilizer in private gardens

Biomass yield and growth performance

The *Saccharina latissima* biomass yields and sporophyte lengths obtained at the 5 sites in Limfjorden were generally low compared to values reported from other cultivation trials in Europe (Peteiro & Freire 2009, 2013b, Edwards & Watson 2011, Forbord et al. 2012, Handå et al. 2013), and in particular in Denmark (Marinho et al. 2015a, Nielsen 2015). The frond lengths were comparable to trials from Norway with a shorter grow-out period (84 vs. 166 d at Færker Vig) (Forbord et al. 2012), and the sporophyte lengths obtained at Lysen Broad were fully comparable to earlier trials at the same site (Wegeberg 2010). Only sporophytes obtained in the German Baltic Sea were smaller, with maximum lengths of 20 cm obtained in a 1 yr grow-out period (Rössner & Krost 2012). The low biomass yields were explained by several factors: (1) light limitation; (2) P-limitation reducing the SGR, and contents of DM and N in the biomass; and (3) the high degree of biofouling forcing an early harvest.

The higher yield observed at Færker Vig was most likely a consequence of the combination of less turbid waters at this site during winter and early spring as well as the earlier deployment (October instead of December), which may have given the juvenile sporophytes there a head start in growth, as has been documented from trials in Spain and Ireland (Peteiro & Freire 2009, Edwards & Watson 2011). The fact that the sporophytes were derived from a different batch of seeded lines and potentially could have been of superior quality was not supported by visual inspection at deployment. Comparing only sporophytes from Batch 1, the growth performance was best at Odby Bay, where light conditions improved markedly from April to June.

The results in general support light as a main controlling factor for growth of cultivated *S. latissima* in Limfjorden in spring. In early spring, light was positively correlated to growth and the average PAR (100–400 µmol photons m^{-2} s^{-1}) reflected a photon flux density within the range reported to saturate photosynthesis of *S. latissima* (E$_{sat}$ 20–500 µmol photons m^{-2} s^{-1}; Bartsch et al. 2008). In contrast, in late spring, the available PAR exceeded E$_{sat}$, and in this period the higher average PAR appeared to have a negative effect on growth since mainly the sporophytes from the deepest cultivation depth showed increased growth rates between April and June. A positive correlation between frond length and salinity has previously been suggested (Nielsen et al. 2014), and reduced frond length at lower salinities may be a consequence of increased allocation of energy to osmoregulation at the expense of growth.

The high turbidity of the waters in Limfjorden generally limited the extent of the vertical production potential. The turbidity in Limfjorden is primarily a consequence of high nutrient loadings supporting a high pelagic primary production (Krause-Jensen et al. 2012). Historically, the primary production in marine waters is considered to be controlled by N availability (Howarth 1988). Limfjorden, however, is an estuary with strong influence of freshwater run-off from agricultural land, and in this study, DIN was available in the water column until late spring/early summer. The limiting nutrient appeared to be P, since the bioavailable P disappeared with the onset of the phytoplankton spring bloom in early spring, and subsequently, P availability appeared to control the primary production. That P rather than N availability controlled the growth performance of *S. latissima* in this study was supported by several observations: tissue N:P ratios were, already in early spring, almost 3 times as high as other reports of kelp N:P ratios (9–25:1) (Atkinson & Smith 1983); the tissue P contents were generally below the P concentration defined as being critical for growth (0.22% P of DM, 69.4 μmol P g^{-1} DM) as suggested by Pedersen et al. (2010), whereas the tissue N concentrations were not below the concentrations critical for growth (N_C) of 1.71, as suggested for brown algae (Pedersen & Borum 1997), and 1.88 specifically for *S. latissima* (Chapman et al. 1978), and finally, SGRs in early spring corresponded positively to the ortho-P concentrations. Phosphorus limitation of macroalgae growth has previously been observed (Pedersen et al. 2010).

Biofouling of the biomass precluded a late summer harvest of the sporophytes and thus, a biomass build-up over summer. Devastation of biomass by biofouling has been reported from cultivation trials in Norway (Handå et al. 2013), Spain (Peteiro & Freire 2013a) as well as from other trials in Denmark (Wegeberg 2010, Marinho et al. 2015a, Nielsen 2015), and the phenomenon appears to be coupled to relatively sheltered locations with established natural or cultured populations of suspension-feeders. Temperature generally exerts positive control on the growth and development of juvenile filter-feeders (Widdows 1991, Nasrolahi et al. 2013), and in this study biofouling was positively correlated to temperature, even within a very narrow range of temperature differences. In the eutrophic environment in Limfjorden, food (phytoplankton) is not a limiting factor for the juvenile filter-feeders, whereas suitable substrate for settling might be. Thus, any substrate introduced in the water column, including macroalgae sporophytes, has the risk of becoming fully overgrown. In this study, the degree of biofouling was most pronounced at the deeper cultivation depths, but did not correspond to the estimated degree of exposure at the individual cultivation sites. The negative correlation between biofouling and length growth may indicate that heavily bio-fouled fronds did not grow well, or that once the fronds were sufficiently long in early spring, they were able to avoid the biofouling, the latter partly being supported by recent findings showing that dense natural kelp canopies tend to be less prone to settling of epiphytic organisms (Bennett et al. 2015).

Biomass quality

If harvested before the onset of biofouling, *S. latissima* cultivated in Færker Vig, Limfjorden, provided a rich source of protein, essential AAs and pigments with bioactive properties suitable for food or feed purposes. Availability of ortho-P influenced the quality of the biomass, significantly increasing tissue DM and N content.

As for growth performance, the biochemical composition of *S. latissima* biomass showed large differences among cultivations sites. Tissue P concentrations were generally in the same range as reported from cultivation trials in Kattegat, Denmark (Marinho et al. 2015a). A doubling of tissue P concentrations in macroalgae cultivated at 2.5 m depth at 2 sites (Odby Bay and Riisgaarde Broad), where the seabed was characterized by soft mud, indicated local differences in resuspension events as also indicated by the poorer light conditions at these sites during winter and early spring. The N content of 3–4.5% of DM in April was high for this time of the year compared to natural populations and cultivated biomass from other locations in Denmark (Nielsen et al. 2014, 2016, Marinho et al. 2015a) and was more comparable to N contents obtained in close proximity to fish farms or in late autumn/winter months where environmental N concentrations are naturally higher (Gevaert et al. 2001, Handå et al. 2013).

The high tissue N concentrations were indicative of high tissue protein concentration in the range of 16.0–17.0%. Compared to other cultivation trials in Denmark, this protein content was high for April (Marinho et al. 2015b), but comparable to what has been reported elsewhere (Black 1950). The ratio of EAAs, and the content of methionine and lysine in the biomass in Færker Vig were higher than described from *S. latissima* biomass cultivated in proximity to fish cages, and thus the *S. latissima* biomass from Limfjorden represented a biomass with an attractive profile for applications within food or feed (Marinho et al. 2015b).

Light availability influenced biomass quality, correlating positively to tissue C content, but negatively to the tissue concentrations of P and all pigments. The pigment contents in *S. latissima* from the 5 sites varied by a factor of 2–5 and were generally high due to the turbid conditions, in particular in the deeper cultivated biomass. The tissue contents of chl *a* and fucoxanthin in the biomass were up to 8 and 5 times higher, respectively, than the tissue contents in *S. latissima* fronds cultivated in autumn under low light conditions in tanks (Boderskov et al. 2016). The antioxidant and other bioactive properties of fucoxanthin have recently drawn attention as being active against obesity and diabetes (Miyashita et al. 2011, D'Orazio et al. 2012). Thus, high contents of this pigment in kelp biomass are attractive for applications in (functional) food and feed.

The positive effects of temperature on DM, N and P tissue contents may in part be explained by increased activity of enzymes involved in nutrient assimilation over the range of temperatures experienced during early spring (Davison & Davison 1987).

Only extreme levels of pollution are considered to cause significant reduction in production of marine plants (Sharp et al. 1988); however, tissue concentrations of specific metals (i.e. As, Cd, Hg and Pb) may prevent the use of the produced biomass for food, food supplement, feed or fertilizer (Miljøstyrelsen 2006, EU 2008b, 2013). In this study, we only had access to sediment concentrations of selected metals from the national environmental monitoring program, as water concentrations are not monitored. For this reason, we had no basis for estimating the environmental metal concentrations experienced by the algae, and the potential direct consequential physiological impacts. However, through the sediment concentrations, we may get an indication of the local level of environmental pollution and an indication of whether this may be a predictive tool in future site selection. The tissue concentrations of As, Pb and Hg fluctuated by a factor of 3–5 between the 5 cultivation sites, whereas the tissue concentrations of Cd were relatively constant. The tissue metal concentrations in this study did not exceed the limit values set for human consumption, and only at one site (Riis-

gaarde Broad) would the tissue Hg concentrations prevent the use of the biomass for food supplements. For use in animal feed, the Pb concentrations in the biomass cultivated at Fur Sund and Odby Bay (2.5 m) exceeded limit values, whereas the Cd concentrations would prevent the use for fertilizer of the biomass cultivated at any of the sites (limit value = 0.8 mg kg DM^{-1}, Danish Ministry of Environment 2006). The As concentrations found in this study did not exceed limit values for use in food or feed, and they were generally lower than what has been found in natural populations in more open Danish waters (Nielsen et al. 2016). Since tissue As concentrations were positively correlated to growth, bioaccumulation may explain the higher As concentrations found in older individuals in natural populations, as compared to the 1-yr-old cultivated individuals in this study. The linear correlation between tissue and sediment Cd concentrations indicated that elevated sediment concentrations of Cd may cause increased availability and hence uptake into the seaweed tissue. At the 2 stations with the highest sediment Cd concentrations (Riisgaarde: 0.44 ppm and Lysen: 0.25 ppm), the seabed sediment and depth, as well as the degree of exposure, differed. At Riisgarde the sediment was soft and muddy, and a high degree of exposure increased the risk of resuspension of the sediment into the water column, potentially increasing the availability of Cd to the seaweed. Below the cultivation structures at the more sheltered and shallow site in Lysen Broad, the seabed consisted of fine sand and clay, and there the depth was lower. Thus, despite different conditions regarding sediment, depth and exposure, the sediment concentration of Cd demonstrated a potential value as an instrument in site selection.

Bio-mitigation

The bio-mitigation capacity of *S. latissima* in this study proved to be relatively poor in comparison with other studies of Laminariales in Denmark (Marinho et al. 2015a) and Scotland (Sanderson et al. 2012), where up to 4 and 1.4 times more N was removed per metre of seeded line, respectively. The low bio-mitigation capacity was primarily a consequence of the low biomass yields obtained due to turbid waters, P-limitation and biofouling. Thus, in highly eutrophic waters such as Limfjorden, the pelagic primary productivity limits the efficiency of kelp cultivation as a tool for bio-mitigation of N. Consequently, care should be taken when extrapolating the bio-mitigation capacities described in the literature to any cultivation site assuming high areal productivity (i.e. Holdt & Edwards 2014). This study highlights the limitations and challenges of kelp production for bio-mitigation purposes in eutrophic waters, where bio-mitigation is needed the most.

Site selection

Even within the relatively homogenous eutrophic Limfjorden, production yields varied by a factor of 10 between different basins. Environmental monitoring data proved useful as predictive instruments for site selection. Regarding the pelagic parameters, generally, highly N-enriched sites with low light availability, high pelagic N:P ratios and high chl *a* concentrations should be avoided, as they supported a lower biomass production and, in conjunction with marginally higher temperatures during spring, also presented a higher risk of biofouling.

Regarding sediment characteristics, 2 recommendations for site selection are suggested: (1) kelp cultivation should be reconsidered in shallow areas dominated by soft muddy seabed, as resuspension events tend to increase turbidity; and (2) sediment Cd concentrations could be investigated as a part of site selection. High sediment Cd concentrations were reflected as high Cd concentrations in seaweed biomass, and depending on the post-harvest use of the biomass, high tissue Cd concentrations may have a strong negative impact on biomass value.

CONCLUSIONS

Basin-scale differences in light and nutrient availability, seabed properties and sediment metal concentrations cause pronounced local differences in the suitability of an area for cultivation of *Saccharina latissima* in terms of biomass yield and quality as well as bio-mitigation, and hence, impact the profitability of potential seaweed production. When selecting sites for cultivation of *S. latissima*, highly N-enriched sites with low light availability, high pelagic N:P ratios and chl *a* concentrations, and high sediment Cd concentration should be avoided. The highly N-enriched waters of Limfjorden appeared less suitable for efficient biomass production of *S. latissima* due to reduced light conditions and P-limitation in early spring, and a high risk of devastating biofouling impairing growth performance, bio-mitigation capacity as well as biomass quality. However, *S. latissima* bio-

mass harvested in spring in Limfjorden had a high content of pigments and protein with a beneficial amino acid composition, and proved highly suitable for food or feed purposes.

Acknowledgements. The work behind this article was supported by 'De Lokale Dyder' (The Market Development Fund), the PEER project on improved resource flows between human and natural systems, the Macroalgae Biorefinery (MAB3) (Danish Council for Strategic Research) and finally a grant supplied by the National Centre for Environment and Energy (DCE). The authors thank Kristian Oddershede Nielsen, Helge Boesen, Finn Bak and Pascal Barreau for the field work, Tanja Quottrup Egholm, Kitte Linding Gerlich, Gitte Jacobsen, Anne Marie Plejdrup and Peter Kofoed for skillful lab work, Ole Manscher and David Rytter for extraction of data from ODAM, Tinna Christensen for graphical assistance and 3 anonymous reviewers for constructive comments improving the manuscript.

LITERATURE CITED

Angell A, Mata L, de Nys R, Paul N (2016) The protein content of seaweeds: a universal nitrogen-to-protein conversion factor of five. J Appl Phycol 28:511

Atkinson MJ, Smith SV (1983) C-N-P ratios of benthic marine plants. Limnol Oceanogr 28:568–574

Bartsch I, Wiencke C, Bischof K, Buchholz CM and others (2008) The genus *Laminaria* sensu lato: recent insights and developments. Eur J Phycol 43:1–86

Bennett S, Wernberg T, de Bettignies T, Kendrick GA and others (2015) Canopy interactions and physical stress gradients in subtidal communities. Ecol Lett 18:677–686

Black WAP (1950) The seasonal variation in weight and chemical composition of the common British Laminariaceae. J Mar Biol Assoc UK 29:45–72

Boderskov T, Schmedes PS, Bruhn A, Rasmussen MB, Nielsen MM, Pedersen MF (2016) The effect of light and nutrient availability on growth, nitrogen, and pigment contents of *Saccharina latissima* (Phaeophyceae) grown in outdoor tanks, under natural variation of sunlight and temperature, during autumn and early winter in Denmark. J Appl Phycol 28:1163–1165

Bruton T, Lyons H, Lerat Y, Stanley M, Rasmussen B (2009) A review of the potential of marine algae as a source of biofuel in Ireland. Sustainable Energy Authority of Ireland, Dublin

Buck BH, Buchholz CM (2004) The offshore-ring: a new system design for the open ocean aquaculture of macroalgae. J Appl Phycol 16:355–368

Buck BH, Buchholz CM (2005) Response of offshore cultivated *Laminaria saccharina* to hydrodynamic forcing in the North Sea. Aquaculture 250:674–691

Buck BH, Krause G, Michler-Cieluch T, Brenner M and others (2008) Meeting the quest for spatial efficiency: progress and prospects of extensive aquaculture within offshore wind farms. Helgol Mar Res 62:269–281

Castine SA, McKinnon AD, Paul NA, Trott LA, De Nys R (2013) Wastewater treatment for land-based aquaculture: improvements and value-adding alternatives in model systems from Australia. Aquacult Environ Interact 4:285–300

Chapman ARO, Markham JW, Lüning K (1978) Effects of nitrate concentrations on the growth and physiology of Laminaria saccharina (Phaeophyta) in culture. J Phycol 14:195–198

Christiansen T, Christensen TJ, Markager S, Petersen JK, Mouritsen LT (2006) Limfjorden i 100 år. Klima, hydrografi, næringsstoftilførsel, bundfauna og fisk i Limfjorden fra 1897 til 2003. Report 578. National Environmental Research Institute, Roskilde

Conley DJ, Kaas H, Møhlenberg F, Rasmussen B, Windolf J (2000) Characteristics of Danish estuaries. Estuaries 23: 820–837

Coquery M, Carvalho FP, Azemard S, Bachelez M, Horvat M (2000) Certification of trace and major elements and methylmercury concentrations in a macroalgae (Fucus sp.) reference material, IAEA-140. Fresenius J Anal Chem 366:792–801

Culmo RF (2010) Methods of organic nitrogen analysis: Kjeldahl and the EA2410 N Analyzer (Dumas Method). PerkinElmer publication EAN-8

D'Orazio N, Gemello E, Gammone MA, de Girolamo M, Ficoneri C, Riccioni G (2012) Fucoxantin: a treasure from the sea. Mar Drugs 10:604–616

Daly HE (1998) The return of Lauderdale's paradox. Ecol Econ 25:21–23

Danish Ministry of Environment (2006) Anvendelse af affald til jordbrugsformål. BEK nr 1650 af 13/12/2006 (Slambekendtgørelsen). Danish Ministry of Environment, Copenhagen

Davison IR, Davison JO (1987) The effect of growth temperature on enzyme activities in the brown alga Laminaria saccharina. Br Phycol J 22:77–87

Edwards M, Watson L (2011) Cultivating Laminaria digitata. BIM Aquaculture Explained 26. Irish Sea Fisheries Board, Dublin

EU (2008a) Marine Strategy Framework Directive. 2008/56/EC. Off J Eur Union L164:19-40

EU (2008b) Commission regulation (EC) No 629/2008 of 2 July 2008 amending Regulation (EC) No 1881/2006 setting maximum levels for certain contaminants in foodstuffs. Off J Eur Union L 173:6–9

EU (2013) Commission regulation (EU) No 1275/2013 of 6 December 2013 -amending Annex I to Directive 2002/32/EC of the European Parliament and of the Council as regards maximum levels for arsenic, cadmium, lead, nitrites, volatile mustard oil and harmful botanical impurities. Off J Eur Union L 328:86–92

EU (2014) Establishing a framework for maritime spatial planning 2014/89/EU. Off J Eur Union L 257:135–145

FAO (2016) The state of world fisheries and aquaculture 2016. Contributing to food security and nutrition for all. FAO, Rome. www.fao.org/3/a-i5555e

Forbord S, Skjermo J, Arff J, Handa A, Reitan KI, Bjerregaard R, Lüning K (2012) Development of Saccharina latissima (Phaeophyceae) kelp hatcheries with year-round production of zoospores and juvenile sporophytes on culture ropes for kelp aquaculture. J Appl Phycol 24: 393–399

Gevaert F, Davoult D, Creach A, Kling R, Janquin MA, Seuront L, Lemoine Y (2001) Carbon and nitrogen content of Laminaria saccharina in the eastern English Channel: biometrics and seasonal variations. J Mar Biol Assoc UK 81:727–734

Grasshoff K, Ehrhardt M, Kremling K (1983) Methods of seawater analysis. Verlag Chemie, Weinheim

Handå A, Forbord S, Wang XX, Broch OJ and others (2013) Seasonal- and depth-dependent growth of cultivated kelp (Saccharina latissima) in close proximity to salmon (Salmo salar) aquaculture in Norway. Aquaculture 414–415:191–201

Holdt SL, Edwards MD (2014) Cost-effective IMTA: a comparison of the production efficiencies of mussels and seaweed. J Appl Phycol 26:933–945

Holmboe N, Jensen HS, Andersen FØ (1999) Nutrient addition bioassays as indicators of nutrient limitation of phytoplankton in an eutrophic estuary. Mar Ecol Prog Ser 186:95–104

Howarth RW (1988) Nutrient limitation of net primary production in marine ecosystems. Annu Rev Ecol Syst 19: 89–110

Kerrison PD, Stanley MS, Edwards MD, Black KD, Hughes AD (2015) The cultivation of European kelp for bioenergy: site and species selection. Biomass Bioenergy 80: 229–242

Kraan S (2013) Mass-cultivation of carbohydrate rich macroalgae, a possible solution for sustainable biofuel production. Mitig Adapt Strategies Glob Change 18:27–46

Krause-Jensen D, Markager S, Dalsgaard T (2012) Benthic and pelagic primary production in different nutrient regimes. Estuaries Coasts 35:527–545

Kronvang B, Jeppesen E, Conley DJ, Søndergaard M, Larsen SE, Ovesen NB, Carstensen J (2005) Nutrient pressures and ecological responses to nutrient loading reductions in Danish streams, lakes and coastal waters. J Hydrol (Amst) 304:274–288

Larsen MM (2013) Environmentally dangerous substances in sediment (NOVANA Technical Instruction for Marine Monitoring). M24. Danish Centre for Environment and Energy, Aarhus University

Lyngby JE (1990) Monitoring of nutrient availability and limitation using the marine macroalgae Ceramium rubrum (Huds) G. Ag. Aquat Bot 38:153–161

Lyngby JE, Mortensen S, Ahrensberg N (1999) Bioassessment techniques for monitoring of eutrophication and nutrient limitation in coastal ecosystems. Mar Pollut Bull 39:212–223

Maar M, Timmermann K, Petersen JK, Gustafsson KE, Storm LM (2010) A model study of the regulation of blue mussels by nutrient loadings and water column stability in a shallow estuary, the Limfjorden. J Sea Res 64: 322–333

Manns D, Deutschle AL, Saake B, Meyer AS (2014) Methodology for quantitative determination of the carbohydrate composition of brown seaweeds (Laminariaceae). RSC Advances 4:25736–25746

Marinho G, Holdt S, Birkeland M, Angelidaki I (2015a) Commercial cultivation and bioremediation potential of sugar kelp, Saccharina latissima, in Danish waters. J Appl Phycol 27:1963–1973

Marinho GS, Holdt SL, Angelidaki I (2015b) Seasonal variations in the amino acid profile and protein nutritional value of Saccharina latissima cultivated in a commercial IMTA system. J Appl Phycol 27:1991–2000

Markager SS (2004) Light extinction (NOVANA technical instruction for marine monitoring). Book 1.3. Danish Centre for Environment and Energy, Aarhus University

Markager SS, Fossing H (2013) Chlorophyll a concentration (NOVANA technical instruction for marine monitoring). Book M07. Danish Centre for Environment and Energy, Aarhus University

Markager S, Storm LM, Stedmon CA (2006) Limfjordens miljøtilstand 1985 til 2003. Sammenhæng mellem næringsstoftilførsler, klima og hydrografi belyst ved empiriske modeller. Report 577 National Environmental Research Institute, Roskilde

Miyashita K, Nishikawa S, Beppu F, Tsukui T, Abe M, Hosokawa M (2011) The allenic carotenoid fucoxanthin, a novel marine nutraceutical from brown seaweeds. J Sci Food Agric 91:1166–1174

Nasrolahi A, Pansch C, Lenz M, Wahl M (2013) Temperature and salinity interactively impact early juvenile development: a bottleneck in barnacle ontogeny. Mar Biol 160: 1109–1117

Neori A, Chopin T, Troell M, Buschmann AH and others (2004) Integrated aquaculture: rationale, evolution and state of the art emphasizing seaweed biofiltration in modern mariculture. Aquaculture 231:361–391

Nielsen MM (2015) Cultivation of kelps for energy, fish feed and bioremediation. PhD thesis, Aarhus University

Nielsen MM, Bruhn A, Rasmussen MB, Olesen B, Larsen MM, Møller HB (2012) Cultivation of Ulva lactuca with manure for simultaneous bioremediation and biomass production. J Appl Phycol 24:449–458

Nielsen MM, Krause-Jensen D, Olesen B, Thinggaard R, Christensen P, Bruhn A (2014) Growth dynamics of Saccharina latissima (Laminariales, Phaeophyceae) in Aarhus Bay, Denmark, and along the species' distribution range. Mar Biol 161:2011

Nielsen MM, Manns D, D'Este M, Krause-Jensen D and others (2016) Variation in biochemical composition of Saccharina latissima and Laminaria digitata along an estuarine salinity gradient in inner Danish waters. Algal Res 13:235–245

Nordic Committee on Food Analysis (2003) Nitrogen. Determination in foods and feeds according to Kjeldahl, 4th edn. NMKL 6, NordVal International, DTU Food, Danish Technical University, Søbord

Pedersen MF, Borum J (1997) Nutrient control of estuarine macroalgae: growth strategy and the balance between nitrogen requirements and uptake. Mar Ecol Prog Ser 161:155–163

Pedersen B, Ærtebjerg G, Larsen MM (2004) Water chemistry parametres (NOVANA Technical Instruction for Marine Monitoring). Book 2.2. Danish Centre for Environment and Energy, Aarhus University

Pedersen MF, Borum J, Fotel FL (2010) Phosphorus dynamics and limitation of fast- and slow-growing temperate seaweeds in Oslofjord, Norway. Mar Ecol Prog Ser 399: 103–115

Peteiro C, Freire O (2009) Effect of outplanting time on commercial cultivation of kelp Laminaria saccharina at the southern limit in the Atlantic coast, NW Spain. Chin J Oceanology Limnol 27:54–60

Peteiro C, Freire O (2013a) Epiphytism on blades of the edible kelps Undaria pinnatifida and Saccharina latissima farmed under different abiotic conditions. J World Aquacult Soc 44:706–715

Peteiro C, Freire Ó (2013b) Biomass yield and morphological features of the seaweed Saccharina latissima cultivated at two different sites in a coastal bay in the Atlantic coast of Spain. J Appl Phycol 25:205–213

Petersen JK, Hasler B, Timmermann K, Nielsen P, Tørring DB, Larsen MM, Holmer M (2014) Mussels as a tool for mitigation of nutrients in the marine environment. Mar Pollut Bull 82:137–143

Rössner Y, Krost P (2012) Verfahrensentwicklung und Anlagenkonzeption für die extraktive Aquakultur von Muscheln und Makroalgen in der Ostsee (Extractive Baltic Aquaculture of Mussels and Algae EBAMA). Abschlussbericht für das Projekt EBAMA AZ 27119-34. Coastal Research & Management, Kiel

Sanderson JC, Dring MJ, Davidson K, Kelly MS (2012) Culture, yield and bioremediation potential of Palmaria palmata (Linnaeus) Weber & Mohr and Saccharina latissima (Linnaeus) C.E. Lane, C. Mayes, Druehl & G.W. Saunders adjacent to fish farm cages in northwest Scotland. Aquaculture 354–355:128–135

Seghetta M, Tørring DB, Bruhn A, Thomsen M (2016) Bioextraction potential of macroalgae in Denmark—an instrument for circular nutrient management. Sci Total Environ 563-564:513–529

Sharp GJ, Samant HS, Vaidya OC (1988) Selected metal levels of commercially valuable seaweeds adjacent to and distant from point sources of contamination in Nova Scotia and New Brunswick. Bull Environ Contam Toxicol 40:724–730

Smale DA, Burrows MT, Moore P, O'Connor N, Hawkins SJ (2013) Threats and knowledge gaps for ecosystem services provided by kelp forests: a northeast Atlantic perspective. Ecol Evol 3:4016–4038

Stephens D, Capuzzo E, Aldrigde J, Forster RM (2014) Potential interactions of seaweed farms with natural nutrient sinks in kelp beds. The Crown Estate, London

Timmermann K, Dinesen GE, Markager S, Ravn-Jonsen L, Bassompierre M, Roth E, Støttrup JG (2014) Development and use of a bio-economic model for management of mussel fisheries under different nutrient regimes in the temperate estuary of the Limfjord, Denmark. Ecol Soc 19:14

Troell M, Rönnbäck P, Halling C, Kautsky N, Buschmann A (1999) Ecological engineering in aquaculture: use of seaweeds for removing nutrients from intensive mariculture. J Appl Phycol 11:89–97

Vang T (2013) CTD measurement (NOVANA technical instruction for marine monitoring). Book M03. Danish Centre for Environment and Energy, Aarhus University

Vang T, Hansen JW (2013) Oxygen in the water column (NOVANA technical instruction for marine monitoring). Book M04. Danish Centre for Environment and Energy, Aarhus University

Wegeberg S (2010) Cultivation of kelp species in the Limfjord, Denmark. Department of Biology, SCIENCE, Copenhagen University

Wegeberg S, Mols-Mortensen A, Engell-Sørensen K (2013) Sustainable production and utilization of marine resources in the Arctic, fish and seaweed (SPUMA). Danish Centre for Environment and Energy rapport, Aarhus University

Wei N, Quarterman J, Jin YS (2013) Marine macroalgae: an untapped resource for producing fuels and chemicals. Trends Biotechnol 31:70–77

Werner A, Edwards M, Mineur F, O'Mahony F, Guiry M, Maggs C, Dring MJ (2009) Development of commercial-scale seaweed aquaculture for selected species in Ireland. Phycologia 48(Suppl):141

Widdows J (1991) Physiological ecology of mussel larvae. Aquaculture 94:147–163

Wiles PJ, van Duren LA, Häse C, Larsen J, Simpson JH (2006) Stratification and mixing in the Limfjorden in relation to mussel culture. J Mar Syst 60:129–143

7

Effect of exposure on salmon lice *Lepeophtheirus salmonis* population dynamics in Faroese salmon farms

Esbern J. Patursson[1,2,*], Knud Simonsen[1], André W. Visser[2], Øystein Patursson[1]

[1]Fiskaaling – Aquaculture Research Station of the Faroes, við Áir, 430 Hvalvík, Faroe Islands
[2]VKR Centre for Ocean Life, National Institute of Aquatic Resources, Technical University of Denmark, Kavalergaarden 6, 2920 Charlottenlund, Denmark

ABSTRACT: We assessed variations in salmon lice *Lepeophtheirus salmonis* population dynamics in Faroese salmon farms in relationship to their physical exposure to local circulation patterns and flushing with adjacent waters. Factors used in this study to quantify physical exposure are estimates of the freshwater exchange rate, the tidal exchange rate and dispersion by tidal currents. Salmon farms were ranked according to the rate of increase in the average numbers of salmon lice per fish. In a multiple linear regression, physical exposure together with temperature were shown to have a significant effect on the rate of lice infection. The sites with low exposure revealed higher rates of self-infection and internally driven outbreak dynamics, while high-exposure sites showed lower rates of self-infection, tending towards externally driven outbreak dynamics. The low-exposure sites also appeared to have a lower threshold of salmon stocking numbers for outbreaks of infection. The study presents a simple method of characterizing salmon farming fjords in terms of their different exposure levels and how they relate to potential self-infection at these sites.

KEY WORDS: Sea lice · Fjord · Freshwater exchange · Tidal exchange · Tidal currents · *Salmo salar* · Self-infection · Faroe Islands

INTRODUCTION

Stocking hundreds of thousands of fish in small areas makes fish farms ideal breeding grounds for sea lice (including salmon lice *Lepeophtheirus salmonis* [Krøyer, 1837]). The lice feed on the protective slime layer, the scales and blood of the salmon, and thereby impair the salmon immune system, reduce its ability to osmo-regulate and cause stress (Pike & Wadsworth 1999). The Faroe Islands require by law that the number of sea lice are monitored in the farming industry (Faroese Food and Veterinary Authority 2009, 2013). The total yearly cost of the sea lice is estimated to €0.19 kg^{-1} fish (Costello 2009), which totals over €15 million for the production of 82 000 t in 2015 in the Faroe Islands alone. High costs, environmental concerns and increased resistance to pharmaceutical treatments (Jimenez et al. 2011) in this rapidly expanding industry provide a strong incentive to develop effective and sustainable methods for the control of sea lice.

Outbreaks of sea lice infections vary considerably between sites. Some farms regularly experience severe sea lice epidemics, while others are barely affected. The reason for such variation is complex, but includes animal husbandry practices such as disease management procedures as well as the physical and biological conditions of the particular location. For example, flushing with adjacent waters may reduce reinfection rates (Pike & Wadsworth 1999).

The life cycle of the salmon louse *L. salmonis* is composed of multiple stages, including 2 planktonic

*Corresponding author: esbern_p@hotmail.com

stages (nauplius I and II) that are followed by a free-swimming infectious copepodid stage, which is mainly restricted to the upper layers of the sea (Hevrøy et al. 2003). Once attached to a host, copepodids eventually develop into adults, and adult females potentially release 26 to 68 nauplii through the protruding egg sacs daily, which then are free to develop to the copepodid stage and to infect other hosts (Heuch et al. 2000, á Norði et al. 2016). The growth of a sea lice population has an exponential nature (Frazer et al. 2012) and is generally modelled as such (Costello 2006, Krkošek et al. 2010, Frazer et al. 2012). Several dispersion models have been developed which assume that the planktonic stages of the sea lice drift freely with prevailing currents close to the surface (Amundrud & Murray 2009, Adams et al. 2012, Salama et al. 2013, Asplin et al. 2014), with recent works including vertical positioning as a response to environmental cues (Johnsen et al. 2014, 2016). Sea temperature is widely accepted as a basic factor influencing the growth rate of sea lice populations, as the generation time decreases with increasing temperature (Tully 1992, Heuch et al. 2000). However, the relationship between temperature and the sea lice abundance is not a simple relationship, and annual peaks and troughs in the abundance of mobile *L. salmonis* may appear delayed compared with maximum and minimum annual temperature (Jansen et al. 2012). Naupliar development is salinity dependent, and complete development from nauplii to copepodid is only achieved at salinities ≥30‰ (Pike & Wadsworth 1999, Brooks 2005). The copepodid has a slightly greater survival rate at lowered salinities than the nauplii, but survives best above 15‰ (Pike & Wadsworth 1999). Other authors have stated that *L. salmonis* tend to avoid salinities below 24–25‰ (Krkošek et al. 2005, Asplin et al. 2014). A key concept in theoretical epidemiology is that increasing host density should promote the population growth of a parasite as the chance of contact increases with increased host density (Anderson & May 1991). High abundances of fish, not only within individual farms but also integrated across farming regions, can affect sea lice outbreaks. Thus, the negative feedback from high densities is not solely due to the density of fish at each site or pen, but also to the density in farming regions (Jansen et al. 2012).

High *L. salmonis* abundance and potential re-infection is often associated with farms situated in areas with weak exchange of waters (Tully & Nolan 2002, Revie et al. 2003). Krkošek et al. (2010) introduced a salmon lice growth model based on simple mathematical host–macroparasite models following the Anderson-May approach (Anderson & May 1978). Growth is split into 2 modes. One is the externally driven mode, which assumes that copepodids enter the farm from the external environment and nauplii released from adult sea lice disperse into the external environment with no re-infection of farmed fish. The other mode is internally driven, and assumes that re-infection and population dynamics are driven from the parasite population inside the farm or the local environment. The internally driven mode will have an exponential salmon lice population growth, whereas the externally driven model will have a constant growth rate reaching a steady-state when the infection pressure equals mortality.

There are 3 basic processes that induce circulation in an estuary: wind, tides and the density-driven flow associated with freshwater inflow (Pritchard 1967). Tides are mainly manifested as oscillatory currents, but also produce residual eddies that can effect a net exchange through inlets (Visser & Bowman 1991), and exchange of suspended material (e.g. planktonic organisms) by tidal dispersion (Geyer & Signell 1992, Nguyen et al. 2008). In the Faroe Islands, the tidal currents are quite moderate in most fjords hosting the majority of the fish farms, but relatively strong with maximum speeds up to 7 knots in the straits between the islands (Simonsen & Niclasen 2011).

Together with temperature and host density, sea lice population dynamics in various Faroese fish farming fjords may be expected to vary in accordance with the flushing rate of these fjords with adjacent waters. The principle factors involved in this exchange of waters are the tidal range, tidal dispersion and freshwater forcing, which we combine and refer to herein as the 'exposure' of a fjord. Even though winds are also a strong influence on the exchange of water in the fjords, they tend to be highly variable in strength and direction. Their net effect on the water exchange is not easily quantified, and thus is not included in present study. We also exclude the effects of salinity. Even though sea lice prefer high salinities (Pike & Wadsworth 1999, Brooks 2005, Krkošek et al. 2005, Asplin et al. 2014), the salinity in the typical Faroese fjord is nowhere near the threshold for any observable impact. The salinity in the Faroe shelf ranges from 35.1 to 35.3‰ (Hansen & Østerhus 2000, Larsen et al. 2008), and the salinity in the fjords is only slightly lower (Gaard et al. 2011).

Observations of sea lice dynamics are obtained from the compulsory sea lice monitoring programme (Faroese Food and Veterinary Authority 2009, 2013). These observations are a valuable resource, lending much needed insight into the epidemiology of sea lice infections (Pike & Wadsworth 1999), not only in their inherent dynamics, but also how self-infection can be influenced by the physical exposure of farming sites.

MATERIALS AND METHODS

Study area

The waters around the Faroe Islands are dominated by the relatively warm and saline North Atlantic Current Water, a derivative of the Gulf Stream. On the shelf is a steady current clockwise around the islands, partly driven by tidal rectification (Larsen et al. 2008), i.e. a mean current generated by the nonlinearities inherent in tidal dynamics and strongly influenced by bathymetry. The tidal wave enters the southwestern shelf and propagates on both sides of the archipelago to be rejoined on the northeastern shelf, with a virtual amphidromic point for the dominant semidiurnal tides on the eastern coast on the central island. This causes strong tidal currents in the straits between the islands (Simonsen & Niclasen 2011). In contrast, tidal currents within fjords tend to be much weaker, decreasing in intensity with distance from the open sea. Tidal currents and wave action induce intense mixing

on the shelf and produce a very homogenous water mass, with temperature varying around 12°C in summer to slightly below 6°C in winter, and salinity from about 35 to 35.3‰ (Larsen et al. 2008, Gaard et al. 2011). The fjords are slightly fresher due to the runoff from land, with some stratification. The salinity in the upper layers of Faroese fjords usually ranges between 34.5 and 34.9‰ in summer and between 33.6 and 34.8‰ in winter (Gaard et al. 2011). Most of the fjords in the Faroe Islands are simple systems, with usually 1 or 2 aquaculture sites in each fjord (Fig. 1) compared, for example, to the larger and more complex fjord systems seen in Norway. The homogeneity of the water in the fjords (Larsen et al. 2008, Gaard et al. 2011), the negligible effect of salinity, the simplicity of the fjord systems and the high variation in exposure make the Faroe Islands ideal for studying the effect of physical exposure on salmon lice population dynamics.

Sea lice data

Sea lice data were obtained from the Fiskaaling – Aquaculture Research Station (ARSF) of the Faroes, with permission from the farming companies in the

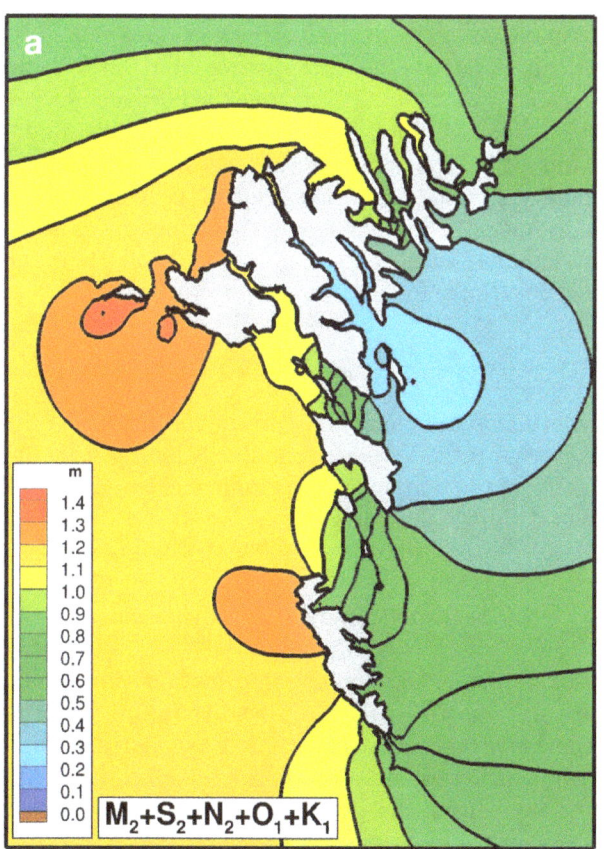

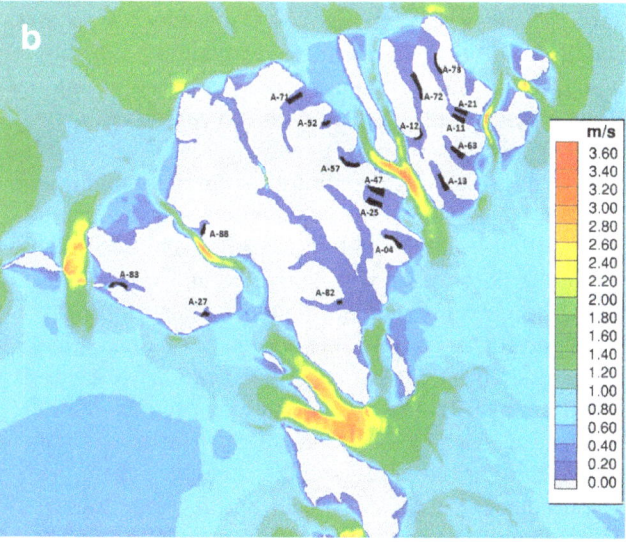

Fig. 1. (a) Calculated amplitude of the maximum tidal height in meters. (b) Calculated maximum tidal currents around the Faroe Islands, obtained from harmonic analysis of numerical model results (Simonsen & Niclasen 2011). The calculation was based on the M2, S2, N2, K2 and O1 constituents of tidal harmonics. The farming areas of the fjords used in this study are marked (shaded black; codes are from Faroese aquaculture authorities and not used in this study)

Faroe Islands. A team at ARSF conducts the monitoring on behalf of the farmers, ensuring standardized monitoring practices. The minimum legal requirement (Faroese Food and Veterinary Authority 2009, 2013), is that 2 fixed and 2 random pens are monitored for sea lice at each site. Here the average numbers of salmon lice (motile *Lepeophtheirus salmonis*) in the 2 fixed pens were used. Non-motile stages (chalimus) were not included, as there is no distinction between *L. salmonis* and *Caligus elongatus* in the monitoring programme. In a few cases, only data from a single pen was available. To avoid the problem of routine treatment and differences in these treatments (Jaworski & Holm 1992), we only used data collected before any treatment was implemented on the farm.

Temperature, freshwater and tidal data

Temperature data were obtained from a measuring station representing the well-mixed shelf water (Larsen et al. 2008). The temperature in the fjords, particularly at the surface, may vary slightly from the shelf temperature, but generally this variation is less than 1°C (Larsen et al. 2008, Gaard et al. 2011).

The average precipitation per unit area over the Faroe Islands was obtained by the isohyet map of Davidsen et al. (1994), which was combined with a topographic map available from an internet map service (www.kortal.fo) to obtain the catchment area and thus average runoff into each fjord.

Tidal data were from a numerical simulation with a barotrophic version of the regional oceanic modeling system (ROMS) (Shchepetkin & McWilliams 2005) set up for the Faroe Shelf with 100 m equidistant resolution. The currents were validated against vertically averaged current profile data in 4 channels and on the western shelf, while sea surface elevation was validated against 21 tidal gauge stations around the islands (Simonsen & Niclasen 2011). The maximum tidal range for the entire shelf is calculated from the amplitudes of the M_2, S_2, N_2, O_1 and K_1 constituents. The calculated maximum tidal range (H) reaches its maximum at 2.8 m at the western islands, and decreases to a minimum of 0.4 m east of the central island (Fig. 1). The minimum amplitudes of the dominating M_2 and S_2 constituents are 0.14 and 0.056, respectively (Farvandsvæsenet 2005). The fjords are therefore all microtidal to mesotidal (Mikhailov 1997).

A theoretical maximum current estimate (U) was estimated as the sum of the semi major axes of the M_2, S_2, N_2, O_1, K_1 and Q_1, respectively. The highest values of U in the straits exceed 3.5 m s^{-1} (Fig. 1b). The simulation underestimates the current near land and in the fjords, but is fairly representative in the straits (Simonsen & Niclasen 2011).

Exchange rates

The freshwater exchange rate (E_f; d^{-1}) is defined as the ratio between the average daily freshwater runoff (R) and the volume of the fjord (V_f):

$$E_f = \frac{R}{V_f} \tag{1}$$

The tidal prism (P) is the area of the fjord (A) multiplied by the tidal range (H):

$$P = H \times A \tag{2}$$

The tidal exchange rate (E_t; d^{-1}) is the tidal prism (P) divided by the volume of the fjord (V_f), divided by the tidal period (t); this may be written as:

$$E_t = P/V_f \times t = H/D \times t \tag{3}$$

where D is the average depth of the fjord.

Energetic currents outside the fjord may drive residual eddies in the fjord, and effect an exchange of waters through tidal dispersion (Geyer & Signell 1992, Nguyen et al. 2008). Several eddies may appear from the mouth to the head of the fjord, where the outermost eddies are the most energetic and the strength decreases with the distance from the fjord mouth. The size and strength of the eddies may be related to the width (W) of the mouth (Nguyen et al. 2008). An estimate of the tidal dispersion effect (E_{tp}) at a fish farm at a distance (L_1) from the mouth of the fjord is estimated by the tidal excursion scaled by the ratio of W to the distance (L_2) to the fjord mouth (Fig. 2), i.e.

$$E_{tp} = \frac{U \times t}{L_1 + L_2} \times \frac{W}{L_1} \tag{4}$$

where U is the maximum current speed outside the fjord, t is the tidal period, and L_2 is length from the mouth of the fjord to the maximum measured current speed outside the fjord.

Salmon lice population dynamics

Examples of the population growth of the salmon lice are shown in Fig. 3. These show the average number of *L. salmonis* (both motile + adult *L. salmonis*) per fish at selected sites. The rate of increase (S), estimated from the slope of the corresponding log-linear plot (Fig. 3) was used as the response variable

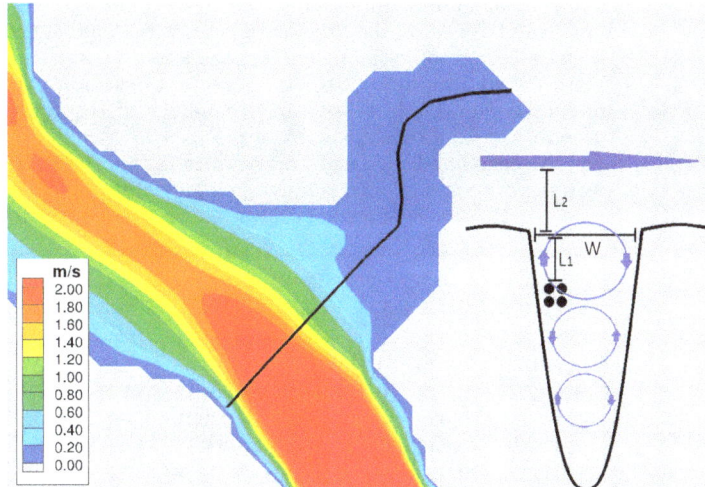

Fig. 2. Enlargement of the calculated maximum tidal currents (color coded m s^{-1}) in a single fjord, with a sketch of the tidal dispersion effect with regards to residual eddies (blue circles). The large arrow at the top of the sketch represents the maximum current measured in the channel. The location of the farm is shown as 4 small circles at the left of the fjord, with distance L_1 from the mouth of the fjord. The other length scales used in the calculations are W (width of the fjord at its mouth) and L_2 (distance from the mouth to the maximum current). Distances are measured along the centerline of the fjord (black line)

in the statistical analysis. We selected only periods where the data showed a clear exponential growth, as illustrated in Fig. 3, to avoid long periods at the start of the monitoring data with no lice, or small outbreaks, which would decrease the slope significantly.

Temperature and host density

In order to isolate the effect of exposure on the growth rate, the effect of temperature (T) can be removed by de-trending the slope with respect to temperature, as shown in Fig. 4.

Host density was quantified as the total number of stocked salmon in the fjord at the time of salmon lice monitoring. However, as this measure did not seem to have a significant effect in the statistical analysis, it was not included.

Statistical analysis

A multiple linear regression was used to examine the effect of exposure on the salmon lice population dynamics. Based on the theory that self-infecting growth follows an exponential trend (Krkošek et al. 2010), we tested the hypothesis that there is a lower rate of self-infection at the more exposed sites. The final analysis of the effect of exposure on salmon lice population dynamics was investigated by a multiple linear regression of the exposure factors E_{tp}, E_f and E_t on the temperature de-trended slopes (S_t) (Fig. 5):

$$S_t = \beta_0 + \beta_1 E_{tp} + \beta_2 E_f + \beta_3 E_t + \varepsilon \qquad (5)$$

The multiple linear regression was used to determine whether the predictor variables have a significant effect on S_t, and to get estimates of the β values to determine their proportional explanatory values. All statistical analyses were made using the software R (R Core Team 2013).

RESULTS

The population growth rates S_t are listed in Table 1, and vary between 0.01 and 0.034 (lice fish^{-1} d^{-1}), and the linear model that best relates these to exposure factors (Eq. 5) is given by the following formula:

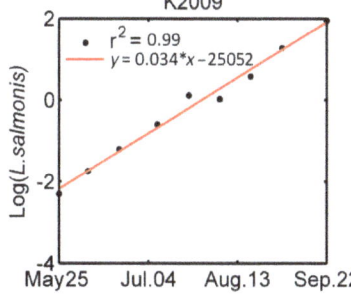
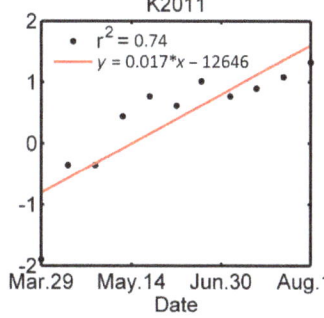
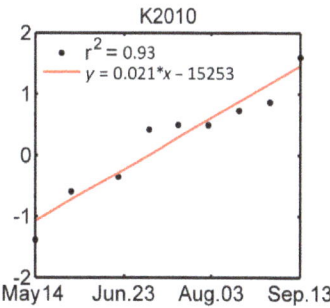

Fig. 3. Log of salmon lice *Lepeophtheirus salmonis* population growth (salmon lice per fish) over time in data sets K2009, K2010 and K2011 (identified by the letter of the fjord and the year of the grow-out phase), and the calculated slopes obtained from the red trend lines using a linear curvefit in Matlab

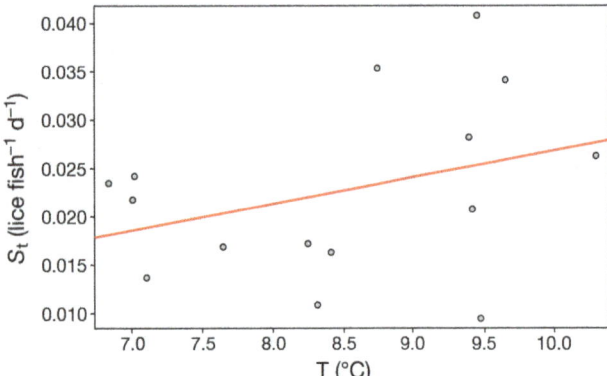

Fig. 4. Correlation between average water temperature (T) in the sampled Faroese fjords and population growth rates (S_t) of salmon lice *Lepeophtheirus salmonis*

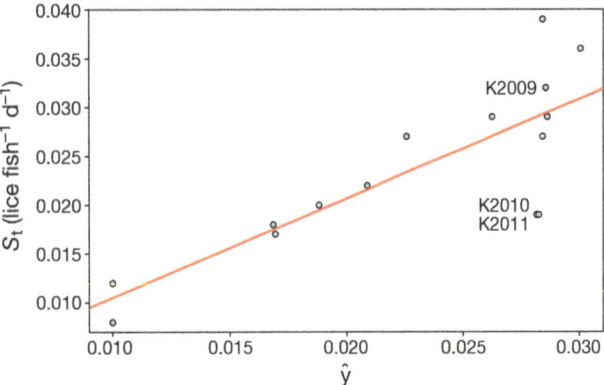

Fig. 5. Observed population growth rate (S_t) of salmon lice *Lepeophtheirus salmonis* and that predicted by exposure effects ($\hat{y}$, Eq. 6). K2009, K2010 and K2011 refer to the different data sets used (identified by the letter of the fjord and the year of the grow-out phase)

$$\hat{y} = 0.0343 - 0.000435\,E_{tp} - 9.36\,E_f - 0.00895\,E_t \quad (6)$$

and is illustrated in Fig. 5. The joint *F*-test for the model was significant (p < 0.01).

The regression coefficients were all negative, indicating that higher exposure factors (i.e. greater tidal exchange, greater freshwater forcing and greater dispersion) resulted in a lower growth rate of salmon lice. Some of these factors were more significant than others. Tidal dispersion (E_{tp}) showed greater significance (p = 0.028) than either freshwater forcing (E_f) or the tidal exchange (E_t) (p = 0.17 and p = 0.87, respectively). This was likely due to the multicollinearity of these factors. However, if either E_f or E_t were removed from the regression, the other was significant, and including both gave the best model fit. The r^2 value of the model was 0.65, meaning that 65% of the variation in S_t can be explained by the exposure variables.

Table 1. Monitoring data sets, identified by a letter for the fjord and the year of the grow-out phase, with exponential growth curves of salmon lice *Lepeophtheirus salmonis*. Fjord D had 2 farms in 2010. The slopes of the salmon lice growth curves (S_t, lice fish^{-1} d^{-1}), r^2 values of the fitted growth lines and the standard errors (SE) for the slopes and the calculated values of the exposure factors E_f (freshwater exchange rate, d^{-1}), E_t (tidal exchange rate, d^{-1}) and E_{tp} (tidal dispersion) (Eqs. 1–4) are given

Data set	S_t	r^2	SE	E_f	E_t	E_{tp}
G2011	0.026	0.92	0.005	0.00076	0.2	10.42
B2010	0.009	0.77	0.003	0.00172	0.19	15.05
B2012	0.011	0.72	0.004	0.00172	0.19	15.05
N2010	0.035	0.86	0.022	0.00021	0.03	4.52
J2009	0.041	0.99	0.003	0.00042	0.04	3.53
J2011	0.022	0.95	0.003	0.00042	0.04	3.53
L2011	0.024	0.87	0.004	0.00035	0.05	4.51
K2009	0.034	0.99	0.002	0.00026	0.04	6.7
K2010	0.021	0.93	0.003	0.00026	0.04	7.58
K2011	0.017	0.74	0.004	0.00026	0.04	7.38
I2009	0.023	0.80	0.011	0.00051	0.08	6.06
D2010_2	0.028	0.93	0.004	0.00027	0.05	19.8
D2010_1	0.016	0.96	0.002	0.00027	0.05	32.89
C2011	0.014	0.86	0.003	0.00063	0.09	24.7
E2010	0.017	0.93	0.004	0.00048	0.09	23.59

The exposure was calculated for all fjords (Fig. 6), including those that did not show exponential growth, or had insufficient data (e.g. due to early implementation of treatments, high *Caligus elongatus* contamination, movement of the farms to different locations). Exposure was estimated according to the best fit regression coefficients:

$$\text{Exposure} = 0.000435\,E_{tp} + 9.36\,E_f + 0.00895\,E_t \quad (7)$$

The exposure of fjord A lies outside the range in the model. Therefore, the exposure and the proportional values of the estimated predictor variables of

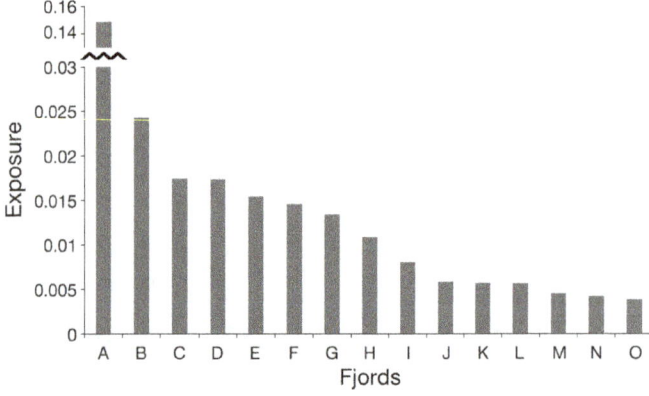

Fig. 6. Estimated exposure for all sampled fjords (Eq. 7). Note the exposure for fjord A is an order of magnitude higher than for the other fjords

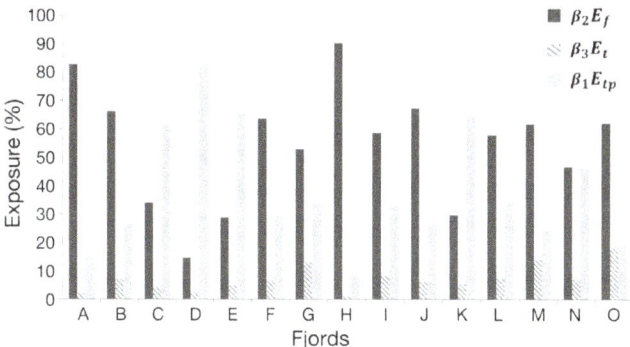

Fig. 7. Percentage of exposure (Eq. 7) explained by each weighed variable, viz. the freshwater exchange rate ($\beta_2 E_f$), the tidal exchange rate ($\beta_3 E_t$) and the tidal dispersion ($\beta_1 E_{tp}$), for all sampled fjords

fjord A are questionable. However, as A has by far the highest values in all exchange rates, it is safe to assume that the level of exposure is highest in this fjord. The relative explanatory values of the exposure factors for all sites are estimated in Fig. 7.

DISCUSSION

We have demonstrated that the physical exposure of a salmon farm site has a highly significant effect on the salmon lice population dynamics, where a higher exposure corresponds to a lower rate of self-infection and vice versa. We have characterized exposure level in terms of relatively easily accessible parameters (tidal range, tidal currents, freshwater inflow), for which we can mechanistically describe a direct influence (E_f, E_t and E_{tp}) on the flushing of water at a farm location.

Nearly all of the salmon farms examined displayed exponential growth of salmon lice, and therefore internally driven population dynamics (Krkošek et

al. 2010). The only clear exception was fjord A, which exhibited externally driven outbreak dynamics (Fig. 8), reaching a relatively uniform level of infection after a rapid growth phase. Fjord B exhibited somewhat of a mixture of internally and externally driven outbreak dynamics (Fig. 8). While fits to an exponential model were significant (Table 1), the infection rates make it difficult to distinguish between exponential and linear growth. These 2 sites displaying signs of externally driven outbreak dynamics were also the most exposed sites (Fig. 6).

The sea temperature on the shelf was used as a reference for the temperature in all sampled fjords and should be a good estimation, as the temperature around the islands is spatially homogenous (Larsen et al. 2008). Including the average sea temperature in the multiple linear regression gave a significant effect on the slope depending on time of year. However, as the aim was to investigate the effect of exposure on salmon lice population dynamics, the temperature effect was removed by de-trending the growth rates.

Both the number and size of salmon in a given area are potentially important parameters in driving the dynamics of sea lice infections. The greater the density of hosts, the more severe an outbreak (Anderson & May 1991). Likewise, large hosts can carry a greater number of parasites (Lees et al. 2008, Heuch et al. 2009, Jansen et al. 2012), increasing contact rates and exposure time (Jackson & Minchin 1992, Tucker et al. 2002). The complexity of host density on sea lice population dynamics is further increased by 'the dilution effect', a negative relationship between host density and infection intensity (Samsing et al. 2014).

We could not find any significant relationship between the total number of salmon in a fjord and infection rate. However, there is an indication that the exposure level of fjords influences the stocking

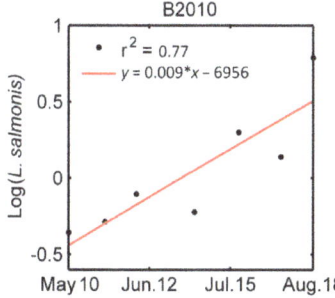

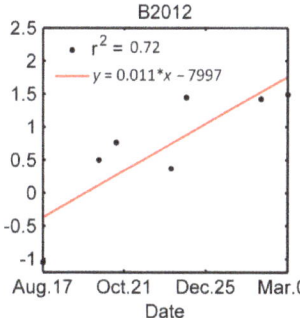

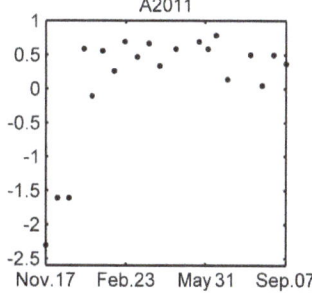

Fig. 8. Log of salmon lice *Lepeophtheirus salmonis* population growth (salmon lice per fish) over time in data sets B2010, B2012 and A2011 (identified by the letter of the fjord and the year of the grow-out phase), and the calculated slopes obtained from the red trend lines

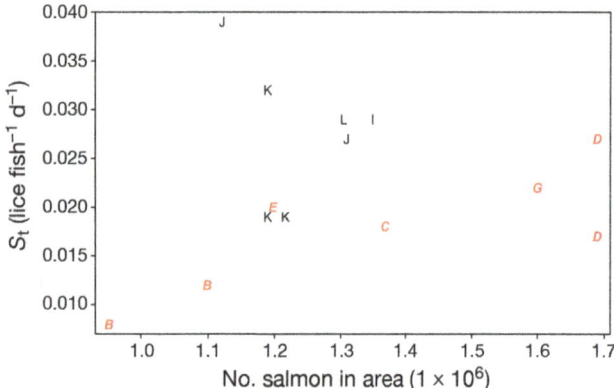

Fig. 9. Correlation between the slopes of the salmon lice *Lepeophtheirus salmonis* growth curves S_t and the total number of stocked Atlantic salmon *Salmo salar* in the sampled fjords in millions. The fjords with the lowest exposure (<0.01) are marked in red *italics*, and the higher-exposure fjords (>0.01) are shown in normal font

density threshold at which outbreaks occur (Fig. 9). Specifically, at low-exposure sites, the infection rate tends to increase faster with the total number of salmon than at high-exposure sites. One final point to note is that current speed can strongly influence the attachment of sea lice to fish (Samsing et al. 2015). Thus the effect of exposure on infection rates is not solely through population dynamics at different sites, but also directly acts on the infestation process.

The simple approach to the estimated exchange rate from freshwater input (E_f) neglects the fact that as fresh water from the river runoff is mixed with saltwater, the volume of the exchange flow out of the fjord may greatly exceed the river flow, easily by as much as a factor of 100 (Hansen 2000). In this study, we did not attempt to calculate an exact exchange rate driven by freshwater input, since it would require salinity observations within the various fjords (Knudsen 1900, as cited by Pickard & Emery 1990). Such salinity profiles were unavailable for all sites. More importantly, actual mixing within the fjord would depend on tidal energy, variables already incorporated in the other exposure factors through tidal exchange and tidal dispersion. An estimate of estuarine-driven exchange as per Knudsen's relationship would exacerbate the correlations of the variables used in the analysis. Freshwater inflow from land drainage therefore provides an independent variable that pertains to exchange circulation and exposure. One possible issue that we have yet to explore is the seasonality in land run-off. In our analysis, we only used a mean annual estimate, but it is known that winter and summer values are quite different. This, coupled with the temperature effects

on salmon lice growth rates, could confound our results somewhat with underestimation of freshwater effects in winter and overestimation in summer.

Despite the simplified physical description of exchange processes, the regression model relating the exposure variables to salmon lice growth rate was highly significant and explained 65% of the observed variation. Two outliers were observed in the model, viz. 2 data sets (K2010 and K2011) from fjord K (Fig. 5). K2009 did not deviate from the model estimate. It seems that the exposure for fjord K was somewhat underestimated in the model. Most of the Faroese fjords, as post-glacial features (Elliott & McLusky 2002), contain a sill at the vicinity of the mouth. Fjord K is distinct as it is not a sill fjord, with a depth of 130 m at the mouth slowly decreasing into the fjord. The tidal effect is likely enhanced on fjord K, as it has a less restricted connection to the open sea compared to other fjords. Further, within fjord K, differences in the salmon lice dynamics observed at K2009 and K2010 can be attributed to their setting within the fjord. In fjord K, the flow is cyclonic with an inflow on one side of the fjord and a corresponding outflow on the other side. K2009, which is on the outflow side, might thus have had an increased rate of infection of salmon lice copepodids from K2010 on the inflow side, and therefore a higher rate of internally driven outbreak than K2010, and a higher rate of internally driven outbreak than K2011, as no other farms were in the fjord simultaneously in 2011.

While our exposure model provides a relatively good explanation of observed differences in infection rates, it still falls short in that 35% of the observed variance remains unexplained, and is likely due to other factors such as host density and wind effects. Weather conditions on the Faroe Islands are notoriously variable, and winds are a powerful influence on the conditions on the shelf as well as within the fjords (Hansen 2000). The nature of wind-driven circulation, however, is largely episodic and random. The ever-changing effect of the wind will have major impacts on short-term self-infectiveness of a site, but over the long term, its effect is to contribute to the general background of mixing that is relatively uniform in space. As with land drainage, there is the possibility of seasonal variations that co-vary with temperature-dependent growth rates.

A range of additional factors can alter sea lice composition and infestation, including functional feeds (Jensen et al. 2015), selective breeding (Gharbi et al. 2015), the use of artificial lights (Hevrøy et al. 2003, Oppedal et al. 2011, Aarseth & Schram 1999) and the physiological status of the fish; for example, salmon stressed by osmoregulation have a higher suscepti-bility to infection by *L. salmonis* (Dawson et al. 1997).

Tidal dispersion (E_{tp}) had a significant effect on the population growth of the salmon lice. It is the only model variable that farmers can influence without too much effort, by relocating the site of the farm. There is, however, a fundamental trade-off to be considered: increasing the exposure of the farm will decrease the level of self-infection, but could also increase the external infection pressure. The exchange of water between sites and the degree of externally derived sea lice is of course highly dependent on the connectivity between sites, a parameter that depends not just on distance, but on circulation patterns. Farms with different rates of self-infection should be managed differently, as low-exposure fjords seem to have a lower threshold of total stocked salmon. Further, at low exposure sites, once the number of sea lice reaches the threshold for treatment, the local environment will also be highly infectious and a single treatment will not be suffi-cient to ensure that re-infection does not occur.

Overall, our attempt to explain the growth rate of salmon lice from estimations of exposure predictor variables produced a robust model, which was highly significant (p = 0.0073). The equation for calculating the combined factor of exposure, although an estima-tion, is believed to be a reasonable parameter for further use in epidemiological studies on sea lice in salmon farms within the range observed in the statistical model.

CONCLUSIONS

As stated by Revie et al. (2005, p. 611): 'No single or simple factor is able to account for the variation in the patterns of infection, and factors interact in a com-plex way'. In this study, we have made an attempt to relate the physical exposure of farming sites with salmon lice population dynamics. The simple fjord systems in the Faroe Islands are well suited to inves-tigate this relationship. The results, although based on simplified proxies, show a significant relationship between infection rate and exposure, with lower rates at high-exposure sites. Our results also indicate that external infestation becomes an increasingly important factor in the most exposed fjords. Knowl-edge of sea lice and their interactions with the envi-ronment is fundamental in the ongoing struggle against the parasites. Characterizing fjords and farms of the Faroe Islands at different exposure levels and how they relate to infection rates is an important step

in an eco-friendly direction for combatting the sea lice. The results can be used to aid in the manage-ment of the different sites, as well as in decision-making processes such as advising on locations that are best suited to mitigate the sea lice problem.

Acknowledgements. We thank Poul Christoffur Thomassen for help and guidance with the statistics. Funding and facilities were provided by Fiskaaling — the Aquaculture Research Station of the Faroes. Sea lice data were provided by the Faroese aquaculture farming companies P/F Bakka-frost, P/F Luna and Marine Harvest Faroes, as well as by the Faroe Marine Research Institute and Landsverk.

LITERATURE CITED

Á Norði G, Simonsen K, Patursson Ø (2016) A method of estimating *in situ* salmon louse nauplii production at fish farms. Aquacult Environ Interact 8:397–405

Aarseth KA, Schram TA (1999) Wavelength-specific behav-iour in *Lepeophtheirus salmonis* and *Calanus finmarchi-cus* to ultraviolet and visible light in laboratory experi-ments (Crustacea: Copepoda). Mar Ecol Prog Ser 186: 211–217

Adams T, Black K, MacIntyre C, MacIntyre I, Dean R (2012) Connectivity modelling and network analysis of sea lice infection in Loch Fyne, west coast of Scotland. Aquacult Environ Interact 3:51–63

Amundrud TL, Murray AG (2009) Modelling sea lice disper-sion under varying environmental forcing in a Scottish sea loch. J Fish Dis 32:27–44

Anderson RM, May RM (1978) Regulation and stability of host-parasite population interactions. J Anim Ecol 47: 219–247

Anderson RM, May RM (1991) Infectious diseases of humans. Oxford University Press, Oxford

Asplin L, Johnsen IA, Sandvik AD, Albretsen J, Sundfjord V, Aure J, Boxaspen KK (2014) Dispersion of salmon lice in the Hardangerfjord. Mar Biol Res 10:216–225

Brooks KM (2005) The effects of water temperature, salinity, and currents on the survival and distribution of the in-fective copepodid stage of sea lice (*Lepeophtheirus salmonis*) originating on Atlantic salmon farms in the Broughton Archipelago of British Columbia, Canada. Rev Fish Sci 13:177–204

Costello MJ (2006) Ecology of sea lice parasitic on farmed and wild fish. Trends Parasitol 22:475–483

Costello MJ (2009) The global economic cost of sea lice to the salmonid farming industry. J Fish Dis 32:115–118

Davidsen E, Førland E, Madsen H (1994) Orographically enhanced precipitation on the Faroe Islands. Nordic Hy-drographical Conference, 2–4 August 1994, Torshavn, p 229–239

Dawson LHJ, Pike AW, Houlihan DF, McVicar AH (1997) Comparison of the susceptibility of sea trout (*Salmo trutta* L.) and Atlantic salmon (*Salmo salar* L.) to sea lice (*Lep-eophtheirus salmonis* (Krøyer, 1837)) infections. ICES J Mar Sci 54:1129–1139

Elliott M, McLusky DS (2002) The need for definitions in understanding estuaries. Estuar Coast Shelf Sci 55: 815–827

Faroese Food and Veterinary Authority (2009) Kunngerð nr. 163. http://logir.fo

Faroese Food and Veterinary Authority (2013) HS mál 13/00854-7. www.hfs.fo

Farvandsvæsenet (2005) Tidevandstabeller 2006 for færøske farvande. Forsvarsministeriets trykkeri, Copenhagen

Frazer LN, Morton A, Krkošek M (2012) Critical thresholds in sea lice epidemics: evidence, sensitivity and sub-critical estimation. Proc R Soc Lond B Biol Sci 279: 1950–1958

Gaard E, á Norði G, Simonsen K (2011) Environmental effects on phytoplankton production in a Northeast Atlantic fjord, Faroe Islands. J Plankton Res 33:947–959

Geyer WR, Signell RP (1992) A reassessment of the role of tidal dispersion in estuaries and bays. Estuaries 15: 97–108

Gharbi K, Matthews L, Bron J, Roberts R, Tinch A, Stear M (2015) The control of sea lice in Atlantic salmon by selective breeding. J R Soc Interface 12:0574

Hansen B (2000) Havið. Føroya skúlabókagrunnur, Tórshavn

Hansen B, Østerhus S (2000) North Atlantic–Nordic Seas exchanges. Prog Oceanogr 45:109–208

Heuch PA, Nordhagen JR, Schram TA (2000) Egg production in the salmon louse [Lepeophtheirus salmonis (Krøyer)] in relation to origin and water temperature. Aquacult Res 31:805–814

Heuch PA, Olsen RS, Malkenes R, Revie CW and others (2009) Temporal and spatial variations in lice numbers on salmon farms in the Hardangerfjord 2004–06. J Fish Dis 32:89–100

Hevrøy EM, Boxaspen K, Oppedal F, Taranger GL, Holm JC (2003) The effect of artificial light treatment and depth on the infestation of the sea louse Lepeophtheirus salmonis on Atlantic salmon (Salmo salar L.) culture. Aquaculture 220:1–14

Jackson D, Minchin D (1992) Aspects of the reproductive output of two caligid copepod species parasitic on cultivated salmon. Invertebr Reprod Dev 22:87–90

Jansen PA, Kristoffersen AB, Viljugrein H, Jimenez D, Aldrin M, Stien A (2012) Sea lice as a density-dependent constraint to salmonid farming. Proc R Soc B 279: 2330–2338

Jaworski A, Holm JC (1992) Distribution and structure of the population of sea lice Lepeoptheirus salmonis Krøyer, on Atlantic salmon Salmo salar L. under typical rearing conditions. Aquacult Fish Manag 23:577–589

Jensen LB, Provan F, Larssen E, Bron JE, Obach A (2015) Reducing sea lice (Lepeophtheirus salmonis) infestation of farmed Atlantic salmon (Salmo salar L.) through functional feeds. Aquacult Nutr 21:983–993

Jimenez D, Heuch PA, Brun E (2011) Evaluering av lusetellingsprotokoll og bioassay for nedsatt følsomhet mot lakselusmidler. Veterinærinst Rapportser 9–2011, Norwegian Veterinary Institute, Oslo

Johnsen IA, Fiksen Ø, Sandvik AD, Asplin LC (2014) Vertical salmon lice behaviour as a response to environmental conditions and its influence on the regional dispersion in a fjord system. Aquacult Environ Interact 5:127–141

Johnsen IA, Asplin LC, Sandvik AD, Serra-Llinares RM (2016) Salmon lice dispersion in a northern Norwegian fjord system and the impact of vertical movements. Aquacult Environ Interact 8:99–116

Knudsen M (1900) Ein hydrographischer Lehrsatz. Ann Hydrogr Mar Meteorol 28:316–320

Krkošek M, Lewis MA, Volpe JP (2005) Transmission dynamics of parasitic sea lice from farm to wild salmon. Proc R Soc Lond B Biol Sci 272:689–696

Krkošek M, Bateman A, Proboszcz S, Orr C (2010) Dynamics of outbreak and control of salmon lice on two salmon farms in the Broughton Archipelago, British Columbia. Aquacult Environ Interact 1:137–146

Larsen KMH, Hansen B, Svendsen H (2008) Faroe shelf water. Cont Shelf Res 28:1754–1768

Lees F, Gettinby G, Revie CW (2008) Changes in epidemiological patterns of sea lice infestation on farmed Atlantic salmon, Salmo salar L., in Scotland between 1996 and 2006. J Fish Dis 31:259–268

Mikhailov VN (1997) Gidrologicheskie prostessy v ust'yakhrek (Hydrological processes in river mouths). GEOS, Moscow

Nguyen AD, Savenije HHG, van der Wegen M, Roelvink D (2008) New analytical equation for dispersion in estuaries with a distinct ebb-flood channel system. Estuar Coast Shelf Sci 79:7–16

Oppedal F, Dempster T, Stien LH (2011) Environmental drivers of Atlantic salmon behaviour in sea-cages. Rev Aquacult 311:1–18

Pickard GL, Emery WJ (1990) Descriptive physical oceanography: an introduction, 5th enlarged edn. Pergamon Press, New York, NY

Pike AW, Wadsworth SL (1999) Sealice on salmonids: their biology and control. In: Baker JR, Muller R, Rollinson D (eds) Advances in parasitology 44. Academic Press, London, p 234–337

Pritchard DW (1967) What is an estuary: physical viewpoint. In: Lauff GH (ed) Estuaries. American Association for the Advancement of Science, Washington, DC, p 3–5

R Core Team (2013) R: a language and environment for statistical computing. R Foundation for Statistical Computing, Vienna. www.r-project.org

Revie CW, Gettinby G, Treasurer JW, Wallace C (2003) Identifying epidemiological factors affecting sea lice Lepeophtheirus salmonis abundance on Scottish salmon farms using general linear models. Dis Aquat Org 57: 85–95

Revie CW, Robbins C, Gettinby G, Kelly L, Treasurer JW (2005) A mathematical model of the growth of sea lice, Lepeophtheirus salmonis, populations on farmed Atlantic salmon, Salmo salar L., in Scotland and its use in the assessment of treatment strategies. J Fish Dis 28: 603–613

Salama NKG, Collins CM, Fraser JG, Dunn J, Pert CC, Murray AG, Rabe B (2013) Development and assessment of a biophysical dispersal model for sea lice. J Fish Dis 36: 323–337

Samsing F, Oppedal F, Johanson D, Bui S, Dempster T (2014) High host densities dilute sea lice Lepeophtheirus salmonis loads on individual Atlantic salmon, but do not reduce lice infection success. Aquacult Environ Interact 6:81–89

Samsing F, Solstorm D, Oppedal F, Solstrom F, Dempster T (2015) Gone with the flow: current velocities mediate parasitic infestation of an aquatic host. Int J Parasitol 45: 559–565

Shchepetkin F, McWilliams JC (2005) The regional oceanic modeling system (ROMS): a split-explicit, free-surface, topography-following-coordinate oceanic model. Ocean Model 9:347–404

Simonsen K, Niclasen BA (2011) On the energy potential in the tidal streams of The Faroe Islands. Tech Rep NVDrit 2011:01. University of the Faroe Islands, Tórshavn

Tucker CS, Sommerville C, Wootten R (2002) Does size really matter? Effects of fish surface area on the settlement and initial survival of Lepeophtheirus salmonis, an ectoparasite of Atlantic salmon Salmo salar. Dis Aquat Org 49:145–152

Tully O (1992) Predicting infestation parameters and impacts of caligid copepods in wild and cultured fish populations. Invertebr Reprod Dev 22:91–102

Tully O, Nolan DT (2002) A review of population biology and host-parasite interactions of the sea lice Lepeophtheirus salmonis (Copepoda: Caligidae). Parasitology 124: 165–182

Visser AW, Bowman MJ (1991) Lagrangian tidal stress and basin-wide residual eddy dynamics in wide coastal sea straits. Geophys Astrophys Fluid Dyn 59:113–145

8

Not too slow, not too fast: water currents affect group structure, aggression and welfare in post-smolt Atlantic salmon *Salmo salar*

Frida Solstorm[1,2,*], David Solstorm[1], Frode Oppedal[1], Rolf Erik Olsen[1,3],
Lars Helge Stien[1], Anders Fernö[1,2]

[1]Institute of Marine Research, 5984 Matredal, Norway
[2]Department of Biology, University of Bergen, PO Box 7803, 5006 Bergen, Norway
[3]Department of Biology, Norwegian University of Science and Technology, 7491 Trondheim, Norway

ABSTRACT: Increased swimming speed of Atlantic salmon is generally considered an improvement to welfare under aquaculture settings, as group structure is improved and agonistic behaviour reduced. As such, establishing fish farms in exposed areas with fast water current velocities should be favourable. However, at some locations, velocities exceed what is known as preferable for salmonids, and this may compromise fish welfare. In this study, behaviour and fin erosion were observed on post-smolt salmon stocked at 39 kg m^{-3} in raceways at 3 water current velocities: fast (1.5 body lengths [BL] s^{-1}), moderate (0.8 BL s^{-1}) and slow (0.2 BL s^{-1}). Movements that affect group structure and interactions between individuals varied by up to 20-fold between velocities. A behavioural change occurred directly after velocities were set. Severe fin erosion decreased over time in all groups, but new injuries increased almost 3-fold in the faster-velocity group. Our results suggest that moderate velocity is ideal from a welfare perspective. At slow velocity, higher frequency of structural movements and between-individual interactions could be stressful for the fish. At faster velocity, the fish have to focus on swimming, which could increase unintentional collisions with obstacles and other individuals and result in new fin erosion. Our results suggest that management of water currents may be an effective way of controlling behaviour and may thereby improve welfare.

KEY WORDS: Exposed farming · Swim speed · Environmental variability · Swimming behaviour

INTRODUCTION

Motivational drivers for swimming in wild Atlantic salmon include migration, feeding, predatory avoidance and mating. For fish in aquaculture systems, few of these drivers are functional. Farmed fish cannot move to another habitat, they do not need to hunt food or avoid predators, and they usually do not mature sexually. In the marine life stages, wild salmonids swim on average at 1 body length (BL) s^{-1} (Drenner et al. 2012). This is similar to observed swimming speeds in salmon farms (Sutterlin et al. 1979, Kadri et al. 1991, Blyth et al. 1993, Juell & Westerberg 1993). In the

wild, this cruising speed is suggested to be the energetic optimum with lowest cost of transport (Drenner et al. 2012). Similarly, a laboratory study by Tudorache et al. (2011) showed that the optimal swimming speed for brook charr *Salvelinus fontinalis* was 1 BL s^{-1} and that this corresponded to the preferred swimming speed (0.8 and 1.0 BL s^{-1}) when the fish was free to choose. Swimming at a preferred speed may be considered positive for welfare. One approach to fish welfare is the Five Freedoms (defined by the UK's Farm Animal Welfare Council [FAWC 1995]). One of the freedoms concludes that the animal should be free to express normal behaviour; swimming at a preferred

*Corresponding author: frida.solstorm@imr.no

speed could be considered a normal behaviour and, hence, positive for welfare.

An increasing number of salmon farms are now located in exposed areas with fast water current velocities, where the fish are forced to swim faster than their preferred speed. Previous laboratory studies have, to our knowledge, only demonstrated positive behavioural effects of increased swimming speeds, which would imply improved welfare (e.g. Korte et al. 2007). Slow current velocities may result in more interactions between individuals and increased aggression that could result in fin erosion (Christiansen & Jobling 1990, Jørgensen & Jobling 1993, Adams et al. 1995, Turnbull et al. 1998) and promote secondary infection (Schneider & Nicholson 1980), reducing production performance and welfare (Stien et al. 2013). Faster current velocities have been reported to reduce agonistic behaviour and create a more ordered group structure with higher production performance (East & Magnan 1987, Christiansen & Jobling 1990, Jørgensen & Jobling 1993, Adams et al. 1995). Based on this, fast current velocities in exposed areas could be considered positive for the fish up to an unknown critical level. However, at water velocities above 0.7 BL s^{-1}, Johansson et al. (2014) observed a breakdown of the circular group structure in Atlantic salmon *Salmo salar* in net cages. When the current increased further to 0.9 BL s^{-1}, all fish abandoned the circular school structure and maintained a position facing the current. As the current shifted, there was a chaotic and challenging transition before a new group structure was established. This indicates that the fish strive for structure to cope with the dynamic environment and the high densities in the cage (see also Ashley 2007). Farms located in areas with strong tidal currents will expose salmon to repeatedly changing current conditions. At more extreme current velocities than Johansson et al. (2014) observed, it is unclear how the group structure and behaviour of the fish change. As the fish approach their maximum sustainable swimming capacity, all available energy would assumedly be allocated to swimming. This may result in decreased energy stores and thus reduced growth (East & Magnan 1987, Farrell et al. 1991, Jørgensen & Jobling 1993) and welfare (e.g. Ashley 2007).

In previous studies, no negative effects of fast water current velocities were found on behaviour and fin erosion, and positive effects of swimming speeds up to 2.5 BL s^{-1} have been reported (East & Magnan 1987). However, negative effects of fast current velocity on physiology and production performance have been demonstrated in Solstorm et al. (2015) and thus are hypothesised to also have impacts at the behavioural level. This study performed a detailed analysis of behaviour and fin erosion when Atlantic salmon post-smolts were exposed to water currents of slower and faster velocities than their preferred range, with the latter prevailing in farms located at exposed sites.

MATERIALS AND METHODS

Facilities and experimental animals

A behavioural study of post-smolt Atlantic salmon (AquaGen strain, hatched March 2011) exposed to water currents was conducted at the Tank Environmental Laboratory at the Institute of Marine Research, Matre (Norway), during 6 wk starting April 2012. The study was part of a larger experiment also investigating physiological effects, where a more detailed description of setup and experimental design can be found (Solstorm et al. 2015). Fish (98.6 ± 20 g, 22.3 ± 1.3 cm, mean ± SD) were transferred to raceways at a mean (±SE) stocking density of 38.7 ± 0.28 kg m^{-3} (n = 80 per raceway) as smolts. This density was chosen as representative for typically observed swimming densities in sea cages, where salmon are known to trade off variable environmental drivers, including temperature and water currents (Oppedal et al. 2011a,b, Johansson et al. 2014). Prior to smoltification, fish had been reared in tanks (Ø 5 m) under natural light and temperature conditions. To finalise smoltification, fish were kept on a constant light regime for 8 wk at 8°C with a flow of 150 l min^{-1} until all fish had smoltified.

Raceways (trans-sectional area 0.10 m^2, Ø 0.36 m, length 2.0 m, giving a volume of 0.20 m^3) were submerged in circular tanks (Ø 3 m, 5.3 m^3). A laminar water current velocity was produced by an electric engine (Minn Kota RT80/EM, Johnson Outdoors Marine Electronics) with adjustable speed followed by a honeycomb (5.0 mm opening, 101.6 mm thickness, PC 5.0 G4, Plascore). Water temperature was 10°C, water exchange was 120 l min^{-1}, salinity was 33 psu and dissolved oxygen levels were above 80%. A constant light regime was maintained during the experiment, and the fish were fed (Skretting Spirit 75) in excess every 15 min throughout the day (24 h) to ensure that food would not be a limiting factor. Feed was distributed by automatic feeding units (Arvo-Tec T Drum 2000, http://www.arvotec.fi) controlled from custom-made computer software (SD Matre, Normatic AS). An underwater video camera (SV27, SeaVision) was mounted in the middle of each raceway, with the field of view covering the posterior area of the chamber. Video recordings were stored on a central PC by a video capture card (GV-800, GeoVision) and multi-camera surveillance system (CV-800, GeoVison).

Experimental design

Fish were kept at a velocity of 0.5 BL s^{-1} in the raceways to acclimatise for 19 d prior to the experiment. This velocity was chosen as it is the velocity typically used for acclimation prior to swimming performance tests in salmonids, and all fish were to experience a change when the trial commenced. During the experiment, fish were kept at 3 different water current velocities for 6 wk in 4 replicate raceways: slow, moderate and fast velocities corresponding to 0.2 ± 0.02, 0.8 ± 0.01 and 1.5 ± 0.02 BL s^{-1} (mean ± SE), respectively. In this setup, water current velocity was considered the same as swimming speed since the fish had to hold station against the current not to be swept back into the netting and had limited space (9 to 7 BL, upstream–downstream direction) to move around freely. Slow current velocity was adjusted to be close to zero without compromising oxygen levels and the transport of faeces and food waste out of the raceway. Moderate current velocity was selected to be in the range of the preferred swimming speed (Tudorache et al. 2011). Fast current velocity was chosen to be twice the amplitude of the moderate but still below the critical swimming speed (Tang & Wardle 1992, Stevens et al. 1998), although actual velocity was slightly lower. After 3 wk, the currents were adjusted to the increased fish length (due to growth) to maintain the same velocity in body length per second.

The experiment was conducted in accordance with laws and regulations of the Norwegian Regulation on Animal Experimentation (application ID 4146).

Behavioural observations

Video recordings of fish behaviour were made without disturbance in Weeks 0, 2, 4 and 6 by the GeoVision system pre-programmed to record for 12 min and 30 s during 08:00 to 12:00 h simultaneously in 2 tanks at a time to increase the video quality. Recordings in Week 0 were done 1 d after the current velocities were set; the other recordings were done at the beginning of the week. Fish behaviour was classified as interactions or movements affecting the group structure (structural movements) and ranked according to the assumed intention (Table 1). Ranking was classified from 1 to 6, with 1 as the highest ranking. Biting was given the highest ranking based on an assumed impact scale. When one type of behaviour was documented, other behaviours were assumed a consequence of the first behaviour, and thus lower-ranked behaviours were not documented. For instance, a biting attempt could be preceded by bursts and collisions, but since these behaviours were of lower ranking, only the biting was documented. Structural behaviours were analyzed according to a Cartesian coordinate system, where X and Y are movements in the cross-sectional plane, and –Z movements are movements with the flow. Behaviours were analysed for 12 min and 30 s, and the number of behavioural occurrences per minute was calculated.

Fin erosion

External injuries and fin erosion were recorded on all fish before and after the experiment. Fin erosion was observed on the pectoral, dorsal, pelvic, anal and caudal fins and divided into 3 categories: fins with 0 to 10 % of the fin missing (eroded) were classified as uninjured due to difficulties in assessing a perfect fin (0 % erosion), fins with 10 to 50 % erosion were classified as moderate fin erosion and fins with > 50 % erosion were classified as severe fin erosion. This

Table 1. Definitions of the behavioural categories used. Behaviours were ranked from 1 to 6, with 1 as the highest ranking

Behaviour	Definition	Ranking
Interaction		
Biting	Active attempt to bite another fish	1
Displacement	A fish forces another out of position, with or without close contact	2
Intentional collision	A fish directs its swimming and bumps into another fish intentionally, not resulting in a bite attack or displacement of the attacked fish	3
Structural movement		
Burst	Fast acceleration against the current	4
Move –Z	Turning and swimming with the current instead of against the current	5
Move XY	Vertical movements in X or Y direction, with a relocation of at least 0.5 body length not followed by other higher-ranked behaviour	6

classification was based on an index used by Swedish governmental agencies and chosen as the index based on the observers' experience. The index is a modified version based on Person-Le Ruyet et al. (2007), where the 0 group has been judged irrelevant for farmed fish and is thus removed from the index. Difficulties distinguishing between their levels 2 and 3 were simplified by adding them to the same group (here, moderate fin erosion). Fin erosions were also classified as either new damage with bleeding or older damage that may have healed.

Statistics

Data analyses were performed using R software Version 3.1.0 (Copyright 2009, The R Foundation for Statistical Computing, Vienna). Counts are reported as mean number of counts $min^{-1} \pm SE$. Deviance in behaviour counts was modelled using generalized linear models with quasi-Poisson errors, as recommended for count data with over-dispersion (Crawley 2012). Week numbers (0, 2, 4 and 6) and treatments were set as explanatory factors in the models (function glm, R). Model simplification was performed in cases where the simpler model was not significantly different from the more complex model (function ANOVA, R, test = Chi; Crawley 2012). Fin erosion data were analyzed in a similar way; however, as data were proportional (percentage fish with damage), the error distribution was set as binomial and, in the case of over-dispersion, quasi-binomial (Crawley 2012). F-tests were used to compare the original to simplified models (function ANOVA, R, test = F; Crawley 2012). Significance level was set at p < 0.05.

During the experiment, 2 replicates were eliminated due to technical problems—one from the fast current velocity in Week 5 and one from the moderate current velocity in Week 4—resulting in triplicate treatments for the aforementioned groups at the final sampling.

RESULTS

Behavioural observations

Group structure

Current velocity had a large influence on structural movements, with the highest frequency of behaviours observed in fish kept at

the slower velocity and decreasing levels of movement with the moderate and faster velocities (Fig. 1). When the treatment velocities were applied, behaviours immediately changed. Similar differences were observed in the following 4 wk. In the final week, there was an increase of structural movements at moderate velocity compared to the second week, due to increased horizontal movements (Move XY). Even so, fish at both moderate and faster velocities had fewer movements than fish at slower velocity. Fish at faster velocity also showed significantly fewer movements than fish at moderate velocity in the final week. Except for an increase in horizontal movements from Week 2 until Week 6, at moderate velocity, there were no changes over time within treatments in the different types of structural behaviours (Fig. 1).

Interactions

More interactions were observed in fish kept at slow velocity compared to both moderate and faster velocities (Fig. 2). Fish at fast velocity did not differ significantly compared to moderate velocity, but a decreasing trend of interactions was observed. At slow velocity, fewer collisions were observed in Week 4 compared to Weeks 0 and 2, resulting in no treatment effect in Week 4. In the other interacting behaviours, no effects were observed over time.

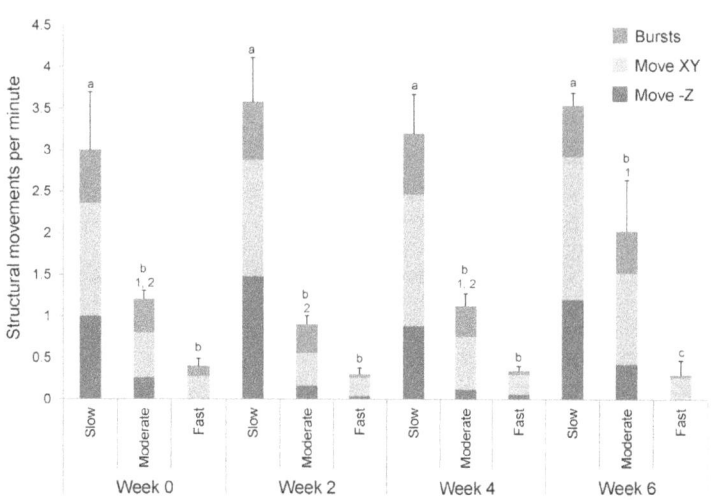

Fig. 1. Mean of structural behaviours per minute in fish at 3 different water current velocities (slow, 0.2 body length [BL] s^{-1}; moderate, 0.8 BL s^{-1}; and fast, 1.5 BL s^{-1}) over 6 wk. Error bars represent SE of total occurrences. Lowercase letters denote significant differences between current velocities, and numbers denote significant differences between weeks based on the cumulated behaviours (p < 0.05)

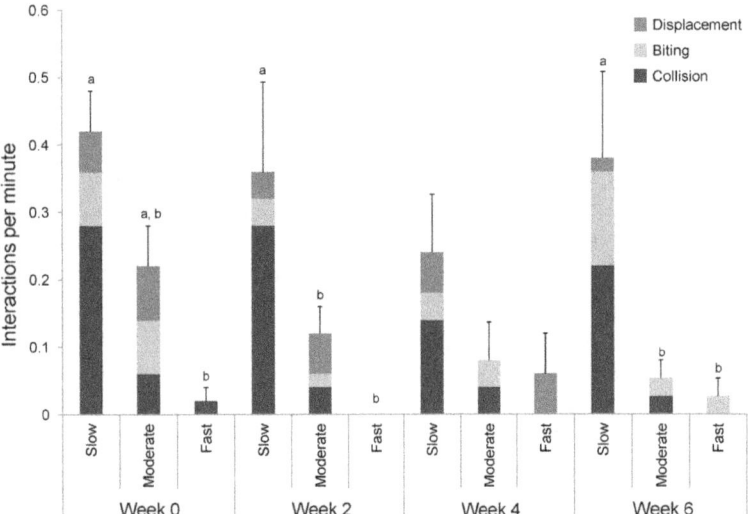

Fig. 2. Mean of interacting behaviours per minute in fish at 3 different water current velocities (slow, 0.2 body length [BL] s^{-1}; moderate, 0.8 BL s^{-1}; and fast, 1.5 BL s^{-1}) over 6 wk. Error bars represent SE of total occurrences. Lowercase letters denote significant difference between current velocities based on cumulated interactions (p < 0.05)

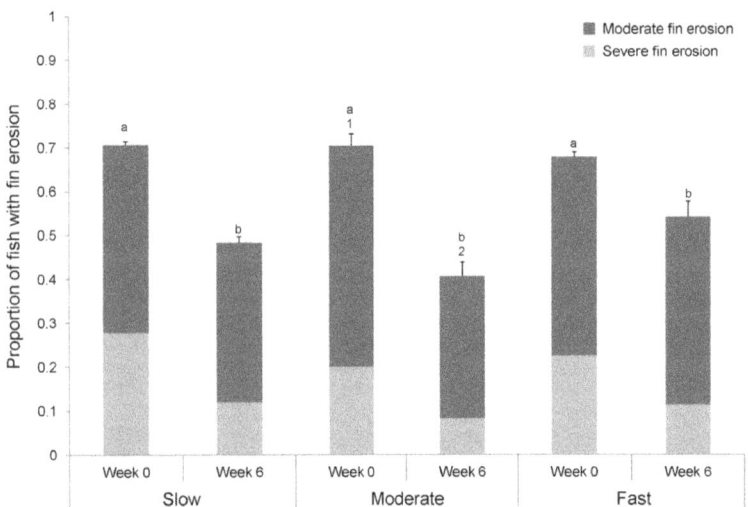

Fig. 3. Mean of cumulated fin erosions at start and end of experiment in fish kept at slow (0.2 body length [BL] s^{-1}), moderate (0.8 BL s^{-1}) and fast (1.5 BL s^{-1}) water current velocities. The columns are separated into fish with moderate fin erosion and fish with severe fin erosion. Error bars show SE of cumulated fin erosions. Lowercase letters denote significant differences between current velocities, and numbers denote significant differences between weeks (p < 0.05)

Fin erosion and other injuries

The proportion of fish with fin erosion decreased from Weeks 0 to 6 in all groups (Fig. 3). Severe fin erosion decreased in fish at all 3 velocities over time, while moderate fin erosion only decreased at the moderate velocity. However, at the faster velocity,

there was an increase in new fin erosion from the start to the end of the experiment (Fig. 4). New fin erosion was also more prevalent in fish at faster velocity than at moderate and slower velocities.

The increase in new fin erosion was caused by an increase in caudal fin erosion (Fig. 5). Damage on the dorsal fin was significantly less frequent at the end, with no effect of velocity. Fish at faster velocity had a significantly higher frequency of caudal fin erosion than fish at slower velocity at the end of the experiment. No effects could be seen for the other fin erosions.

DISCUSSION

The water current velocities had an effect on both behaviour and fin erosion in post-smolt salmon in raceways. Increasing current velocities induced a higher level of organised swimming in salmon, while agonistic behaviour decreased. However, the prevalence of new fin erosion increased at the fast current velocity. These findings suggest that increased current is positive but that welfare may be compromised above a certain velocity.

When the fish are not forced to swim against a current, they may choose various behavioural strategies within the tank. Some fish move at random, while others are stationary. Some may display territorial behaviour, while others are subordinate (Fernö & Holm 1986, Adams et al. 1995). The higher frequency of movements affecting group structure (structural movements) and agonistic behaviour in fish kept at slow current velocity has also been observed in previous studies on salmonids, although these studies did not observe the group structure in detail (East & Magnan 1987, Christiansen & Jobling 1990, Christiansen et al. 1992, Jørgensen & Jobling 1993, Adams et al. 1995). It is assumed that aggression is energetically costly and will result in decreased growth (East & Magnan 1987, Christiansen & Jobling 1990). Solstorm et al. (2015) found, based on physiological parameters, that some individuals at slow velocity experienced elevated

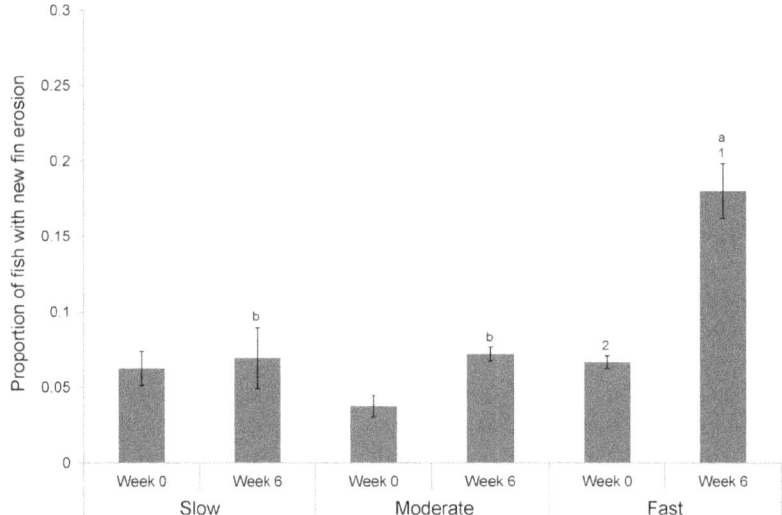

Fig. 4. Mean proportion of fish with new fin erosion at start and end of experiment in fish kept at slow (0.2 body length [BL] s^{-1}), moderate (0.8 BL s^{-1}) and fast (1.5 BL s^{-1}) water current velocities. Error bars represent SE. Lowercase letters denote significant differences between current velocities, and numbers denote significant differences between weeks (p < 0.05)

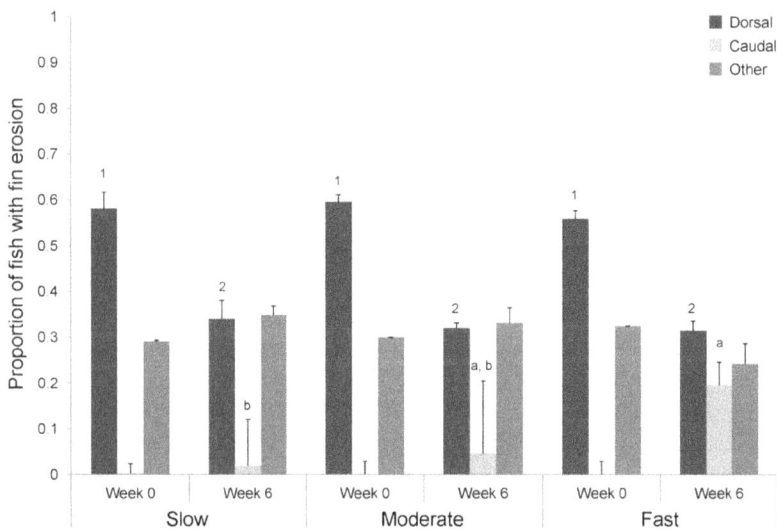

Fig. 5. Mean proportion of fin erosions separated into dorsal, caudal or other, with other including pectoral, pelvic and anal fins. Columns are presented for fish kept at slow (0.2 body length [BL] s^{-1}), moderate (0.8 BL s^{-1}) and fast (1.5 BL s^{-1}) water current velocities at the start and end of a 6 wk experiment. Error bars represent SE. Lowercase letters denote significant differences between water current velocities, and numbers denote significant differences between weeks (p < 0.05)

moderate velocity (Solstorm et al. 2015), indicating that the level of interactions at slow velocity was not severe enough to result in decreased growth. Also, when relating the level of agonistic behaviour to previous studies, these values are low (East & Magnan 1987, Adams et al. 1995). Adams et al. (1995) observed approximately the same number of agonistic behaviours in exercised Arctic charr as was seen at slow velocity in this study, but in their setup, 10 fish were observed, whereas in this study, up to 80 fish were studied. In our setup, there is a possible risk that the slow velocity induced structural movements and interactions as an effect of the possibility to swim at voluntary speed and thereby create an organised group structure. In contrast to raceways, salmon in cages are largely free to choose their swimming speed and create an organized structure, making results from our slow velocity difficult to relate to commercial situations, where fish are commonly held in sea cages. Farmed salmon are raised in high densities and are thus habituated to social organisation; rapid changes in current velocity can break down their schooling structure, and this could cause brief disorder. Johansson et al. (2014) observed that after this type of disorder, salmon strive to organise a new structure.

Fin erosion is a welfare problem that can arise from increased agonistic behaviour. Damage to the tissue is a direct violation of one of the Five Freedoms, i.e. freedom from pain and injury (defined by FAWC [1995]) and may also lead to health problems, with increased susceptibility to pathogens due to damaged tissue (Turnbull et al. 1996). Even though a fish's ability to experience pain is widely debated, the fins contain nociceptors, and adverse behaviour after fin clipping has been demonstrated, suggesting that fin erosion contributes to negative welfare (Roques et al. 2010, Noble et al. 2012). Causes for fin erosion are nipping from others, abrasion and bacterial infection (Latremouille 2003). Fin nipping mostly targets the dorsal fin in salmonid parr (Turnbull et al. 1998). The high frequency of dorsal fin erosions at the start of our experiment

stress levels. This may suggest that agonistic behaviour and structural movements displayed at slow velocity are stressful for the individual, resulting in compromised welfare for the individual. However, fish at slow velocity had the same growth as fish at could be explained by the fish coming directly from the freshwater life stage potentially combined with periodic restrictions in feed availability, as often seen under commercial parr production. At the parr stage in fresh water, wild salmon are territorial and frequently show agonistic behaviours (Keenleyside & Yamamoto 1962), while fish stressed by feed restrictions in aquaculture production units may show increased fin nipping and thus erosion (Noble et al. 2008). The reduction of cumulated fin erosions over time in our experiment may therefore be explained by a life stage-related decrease in agonistic behaviour, where salmon go from being territorial on riverbeds to becoming non-territorial pelagic swimmers in the open ocean (e.g. McCormick et al. 1998). Lending evidence to this is the fact that we observed the highest rate of agonistic behaviours at slower velocity, yet fin erosion declined over time. This suggests that the conditions before the experiment promoted fin erosion.

Previously documented behavioural effects with respect to reduced aggression, increased group structure and decreased fin erosion (East & Magnan 1987, Christiansen & Jobling 1990, Christiansen et al. 1992, Jørgensen & Jobling 1993, Adams et al. 1995) are in line with the results of the present study. Moderate current velocity forced the fish to swim against the current to prevent being swept back into the netting or colliding with other fish, which reduced interactions and structural movements. However, even though agonistic behaviours decreased at moderate compared to slow velocity, the fish displayed the same low level of fin erosion at the end of the experiment.

In previous studies, current velocities far exceeding our fastest velocity caused reduced aggression and improved fin quality in juvenile Arctic charr, brook charr and Atlantic salmon (East & Magnan 1987, Christiansen & Jobling 1990, Jørgensen & Jobling 1993, Adams et al. 1995). At our fast velocity (1.5 BL s^{-1}), fish focused on swimming against the current to avoid being swept back into the netting; thus, structural movements and agonistic behaviour were almost absent. Yet, caudal fin erosion increased at the end of the experiment in the fast velocity. During daily maintenance, fish were observed to have difficulties in maintaining their position without unintentionally colliding with each other and the back netting. Hence, abrasion is the most likely cause for the observed caudal fin erosion, and this is supported by the increase in new fin erosion over time at the fast current velocity. Our experimental setup differed from previous studies and could explain the different

outcome that we observed. As described in Solstorm et al. 2015, it is difficult to compare swimming speed to earlier studies where homogeneity of the current and absolute swimming speed is difficult to evaluate. Previous studies have mostly been conducted in open circular tanks (East & Magnan 1987, Christiansen & Jobling 1990, Jørgensen & Jobling 1993, Adams et al. 1995) with no back netting where the fish can attain abrasive fin erosion, and therefore our experimental setup may have unintentionally caused further fin erosion. Yet, if this is the case, our setup is relevant for fish in sea cages, where fast current velocities could force the fish into the cage netting and thereby cause the same abrasion. With the trend of having land-based post-smolt production, these aspects should not constitute a problem. Abrasion to other fins could also arise from the use of raceways (Arndt et al. 2001), but in our setup, the rough fin index could not detect any clear differences. It is possible that the rounded bottom in our raceways had a positive effect, since no fish were observed to stand on the bottom.

Previous studies have assumed that reduced interactions between fish are always positive and have not focused on the potential need for fish to express different types of behaviour. It could be that fast current velocity does not permit the fish to move around and interact in the environment, which could be seen as a violation of one of the Five Freedoms (defined by FAWC [1995]), i.e. the possibility to perform normal behaviour. In salmon feeding areas in the wild, such as the Nordic seas, individuals experience mean water velocities of 20 to 35 cm s^{-1}, with considerably higher maximum velocities exceeding 100 cm s^{-1} (Orvik et al. 2001). From an evolutionary perspective, salmon should therefore be adapted to a range of velocities, and normal behaviour therefore needs to have a wide range. Even so, wild salmon may drift along when velocities exceed limits in swimming performance, while farmed salmon are forced to maintain position within the cage. This may lead to welfare acceptable limits being breached in aquaculture settings. At the slower velocity, frequent movements and interactions could create a stressful environment for some individuals (Solstorm et al. 2015). Yet, this does not apply to wild salmon, as they do not typically swim in schools in the open sea, and even if interactions occur, they may quickly escape each other in an infinite water volume. As mentioned above, farmed salmon in cages with slow water velocities choose their own swimming speed, making results from slow velocity difficult to relate to fish held in sea cages and therefore welfare evaluation from this group is not commercially relevant. Fish at moderate velocity display a lower degree of interactions and movements, but the current does not seem to be restrictive. The moderate velocity in our study is also within the range considered as the preferable swimming speed for brook charr (Tudorache et al. 2011), as well as in the range of swimming speeds for salmonids in the wild (Drenner et al. 2012). Swimming speeds that are preferred or chosen should be expected to optimise welfare. In view of the welfare concept based on allostasis (Korte et al. 2007), the fish at moderate velocity seemed to uphold stability through change the best (see also Solstorm et al. 2015), resulting in the highest welfare. Altogether, from a welfare perspective, our study showed that a moderate water current velocity likely contributed to the highest welfare with regard to social interactions, swimming speed and fin erosion.

CONCLUSIONS

Our results indicate that there is an upper limit in current velocity where positive effects are gained. An increase in water current velocity and the resulting swimming speed is initially positive by reducing structural movements and agonistic behaviour, but if the current is too strong, it can result in negative effects like increases in new fin damage and a reduced possibility to express different behaviours.

The fish in the present study displayed a behavioural plasticity when rapidly adapting their behaviour to different water current velocities at the start of the experiment, analogous to the response to changing currents in the field (Johansson et al. 2014). This suggests that currents are an effective way of controlling behaviours that may compromise fish welfare. If the fish experience water currents that are too fast, the netting could be modified to decrease the currents (Klebert et al. 2013), or more sheltered areas may be selected. Further studies on both fish behaviour and physiology under different water current velocities are needed to identify optimal current velocities for salmon net cages.

Acknowledgements. We thank the staff at IMR Matre for their assistance during the experiments. The work was funded by the Research Council of Norway Grant No. 207116/S40, Exposed Farming. All experimental work was conducted in accordance with the laws and regulations for experiments and procedures on live animals in Norway, following the Norwegian Regulation on Animal Experimentation 1996. The experiment was approved by Forsøksdyrutvalget (FOTS ID 4146).

LITERATURE CITED

ä Adams CE, Huntingford FA, Krpal J, Jobling M, Burnett SJ (1995) Exercise, agonistic behavior and food acquisition in Arctic charr, *Salvelinus alpinus*. Environ Biol Fishes 43:213–218

ä Arndt RE, Routledge MD, Wagner EJ, Mellenthin RF (2001) Influence of raceway substrate and design on fin erosion and hatchery performance of rainbow trout. N Am J Aquacult 63:312–320

ä Ashley PJ (2007) Fish welfare: current issues in aquaculture. Appl Anim Behav Sci 104:199–235

Blyth PJ, Purser GJ, Russel JF (1993) Detection of feeding rhythms in seacaged Atlantic salmon using a new feeder technology. In: Reinertsen H, Dahle LA, Jørgensen L, Tvinnereim K (eds) Fish farming technology. Balkema, Rotterdam, p 209–216

ä Christiansen JS, Jobling M (1990) The behavior and the relationship between food intake and growth of juvenile Arctic charr, *Salvelinus alpinus* L., subjected to sustained exercise. Can J Zool 68:2185–2191

ä Christiansen JS, Svendsen YS, Jobling M (1992) The combined effects of stocking density and sustained exercise on the behaviour, food intake, and growth of juvenile Arctic charr (*Salvelinus alpinus* L.). Can J Zool 70: 115–122

Crawley MJ (2012) The R book. John Wiley & Sons, Chichester

ä Drenner SM, Clark TD, Whitney CK, Martins EG, Cooke SJ, Hinch SG (2012) A synthesis of tagging studies examining the behaviour and survival of anadromous salmonids in marine environments. PLoS ONE 7:e31311

ä East P, Magnan P (1987) The effect of locomotor activity on the growth of brook charr, *Salvelinus fontinalis* Mitchill. Can J Zool 65:843–846

ä Farrell AP, Johansen JA, Suarez RK (1991) Effects of exercise training on cardiac performance and muscle enzymes in rainbow trout, *Oncorhynchus mykiss*. Fish Physiol Biochem 9:303–312

Fernö A, Holm M (1986) Aggression and growth of Atlantic salmon parr. I. Different stocking densities and size groups. Fiskeridir Skr Ser Havunders 18:113–122

ä Johansson D, Laursen F, Fernö A, Fosseidengen JE and others (2014) The interaction between water currents and salmon swimming behaviour in sea cages. PLoS ONE 9:e97635

ä Jørgensen EH, Jobling M (1993) The effects of exercise on growth, food utilization and osmoregulatory capacity of juvenile Atlantic salmon, *Salmo salar*. Aquaculture 116: 233–246

ä Juell JE, Westerberg H (1993) An ultrasonic telemetric system for automatic positioning of individual fish used to track Atlantic salmon (*Salmo salar* L.) in a sea cage. Aquacult Eng 12:1–18

ä Kadri S, Metcalfe NB, Huntingford FA, Thorpe JE (1991) Daily feeding rhythms in Atlantic salmon in sea cages. Aquaculture 92:219–224

ä Keenleyside MHA, Yamamoto FT (1962) Territorial behaviour of juvenile Atlantic salmon (*Salmo salar* L.). Behaviour 19:139–168

ä Klebert P, Lader P, Gansel L, Oppedal F (2013) Hydrodynamic interactions on net panel and aquaculture fish cages: a review. Ocean Eng 58:260–274

ä Korte SM, Olivier B, Koolhaas JM (2007) A new animal welfare concept based on allostasis. Physiol Behav 92:

422–428

▶ Latremouille DN (2003) Fin erosion in aquaculture and natural environments. Rev Fish Sci 11:315–335

▶ McCormick SD, Hansen LP, Quinn TP, Saunders RL (1998) Movement, migration, and smolting of Atlantic salmon (*Salmo salar*). Can J Fish Aquat Sci 55:77–92

▶ Noble C, Kadri S, Mitchell DF, Huntingford FA (2008) Growth, production and fin damage in cage-held 0+ Atlantic salmon pre-smolts (*Salmo salar* L.) fed either a) on-demand, or b) to a fixed satiation–restriction regime: data from a commercial farm. Aquaculture 275:163–168

▶ Noble C, Jones HAC, Damsgård B, Flood MJ and others (2012) Injuries and deformities in fish: their potential impacts upon aquacultural production and welfare. Fish Physiol Biochem 38:61–83

▶ Oppedal F, Dempster T, Stien LH (2011a) Environmental drivers of Atlantic salmon behaviour in sea-cages. Rev Aquacult 311:1–18

▶ Oppedal F, Vågseth T, Dempster T, Juell JE, Johansson D (2011b) Fluctuating sea-cage environments modify the effects of stocking densities on production and welfare parameters of Atlantic salmon (*Salmo salar* L.). Aquaculture 315:361–368

▶ Orvik KA, Skagseth Ø, Mork M (2001) Atlantic inflow to the Nordic seas: current structures and volume fluxes from moored current meters, VM-ADCP and SeaSoar-CTD observations, 1995–1999. Deep-Sea Res I 48:937–957

▶ Person-Le Ruyet J, Le Bayon N, Gros S (2007) How to assess fin damage in rainbow trout, *Oncorhynchus mykiss?* Aquat Living Resour 20:191–195

▶ Roques JAC, Abbink W, Geurds F, van de Vis H, Flik G (2010) Tailfin clipping, a painful procedure: studies on Nile tilapia and common carp. Physiol Behav 101:533–540

▶ Schneider R, Nicholson BL (1980) Bacteria associated with fin rot disease in hatchery-reared Atlantic salmon (*Salmo salar*). Can J Fish Aquat Sci 37:1505–1513

▶ Solstorm F, Solstorm D, Oppedal F, Fernö A, Fraser TWK, Olsen RE (2015) Fast water currents reduce production performance of post-smolt Atlantic salmon *Salmo salar.* Aquacult Environ Interact 7:125–134

▶ Stevens ED, Sutterlin A, Cook T (1998) Respiratory metabolism and swimming performance in growth hormone transgenic Atlantic salmon. Can J Fish Aquat Sci 55:2028–2035

▶ Stien LH, Bracke MBM, Folkedal O, Nilsson J and others (2013) Salmon Welfare Index Model (SWIM 1.0): a semantic model for overall welfare assessment of caged Atlantic salmon: review of the selected welfare indicators and model presentation. Rev Aquacult 5:33–57

▶ Sutterlin AM, Jokola KJ, Holte B (1979) Swimming behaviour of salmonid fish in ocean pens. J Fish Res Board Can 36:948–954

Tang J, Wardle CS (1992) Power output of two sizes of Atlantic salmon (*Salmo salar*) at their maximum sustained swimming speeds. J Exp Biol 166:33–46

▶ Tudorache C, O'Keefe RA, Benfey TJ (2011) Optimal swimming speeds reflect preferred swimming speeds of brook charr (*Salvelinus fontinalis* Mitchill, 1874). Fish Physiol Biochem 37:307–315

▶ Turnbull JF, Richards RH, Robertson DA (1996) Gross, histological and scanning electron microscopic appearance of dorsal fin rot in farmed Atlantic salmon, *Salmo salar* L., parr. J Fish Dis 19:415–427

▶ Turnbull JF, Adams CE, Richards RH, Robertson DA (1998) Attack site and resultant damage during aggressive encounters in Atlantic salmon (*Salmo salar* L.) parr. Aquaculture 159:345–353

Farm Animal Welfare Council (FAWC) (1995) Five Freedoms of the Farm Animal Welfare Council. http://webarchive.nationalarchives.gov.uk/20121007104210/http:/www.fawc.org.uk/freedoms.htm (accessed on 20 September 2015)

9

Oxygen gradients affect behaviour of caged Atlantic salmon *Salmo salar*

Tina Oldham[1,*], Tim Dempster[2], Jan Olav Fosse[3], Frode Oppedal[3]

[1]Aquatic Animal Health Group, Institute for Marine and Antarctic Studies, University of Tasmania, Launceston, Tasmania 7250, Australia

[2]Sustainable Aquaculture Laboratory – Temperate and Tropical (SALTT), School of BioSciences, University of Melbourne, Parkville, Victoria 3052, Australia

[3]Institute of Marine Research, Matredal 5984, Norway

ABSTRACT: Dissolved oxygen (DO) conditions in marine aquaculture cages are heterogeneous and fluctuate rapidly. Here, by temporarily wrapping a tarpaulin around the top 0 to 6 m of a marine cage (~2000 m^3), we manipulated DO to evaluate the behavioural response of Atlantic salmon *Salmo salar* to hypoxia. Videos were recorded before, during and after DO manipulation at 3 m depth while vertical profiles of temperature, salinity, DO and fish density were continuously measured. The trial was repeated 4 times over a 2 wk period. Temperature and salinity profiles varied little across treatment periods; however, DO saturation was reduced at all depths in all replicate trials during the tarpaulin treatment compared to the periods before or after. In 3 out of 4 trials, swim speeds were 1.5 to 2.7 times slower during the tarpaulin treatment than the before or after periods. Significant changes in vertical distribution of fish density and DO were observed between treatment periods in all replicate trials; salmon swam either above or below the most hypoxic depth layer (59 to 62% DO saturation). In a regression tree analysis, the relative influence of DO in determining fish distribution was 17%, while temperature (39%) and salinity (44%) explained the majority of variation. Our results demonstrate that salmon are capable of modifying their distribution and possibly activity levels in response to intermediate DO levels, but that DO is not a primary driver of behaviour at the saturation levels examined in this study.

KEY WORDS: Hypoxia · Dissolved oxygen · Behaviour · *Salmo salar* · Fish distribution · Aquaculture

INTRODUCTION

The energy yield of anaerobic glycolysis is only 10% that of aerobic metabolism (Hochachka & Somero 2014), thus animals depend on a consistent supply of oxygen from their surroundings to achieve optimal performance. In the aquatic environment however, where diffusion happens slowly and photosynthesis can only partially meet the metabolic demands of organisms in the surface waters (Richards et al. 2009), dissolved oxygen (DO) concentration varies both vertically and horizontally with changing light, temperature, currents (Johansson et al. 2007), wind and rainfall (Diaz 2001). For these reasons, DO is a major limiting factor affecting the growth, distribution and survival of fishes.

Hypoxia, defined in the marine environment as a drop in DO saturation which reduces metabolic scope (Pollock et al. 2007, Richards et al. 2009), occurs naturally in coastal environments (e.g. Johannessen & Macdonald 2009, Silva & Vargas 2014, Brown et al. 2015). Poor DO conditions are exacerbated in aquaculture cages due to restricted water movement, nutrient loading and locally increased biomass (Johansson et al. 2006, 2007, Oppedal et al. 2011b, Burt et al. 2012), and are becoming more common as global temperatures rise (Gruber 2011). In extreme cases, acute hypoxia results in mass mortal-

*Corresponding author: tina.oldham@utas.edu.au

ity (Thronson & Quigg 2008, Stauffer et al. 2012); in less extreme cases, sub-optimal DO concentrations result in decreased growth, appetite, immune function, swimming performance and fish welfare (Oppedal et al. 2011a, Remen et al. 2012, 2014, Burt et al. 2013, Kvamme et al. 2013).

Fish utilize numerous strategies to mitigate the impacts of hypoxia, including physiological adjustments, morphological adaptations, molecular defences and behavioural modifications (Richards et al. 2009). The most immediate strategies to minimize acute hypoxic stress are behavioural adaptations such as avoidance, aquatic surface respiration, air breathing and altered activity levels (Kramer 1987). Many species of fish can actively avoid hypoxic conditions (Pihl et al. 1991, Wannamaker & Rice 2000, Brown et al. 2015), but not all (Kramer 1987, Butler et al. 2001). For example, Atlantic cod *Gadus morhua* L. did not avoid extremely hypoxic conditions when a normoxic refuge was available (Herbert et al. 2011), whereas in a similar trial, rainbow trout *Onchorhynchus mykiss* displayed avoidance behaviour beginning at 80% DO saturation (17 to 19°C) (Poulsen et al. 2011). Even among fish that avoid reduced DO concentrations, the point at which behavioural responses are initiated varies greatly with species, lifestyle and habitat (Whitmore et al. 1960, Richards et al. 2009).

Uncertainty exists regarding the extent to which the world's most farmed marine fish, Atlantic salmon *Salmo salar*, respond behaviourally to hypoxia. One field trial which investigated the relationship between environmental parameters and vertical salmon distribution observed no consistent response to DO, despite reaching levels as low as 57% saturation (Johansson et al. 2006). However, an alternate study at 4 commercial farm sites found that salmon avoided specific depth ranges in the water column where lowest DO concentrations (60% saturation at 15°C) occurred (Johansson et al. 2007). Given that DO concentrations reach severely hypoxic levels in commercial cages for extended periods of time (Oppedal et al. 2011a, Burt et al. 2012, Stien et al. 2012, Dempster et al. 2016), and that commercially viable mitigation options are limited (Bergheim et al. 2006, Srithongouthai et al. 2006, Oppedal et al. 2011b), it is critical to know whether salmon avoid hypoxic areas for farmers to maximize welfare and production performance.

Salmon in sea cages alter their distribution and behaviour in response to numerous stimuli, including light (Oppedal et al. 2007), temperature, salinity, feeding (Oppedal et al. 2011a), water current velocity (Johansson et al. 2014), sound (Bui et al. 2013) and sea lice infestation level (Bui et al. 2016). Given the myriad factors affecting salmon in sea cages, experimental testing within the marine cage environment is required to understand the behavioural trade-offs made by salmon in response to hypoxia and other environmental factors. Here, we manipulated DO levels within a sea cage to determine if salmon altered their behaviour to avoid areas of low DO concentration among the other environmental factors which control their vertical distribution.

MATERIALS AND METHODS

Experimental setup

The experiment was performed in a research scale 12 × 12 × 29 m marine cage at the Institute of Marine Research's Solheim cage environment laboratory, stocked at a density of 7.67 kg m^{-3} with 13 428 (mean mass = 2.4 kg) Atlantic salmon *Salmo salar*. At the 15 to 21 m depth band, below which the majority of fish were observed to aggregate during preliminary observations, a tarpaulin was permanently attached to the cage net to create a barrier to oxygen replenishment. To reduce DO below ambient levels and create a gradient, the entire net cage was raised by crane until the tarpaulin surrounded the upper 0 to 6 m depth band where fish schooled at high densities. During DO reduction periods, the total available cage depth was 14 m (Fig. 1) with a stocking density of 15.34 kg m^{-3}. For each trial, data was recorded during 3 periods: 40 min prior to raising the tarpaulin (before), 60 min with the tarpaulin secured at the sur-

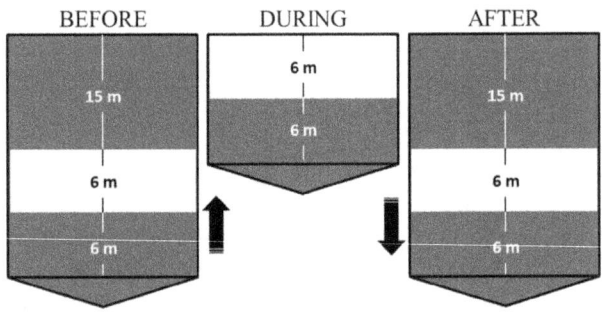

Fig. 1. Cage setup during each of the 3 dissolved oxygen reduction treatment periods: 'before' (control), 'during' (treatment) and 'after' (control). Throughout all trials and treatment periods, temperature, salinity and dissolved oxygen measurements were collected within the top 0 to 6 m of the cage. Vertical distribution of Atlantic salmon *Salmo salar* was estimated by calculating relative echo intensities within the uppermost 6 m depth band during each treatment period. The white band represents the position of the 6 m deep tarpaulin

face (during) and 40 min after the tarpaulin was dropped (after). Four replicate trials were conducted between 30 October and 9 November 2015. All trials were performed in daylight between 09:00 and 16:00 h and were timed to co-occur with slack tide to avoid deformation of the cage.

Though data were collected throughout the entirety of the cage, all analyses were confined to the top 0 to 6 m depth band where the tarpaulin was located during treatment. This strategy was chosen to minimize the potentially confounding effect of the tarpaulin on fish behaviour. If the entire cage area was considered it would have been impossible to distinguish between a response to the reduced DO concentrations within the tarpaulin and a response to the tarpaulin itself. By focusing only on fish which were within the tarpaulin area, we were assured that any changes in distribution could be attributed to environmental variations and not as a result of a response to the tarpaulin.

During each replicate trial, water temperature, DO and salinity were continuously recorded by a CTD (SD204, SAIV AS) vertically profiling between 0 and 13 m at 0.6 m min^{-1} on an automated Belitronics winch. The CTD DO probe was calibrated prior to the start of each trial.

Fish density measurements

Vertical fish distribution was quantified as described by Bjordal et al. (1993) using an echo-integration system (Lindem Data Acquisition) connected to an upward-facing transducer with a 15° acoustic beam positioned beneath the cage at a fixed depth of 30 m. Echo intensity was recorded once per minute in 7 cm depth intervals from 0 to 29 m. The sum of all echo intensity measurements between 0 and 6 m for each minute was then calculated and mean values of every 7 cm depth band were used to calculate relative echo intensity as a measure of percent fish biomass. Mean echo intensity values for each 0.21 m depth band between 0 and 6 m were calculated for all treatment periods (before, during, after).

Swimming speeds

Fish swimming speed was monitored using an underwater 360° pan/tilt Orbit Subsea camera controlled from the surface by a winch. Videos during each treatment period were recorded at a depth of 3 m. Instantaneous swimming speeds were calculated as body lengths per second (BL s^{-1}) based on the time required for the snout and tail of an individual to pass a vertical reference line within the cage (Dempster et al. 2008). Swimming speed was calculated for 20 individuals haphazardly chosen in each treatment period, totalling 240 individuals.

Data analyses

For each trial, differences in vertical distribution between treatment periods of temperature, salinity, DO saturation and fish density were tested for with 2-sample Kolmogorov-Smirnov tests. To correct for multiple comparisons, statistical significance (α = 0.05) was determined at a Bonferroni-corrected p-value of 0.007. Instantaneous swimming speeds were compared using repeated measures ANOVAs. Significant ANOVA results were further analysed using Tukey's HSD (honest significant difference) test for specific pair-wise comparisons.

To determine the relative influence of each environmental factor in explaining vertical fish distribution, data from all trials during the oxygen reduction period were pooled and fish density was modelled as a function of temperature, salinity and DO using a non-parametric regression tree method (Therneau & Atkinson 1997, Johansson et al. 2006, 2007). Briefly, this involves identifying a single variable which best divides the data into 2 groups based on reduction of relative error. The data is separated and the same process repeated, separately, for each sub-group until no further improvements can be made. Cross-validation is then used to 'prune' the tree to the final model. In a graphical presentation, each split is seen as 1 stem dividing into 2 branches. The branch to the left is written out in the split, and the one to the right the opposite. Branch length is proportional to reduction in relative error. The 'leaves' at the end of each terminal branch are predicted fish density, scaled between 0 and 1.

RESULTS

Throughout all trials, a consistent pycnocline was observed with a cool, brackish (~10°C, 20 ppt salinity) surface layer which transitioned to warmer seawater (~13°C, 30 ppt) at a depth between 2 and 4 m. Temperature and salinity varied little between treatment periods, whereas DO saturation was reduced by 10% on average during the tarpaulin treatment compared to the before and after periods in all replicate trials (Table 1). Fish density observations of the

Table 1. Range and variation of environmental conditions experienced by Atlantic salmon *Salmo salar* throughout the dissolved oxygen reduction experiment. Before, during and after refer to the measurement period in relation to dissolved oxygen reduction treatment (see Fig. 1 and 'Materials and methods: Experimental setup' for details)

Period	Sample size	Mean	Standard deviation	Range	Median	Coefficient of variation
Temperature (°C)						
Before	87	12.2	1.1	10.2–13.5	12.7	0.09
During	116	11.6	1.1	9.9–13.4	11.4	0.10
After	116	11.6	1.4	9.6–13.6	11.6	0.12
Salinity (ppt)						
Before	87	25.7	5.0	15.9–31.2	27.8	0.20
During	116	24.5	4.4	16.3–30.9	23.4	0.18
After	115	24.1	5.1	14.8–30.8	24.1	0.21
Dissolved oxygen (% saturation)						
Before	87	75.2	5.4	68.1–86.6	72.9	0.07
During	116	65.3	4.7	59.4–78.4	64.4	0.07
After	116	75.9	5.6	67.2–85.7	75.2	0.07

entire 29 m deep cage area found that, on average, 81% of the fish biomass swam shallower than 14 m prior to the tarpaulin being raised to the surface, and that total fish density was lower in the top 0 to 6 m during treatment than either before or after in all 4 trials. However, during all 4 trials, fish density in the 0 to 2 m surface band was higher during tarpaulin treatment than in either the before or after periods.

Analysis of variance tests on instantaneous swimming speeds detected significant variation between treatment periods in 3 of the 4 replicate trials (F > 21.4, p < 0.001). In all 3 cases, swimming speeds were 1.5 to 2.7 times slower (p < 0.05) during the tarpaulin treatment (range: 0.35 to 0.36 BL s^{-1}) than in the before or after periods (0.53 to 0.95 BL s^{-1}; (Fig. 2).

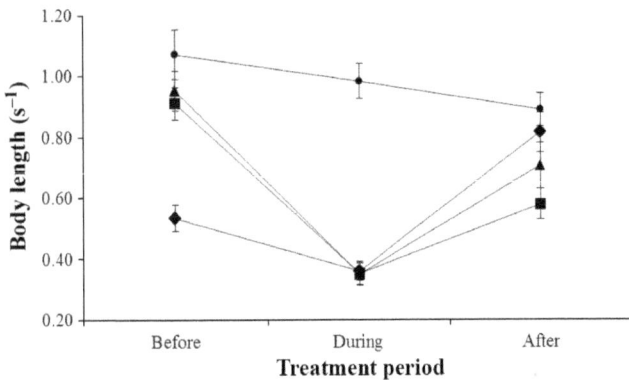

Fig. 2. Mean ± SE instantaneous swimming speeds (body lengths per second; BL s^{-1}) of Atlantic salmon *Salmo salar* before, during and after dissolved oxygen reduction treatment. The 4 replicate trials are represented by (◆) 30 Oct, (■) 2 Nov, (▲) 3 Nov and (●) 9 Nov 2015

In the regression tree model, salinity and temperature had the largest relative influences on fish density, at 44 and 39% respectively, while DO was only 17%. Node 1 of the tree, salinity <28.47 ppt, explained the largest amount of variance; however, the surrogate split of temperature <12.58°C had 97% agreement with the primary split, suggesting that both variables were critical drivers of salmon distribution within the cage. Of the 11 nodes in the tree, 4 splits were attributed to salinity, 5 to temperature and 2 to DO saturation (Fig. 3). The most preferred environment was salinity > 30.43 ppt, temperature < 13.14°C and DO saturation > 65.17%. In both cases where DO saturation was attributed a split, higher levels of DO were the preferred condition.

Trial 1

Environmental conditions between periods were the most variable during Trial 1, with significantly different distributions in both salinity and temperature (Fig. 4). Vertical distribution of fish density also differed significantly during each of the measurement periods. In the period before the tarpaulin treatment, lowest fish density was observed at the surface and increased with depth. During the tarpaulin treatment, minimum fish density occurred at 1.8 m and markedly increased with depth to a maximum at 6 m. In the period after the tarpaulin treatment, fish density distribution was bimodal with minimum density at the surface and maximum density peaks at 2 and 5.4 m (Fig. 5).

Trial 2

Temperature and salinity distributions did not differ significantly between any of the treatment periods (Fig. 4). Vertical distribution of fish biomass was similar during the before and after periods, with minimum fish densities occurring at the surface and increasing with depth. During the tarpaulin treatment, vertical fish distribution differed significantly from the period after, with minimum fish density occurring at 3 m and bimodal peak densities at 1 and 6 m (Fig. 5).

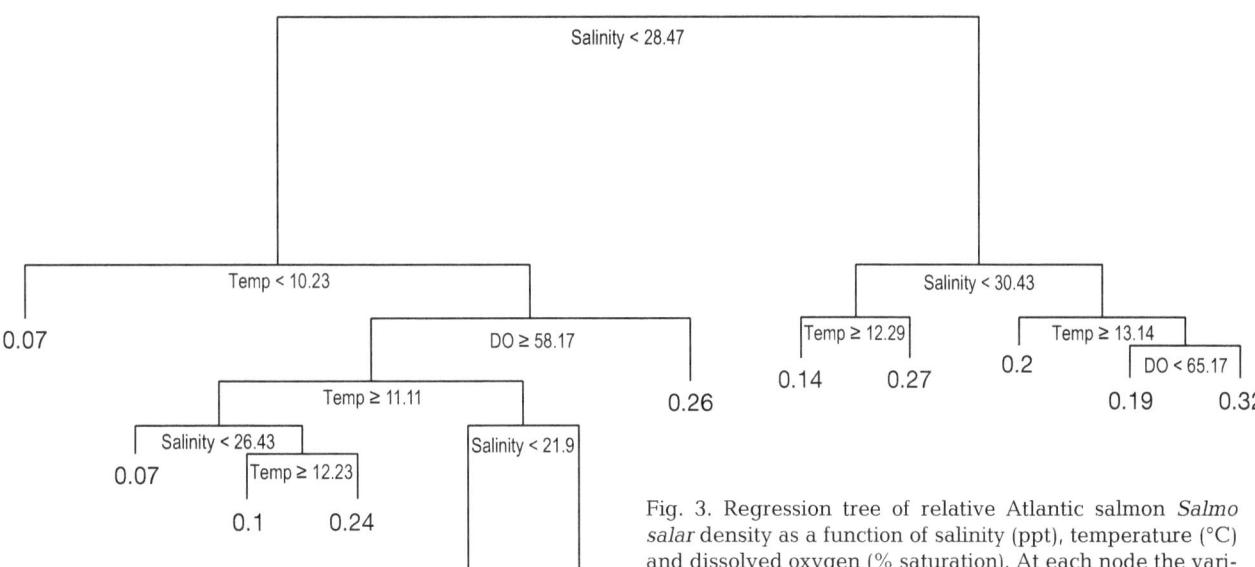

Fig. 3. Regression tree of relative Atlantic salmon *Salmo salar* density as a function of salinity (ppt), temperature (°C) and dissolved oxygen (% saturation). At each node the variable/value causing the split is identified. Predicted relative fish density is noted at the end of each terminal branch. Branch length illustrates the reduction in relative error as a result of the previous split

Trial 3

Temperature and salinity distributions were similar throughout all treatment periods (Fig. 4). Vertical distribution of fish biomass did not differ significantly during the before and after periods, with minimum fish densities occurring at the surface and maximum densities near 4 m. During the tarpaulin treatment, vertical fish density distribution differed significantly from both before and after periods, with minimum fish density occurring at 1.8 m and gradually increasing with depth (Fig. 5).

Trial 4

Fish densities were only recorded during 2 treatment periods in the fourth trial due to equipment malfunction. Temperature and salinity distributions differed significantly between the tarpaulin treatment period and the period after (Fig. 4). During the tarpaulin treatment, minimum fish densities occurred in the top 4 m and increased sharply to maximum density at 6 m. After the tarpaulin treatment, minimum fish density occurred at the surface and increased with depth (Fig. 5).

DISCUSSION

Behavioural responses

When at the surface, surrounding the cage perimeter with a tarpaulin quickly and consistently reduced DO saturations by as much as 20% within a 60 min period. Using this technique, our results provide evidence that salmon have some capacity to modify their behaviour in response to intermediate DO levels (59 to 78% saturation) well above the limiting oxygen saturation (39 ± 1% at 12°C; Remen et al. 2013) in a marine cage environment. In all 4 trials, vertical fish distribution shifted during the DO reduction treatment, with movement away from the depths with the lowest DO concentrations and an increase in fish density in surface waters.

However, whether salmon avoid depths within sea cages with lowest DO appears to be determined by whether a DO gradient is available within their preferred depth band based on other environmental cues, such as temperature, which override a response to intermediate DO concentrations (Oppedal et al. 2011a). With a regression tree model that included temperature, salinity and DO as predictors, the relative importance of DO in determining fish density was only 17%, compared to 44% for salinity and 39% for temperature. In a more holistic model which considered all known determinants of fish distribution within sea cages, such as artificial and natural light, hunger, water current velocity and social cues, the relative influence of DO would be reduced even further.

The results of our manipulative experiment align with previous observations that salmon remained in the warm surface waters of a cage despite DO saturation being 20 to 30% lower than in the deeper, cooler water (Stien et al. 2012). During our trial, the heterogeneity of the cage environment meant the width of preferred depth bands was quite small, however in environments more homogeneous in salinity and temperature, as is typical of coastal salmon farms, responses to DO may be more pronounced as they will seldom be overruled by a pycnocline.

Given that previous work on healthy Atlantic salmon suggested 70 % DO saturation at 16°C as a threshold for reduced growth, and 60 % DO saturation a threshold for fish welfare (Remen et al. 2012), we conclude that salmon have a limited capacity to align their swimming depth with DO conditions

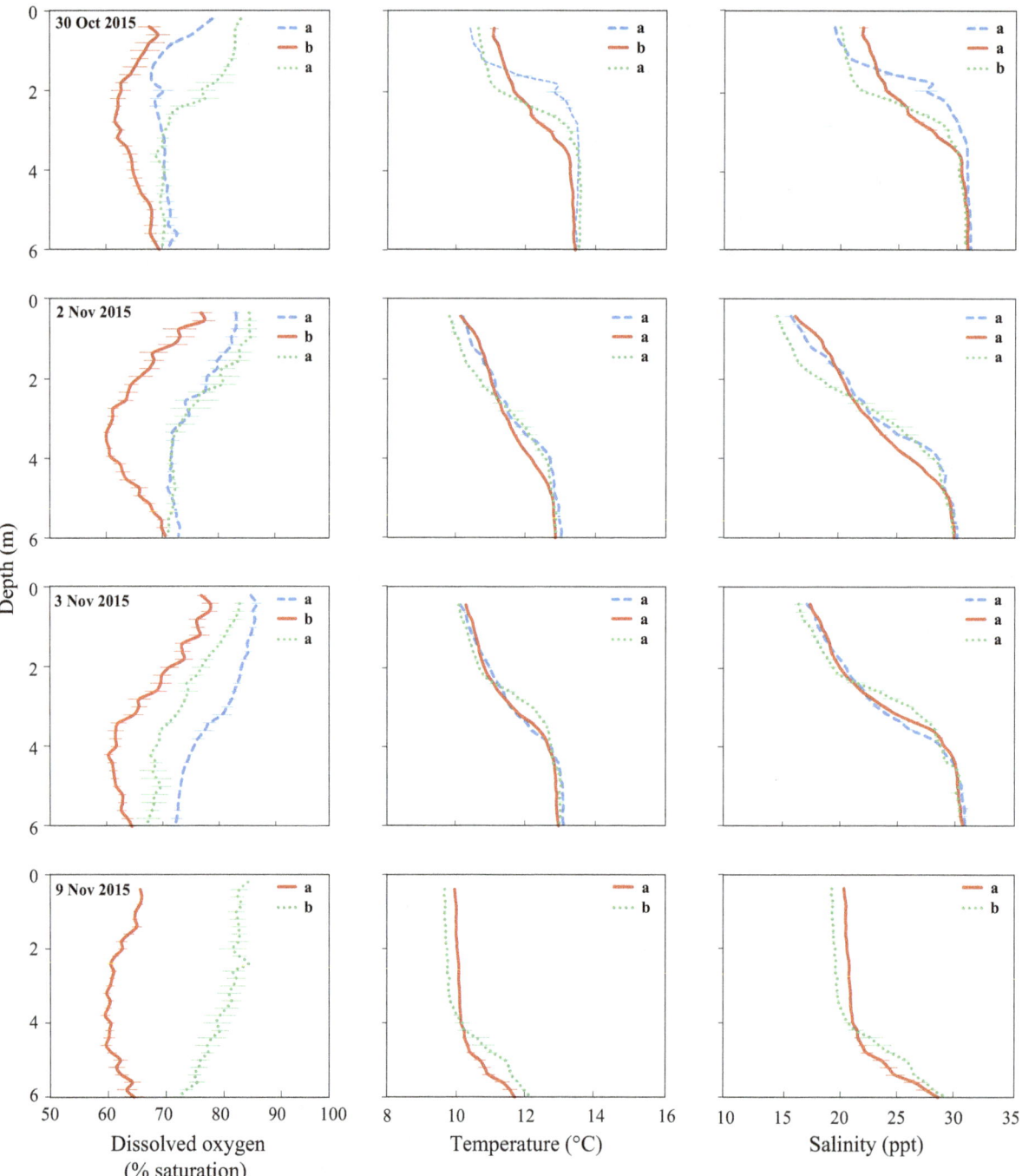

Fig. 4. Vertical profiles of dissolved oxygen (% saturation), temperature (°C) and salinity (ppt) in a marine aquaculture cage before tarpaulin treatment (blue line), during reduced oxygen treatment (red line) and after returning to normal conditions (green line) for each of 4 replicate trials. Values are mean ± SE. Significant differences between treatment periods in each plot are indicated by different letters (p < 0.007)

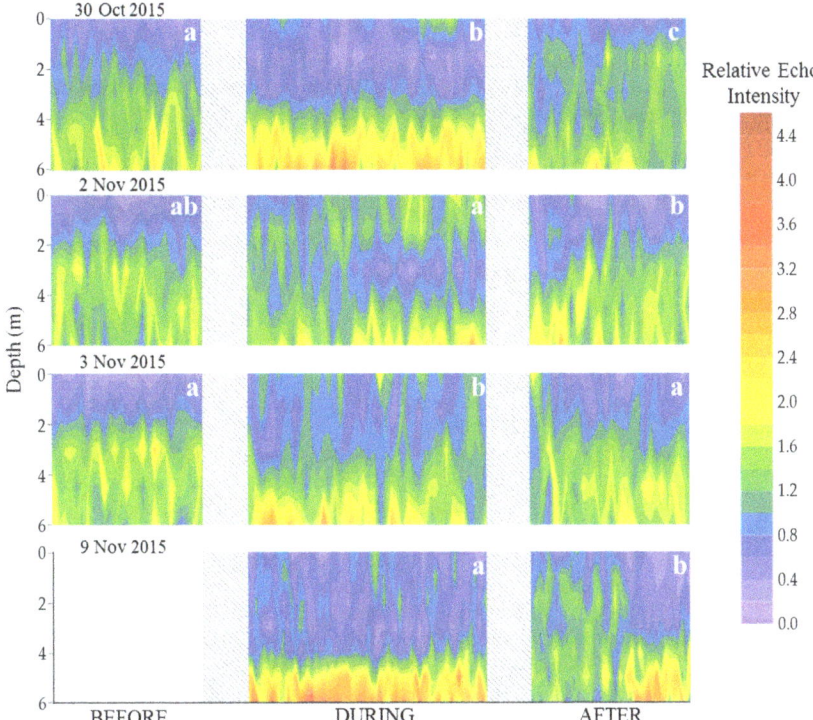

Fig. 5. Depth distribution of Atlantic salmon *Salmo salar* (relative echo intensity) in a marine aquaculture cage before oxygen reduction (40 min), during reduced oxygen treatment (60 min) and after returning to normal conditions (40 min) for each of 4 replicate trials. Significantly different distributions between treatment periods within each trial are indicated by different letters (p < 0.007). Hatching indicates transitional periods during tarp movement. On 9 Nov 2015 fish densities were only recorded during 2 treatment periods due to equipment malfunction

the 4 replicate trials. Though the observed reduction in swim speed cannot be conclusively attributed to the change in DO, as it could also be related to the presence of the tarpaulin or altered social interactions as a result of the reduced fish densities, it is an interesting result for further investigation. The anadromous lifecycle of salmonids means that during some portions of their lives the fish may find themselves in rivers (Elliott et al. 1998) and estuaries (Priede et al. 1988) with low DO, little vertical stratification and no choice but to carry on. In such conditions, an increase in activity level could be lethal, whereas a reduction may allow them to survive until better conditions occur. Such an adaptation would likely contribute to the success of fish in aquaculture given that the cage environment periodically limits their ability to escape hypoxic conditions which will often improve with a changing tide (Johansson et al. 2007).

which would maximize production performance. Testing responses at lower DO concentrations, such as the sustained low saturations (26 to 52%) recently recorded on a commercial farm in Macquarie Harbour, Tasmania (Dempster et al. 2016), is required to determine if the relative importance of DO would increase with more extreme reductions in DO concentration.

With regards to swim speed, behavioural reactions of fish that encounter hypoxic conditions vary from no response to burst swimming, depending on the species and extent of hypoxia (Schurmann & Steffensen 1994, Richards et al. 2009). Increased swim speed improves an individual's likelihood of encountering better conditions, but also increases its oxygen requirements. Alternatively, reduced swim speeds in response to hypoxia minimize the fish's oxygen requirements, but also reduces its chance of reaching more oxygenated water. In this study, a marked decrease of instantaneous swim speed was observed during the reduced DO treatment compared to both the before and after periods in 3 of

Practical implications

As global sea surface temperatures continue to rise and oxygen solubility decreases, hypoxia is expected to become a more frequent occurrence globally (IPCC 2014). The knowledge gained from our experiment stresses the importance for the aquaculture industry to continue developing mitigation and management practices which minimize the occurrence and impacts of hypoxia on farmed salmon.

Potential mitigation measures include site selection to prioritize water movement so that DO replenishment within cages is maximized (Johansson et al. 2007), and farming in deeper areas where there is increased distance between the cage bottom and decomposing organic matter in benthic sediments (Bannister et al. 2014). Frequent fallowing, which minimizes organic enrichment beneath cages, will also reduce biological oxygen demand and thus formation of deep water hypoxia (Valdemarsen et al. 2010). Further, if future research detects that salmon display more pronounced avoidance of depths with poor DO conditions when other environmental factors are uniform, then selection of locations with more vertically homogenous temperatures and salinities could minimize the need for intervention.

Preliminary work has partially tested the benefits of supplemental aeration (Srithongouthai et al. 2006) and oxygenation (Bergheim et al. 2006) in marine net cages. While these techniques improve DO conditions within cages at some depths and in some conditions, further study and cost–benefit analyses are required to optimize performance and assess feasibility at full commercial scale.

Finally, the use of environmental stimuli to alter fish distribution within cages has proven very successful (Oppedal et al. 2007, Bui et al. 2013, Stien et al. 2014). Underwater lighting is commonly used to delay maturation in salmon aquaculture. Recent studies have exploited lights for the secondary purpose of attracting the school to cage depths with reduced parasite load (Frenzl et al. 2014). The same technique could be used to attract salmon away from hypoxic depth layers, or through continuous movement of lights vertically at 1 m min^{-1} to prevent the formation of hypoxic layers by minimizing prolonged schooling at any one depth (Wright et al. 2015).

CONCLUSIONS

Fish in marine aquaculture cages are exposed to substantial environmental variation, but are spatially restricted in their ability to adapt and respond to suboptimal conditions. The impact of hypoxia depends critically on which, if any, response is undertaken. Our manipulative, field-based experiment provides evidence that Atlantic salmon are capable of altering their behaviour in response to intermediate DO concentrations by seeking out water layers with higher DO levels, and possibly with reduced activity levels, but that such responses can be overridden by other factors. These results confirm previous observation-based studies that DO is not a primary driver of Atlantic salmon distribution within marine cages (Johansson et al. 2006, 2007).

Acknowledgements. Many thanks to Tone Vågseth for technical support and Jan Harald Nordahl, Bjørn Frode Grønevik and Marita Laupsa for the heavy lifting and research farm upkeep. This experiment was conducted in accordance with the Norwegian Regulation on Animal Experimentation. The protocol was approved by the Norwegian Animal Research Authority (permit number 8260).

LITERATURE CITED

Bannister RJ, Valdemarsen T, Hansen PK, Holmer M, Ervik A (2014) Changes in benthic sediment conditions under an Atlantic salmon farm at a deep, well-flushed coastal site. Aquacult Environ Interact 5:29–47

Bergheim A, Gausen M, Næss A, Hølland PM, Krogedal P, Crampton V (2006) A newly developed oxygen injection system for cage farms. Aquacult Eng 34:40–46

Bjordal Å, Juell JE, Lindem T, Femö A (1993) Hydroacoustic monitoring and feeding control in cage rearing of Atlantic salmon (*Salmo salar* L.) In: Reinertsen H, Dahle LA, Jørgensen L, Tvinnereim K (eds) Fish farming technology. AA Balkema, Rotterdam, p 203–208

Brown DT, Aday DD, Rice JA (2015) Responses of coastal largemouth bass to episodic hypoxia. Trans Am Fish Soc 144:655–666

Bui S, Oppedal F, Korsøen ØJ, Sonny D, Dempster T (2013) Group behavioural responses of Atlantic salmon (*Salmo salar* L.) to light, infrasound and sound stimuli. PLOS ONE 8:e63696

Bui S, Oppedal F, Stien L, Dempster T (2016) Sea lice infestation level alters salmon swimming depth in sea-cages. Aquacult Environ Interact 8:429–435

Burt K, Hamoutene D, Mabrouk G, Lang C and others (2012) Environmental conditions and occurrence of hypoxia within production cages of Atlantic salmon on the south coast of Newfoundland. Aquacult Res 43:607–620

Burt K, Hamoutene D, Perez-Casanova J, Kurt Gamperl A, Volkoff H (2013) The effect of intermittent hypoxia on growth, appetite and some aspects of the immune response of Atlantic salmon (*Salmo salar*). Aquacult Res 45:124–137

Butler M, Bollens SM, Burkhalter B, Madin LP, Horgan E (2001) Mesopelagic fishes of the Arabian Sea: distribution, abundance and diet of *Chauliodus pammelas*, *Chauliodus sloani*, *Stomias affinis*, and *Stomias nebulosus*. Deep Sea Res II 48:1369–1383

Dempster T, Juell JE, Fosseidengen JE, Fredheim A, Lader P (2008) Behaviour and growth of Atlantic salmon (*Salmo salar* L.) subjected to short-term submergence in commercial scale sea-cages. Aquaculture 276:103–111

Dempster T, Wright D, Oppedal F (2016) Identifying the nature, extent and duration of critical production periods for Atlantic salmon in Macquarie Harbour, Tasmania, during summer. Fisheries Research and Development Corporation report no. 2016-229-DLD. FRDC, Deakin

Diaz RJ (2001) Overview of hypoxia around the world. J Environ Qual 30:275–281

Elliott SR, Coe TA, Helfield JM, Naiman RJ (1998) Spatial variation in environmental characteristics of Atlantic salmon (*Salmo salar*) rivers. Can J Fish Aquat Sci 55:267–280

Frenzl B, Stien L, Cockerill D, Oppedal F and others (2014) Manipulation of farmed Atlantic salmon swimming behaviour through the adjustment of lighting and feeding regimes as a tool for salmon lice control. Aquaculture 424–425:183–188

Gruber N (2011) Warming up, turning sour, losing breath: ocean biogeochemistry under global change. Philos Trans R Soc A 369:1980–1996

Herbert NA, Skjæraasen JE, Nilsen T, Salvanes AG, Steffensen JF (2011) The hypoxia avoidance behaviour of juvenile Atlantic cod (*Gadus morhua* L.) depends on the provision and pressure level of an O_2 refuge.

Mar Biol 158:737–746

Hochachka PW, Somero GN (2014) Biochemical adaptation. Princeton University Press, Princeton, NJ

IPCC (2014) Climate change 2014: synthesis report. Contribution of Working Groups I, II and III to the fifth assessment report of the Intergovernmental Panel On Climate Change. IPCC, Geneva

Johannessen S, Macdonald R (2009) Effects of local and global change on an inland sea: the Strait of Georgia, British Columbia, Canada. Clim Res 40:1–21

Johansson D, Ruohonen K, Kiessling A, Oppedal F, Stiansen JE, Kelly M, Juell JE (2006) Effect of environmental factors on swimming depth preferences of Atlantic salmon (Salmo salar L.) and temporal and spatial variations in oxygen levels in sea cages at a fjord site. Aquaculture 254:594–605

Johansson D, Juell JE, Oppedal F, Stiansen JE, Ruohonen K (2007) The influence of the pycnocline and cage resistance on current flow, oxygen flux and swimming behaviour of Atlantic salmon (Salmo salar L.) in production cages. Aquaculture 265:271–287

Johansson D, Laursen F, Fern A, Fosseidengen JE and others (2014) The interaction between water currents and salmon swimming behaviour in sea cages. PLOS ONE 9: e97635

Kramer DL (1987) Dissolved oxygen and fish behavior. Environ Biol Fishes 18:81–92

Kvamme BO, Gadan K, Finne-Fridell F, Niklasson L and others (2013) Modulation of innate immune responses in Atlantic salmon by chronic hypoxia-induced stress. Fish Shellfish Immunol 34:55–65

Oppedal F, Juell JE, Johansson D (2007) Thermo-and photoregulatory swimming behaviour of caged Atlantic salmon: implications for photoperiod management and fish welfare. Aquaculture 265:70–81

Oppedal F, Dempster T, Stien LH (2011a) Environmental drivers of Atlantic salmon behaviour in sea-cages: a review. Aquaculture 311:1–18

Oppedal F, Vågseth T, Dempster T, Juell JE, Johansson D (2011b) Fluctuating sea-cage environments modify the effects of stocking densities on production and welfare parameters of Atlantic salmon (Salmo salar L.). Aquaculture 315:361–368

Pihl L, Baden S, Diaz R (1991) Effects of periodic hypoxia on distribution of demersal fish and crustaceans. Mar Biol 108:349–360

Pollock M, Clarke L, Dube M (2007) The effects of hypoxia on fishes: from ecological relevance to physiological effects. Environ Rev 15:1–14

Poulsen SB, Jensen LF, Nielsen KS, Malte H, Aarestrup K, Svendsen JC (2011) Behaviour of rainbow trout Oncorhynchus mykiss presented with a choice of normoxia and stepwise progressive hypoxia. J Fish Biol 79:969–979

Priede IG, Solbé JFLG, Nott JE, O'Grady KT, Cragg-Hine D (1988) Behaviour of adult Atlantic salmon, Salmo salar L., in the estuary of the River Ribble in relation to variations in dissolved oxygen and tidal flow. J Fish Biol 33: 133–139

Remen M, Oppedal F, Torgersen T, Imsland AK, Olsen RE (2012) Effects of cyclic environmental hypoxia on physiology and feed intake of post-smolt Atlantic salmon: initial responses and acclimation. Aquaculture 326–339: 148–155

Remen M, Oppedal F, Imsland AK, Olsen RE, Torgersen T (2013) Hypoxia tolerance thresholds for post-smolt Atlantic salmon: dependency of temperature and hypoxia acclimation. Aquaculture 416–417:41–47

Remen M, Aas TS, Vågseth T, Torgersen T, Olsen RE, Imsland A, Oppedal F (2014) Production performance of Atlantic salmon (Salmo salar L.) postsmolts in cyclic hypoxia, and following compensatory growth. Aquacult Res 45:1355–1366

Richards JG, Farrell AP, Brauner CJ (eds) (2009) Hypoxia. Academic Press, London

Schurmann H, Steffensen JF (1994) Spontaneous swimming activity of Atlantic cod Gadus morhua exposed to graded hypoxia at three temperatures. J Exp Biol 197:129–142

Silva N, Vargas CA (2014) Hypoxia in Chilean Patagonian fjords. Prog Oceanogr 129:62–74

Srithongouthai S, Endo A, Inoue A, Kinoshita K and others (2006) Control of dissolved oxygen levels of water in net pens for fish farming by a microscopic bubble generating system. Fish Sci 72:485–493

Stauffer BA, Gellene AG, Schnetzer A, Seubert EL, Oberg C, Sukhatme GS, Caron DA (2012) An oceanographic, meteorological, and biological 'perfect storm' yields a massive fish kill. Mar Ecol Prog Ser 468:231–243

Stien LH, Nilsson J, Hevrøy EM, Oppedal F, Kristiansen TS, Lien AM, Folkedal O (2012) Skirt around a salmon sea cage to reduce infestation of salmon lice resulted in low oxygen levels. Aquacult Eng 51:21–25

Stien LH, Fosseidengen JE, Malm ME, Sveier H, Torgersen T, Wright DW, Oppedal F (2014) Low intensity light of different colours modifies Atlantic salmon depth use. Aquacult Eng 62:42–48

Therneau TM, Atkinson EJ (1997) An introduction to recursive partitioning using the RPART routines. Mayo Foundation Technical Report 61. Rochester, MN. www.mayo. edu/research/documents/biostat-61pdf/doc-10026699

Thronson A, Quigg A (2008) Fifty-five years of fish kills in coastal Texas. Estuaries Coasts 31:802–813

Valdemarsen T, Kristensen E, Holmer M (2010) Sulfur, carbon, and nitrogen cycling in faunated marine sediments impacted by repeated organic enrichment. Mar Ecol Prog Ser 400:37–53

Wannamaker CM, Rice JA (2000) Effects of hypoxia on movements and behavior of selected estuarine organisms from the southeastern United States. J Exp Mar Biol Ecol 249:145–163

Whitmore CM, Warren CE, Doudoroff P (1960) Avoidance reactions of salmonid and centrarchid fishes to low oxygen concentrations. Trans Am Fish Soc 89:17–26

Wright DW, Glaropoulos A, Solstorm D, Stien LH, Oppedal F (2015) Atlantic salmon Salmo salar instantaneously follow vertical light movements in sea cages. Aquacult Environ Interact 7:61–65

10

Intertidal rack-and-bag oyster farms have limited interaction with horseshoe crab activity in New Jersey, USA

Daphne Munroe*, David Bushek, Patricia Woodruff, Lisa Calvo

Haskin Shellfish Research Laboratory, Rutgers University, 6959 Miller Ave., Port Norris, NJ 08349, USA

ABSTRACT: Concern has been raised about the ability of horseshoe crabs *Limulus polyphemus* to traverse intertidal rack-and-bag oyster farms, and how farms may change shorebird foraging activity. During the 2016 horseshoe crab spawning season, experiments conducted in Delaware Bay (New Jersey, USA) assessed the ability of crabs to move among oyster farms and access landward nesting grounds, and surveyed the distribution of dislodged eggs upon which many shorebirds feed. Experiments included testing (1) for impairment of crab passage by oyster racks, (2) for differences in crab abundance among paired farm/control transects, (3) whether farms affect crab stranding rates on nesting beaches, and (4) assessing the spatial distribution of dislodged eggs along the wrack zone among farm and non-farm areas. All crabs, regardless of size, passed beneath racks ≥10 cm tall, indicating that the regulated rack height of 30.5 cm is abundantly precautious to allow crab movement beneath racks. Farm/control census observed 853 crabs in total, with no evidence of differing crab numbers among farmed and control transects. Only 2 of 853 (<0.5%) crabs were obstructed by farm gear, and more crabs were present on nesting beaches inshore of farms compared to adjacent farm-free areas. The proportion of crabs flipped (stranded) at low tide within nesting habitats was constant regardless of farm presence. Dislodged eggs in the wrack zone were observed most frequently in the center of the survey area, and were not concentrated near farms, suggesting that in 2016, shorebird foraging opportunities were not coincident with farm locations.

KEY WORDS: Oyster aquaculture · *Crassostrea virginica* · Rack-and-bag · *Limulus polyphemus* · Ecological interactions · Shorebird foraging

INTRODUCTION

For the first time since records have been kept, the contribution of aquaculture to global seafood supply has exceeded that of wild capture fisheries (FAO 2016). Although shellfish culture is largely viewed as an environmentally and ecologically sound industry (Shumway et al. 2003), as global production expands, the industry is faced with key challenges to ensure ecological sustainability and social acceptance. The

ways in which molluscan aquaculture interacts with fundamental ecosystem processes such as particle depletion, nutrient cycling, and benthic–pelagic coupling has been relatively well studied (Newell 2004, Dumbauld et al. 2009, Rose et al. 2015); however, central to sustainability and acceptance is also understanding the nature of the interaction among farms and wildlife such as birds and mammals that may use habitat near to or occupied by farms. Studies that have examined the ways that molluscan farms inter-

*Corresponding author: dmunroe@hsrl.rutgers.edu

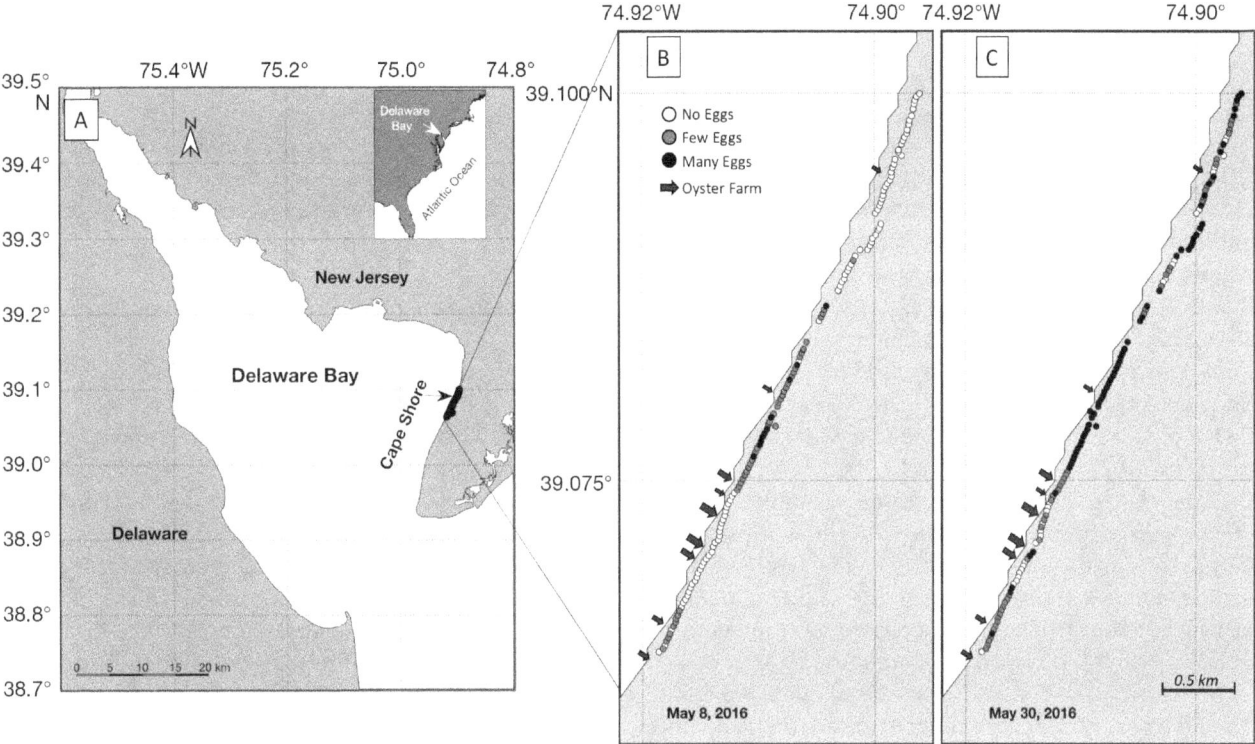

Fig. 1. (A) Study area in Delaware Bay, USA, with the Cape Shore region noted in black. Zoomed-in panels show the distributions of washed up horseshoe crab *Limulus polyphemus* eggs on (B) 8 May and (C) 30 May 2016 within the Cape Shore region relative to farm locations; arrows denote farm locations, and arrow size scales with the relative farm footprint

act with local avian populations provide examples of mixed interactions (Roycroft et al. 2004, Godet et al. 2009, Žydelis et al. 2009), leaving the question unable to be generalized broadly across species, habitats, and culture practices.

Shellfish aquaculture in the Cape Shore region of New Jersey (USA) has a long history as a local food production system (Ford 1997). Eastern oyster *Crassostrea virginica* production declined sharply following the appearance of Multinucleated Sphere X, more commonly known as MSX disease (caused by *Haplosporidium nelsoni*) in 1957 and the more recent establishment of perkinsosis (caused by *Perkinsus marinus*) (Ford 1997). Advances in hatchery and production methods as well as the development of disease-resistant stocks and triploid technology (Dégremont et al. 2015) have revived the industry. Today, an annual production of approximately 1.8 million market-sized aquacultured oysters is primarily carried out on intertidal farms in the Cape Shore, encompassing 10 acres (~4.05 ha) of actively farmed bottom (Calvo 2017).

Horseshoe crabs *Limulus polyphemus* are an iconic and ecologically important species in the Delaware Bay area. Hundreds of thousands come ashore during spring to mate and lay eggs along sandy beaches of the New Jersey and Delaware coastline (Smith et al. 2002, 2017). Approximately 5% of the total shoreline baywide that has been categorized as suitable for crab spawning (Lathrop et al. 2006) is also home to rack-and-bag oyster farming. These activities occur in the Cape Shore region, where farmers grow oysters in specialized cultivation bags on top of short metal racks. Recently, horseshoe crab biologists with a history of studying this region have indicated that most of this area is no longer suitable habitat due to shoreline erosion following recent changes in sea level (Loveland & Botton 2015). Variations of this cultivation method have occurred along this region for more than a century (Ford 1997). An important migratory food source for red knots are lipid-rich horseshoe crab eggs (Haramis et al. 2007) that are deposited on beaches along the migratory route by mating crabs during the birds' spring northward migration, typically in May of each year (Castro & Myers 1993, Botton et al. 1994). The eggs become available to the transitory bird flocks when they are exhumed from nests by sediment disturbance such as crab burrowing and wave action (Kraeuter & Fegley 1994, Smith 2007) and become concentrated in the upper intertidal zone by wave action in the swash zone as the tide rises (Nordstrom et al. 2006).

The *rufa* subspecies of the red knot *Calidris canutus* is a medium-bodied shorebird that breeds in the Canadian Arctic and winters in parts of the southern USA, the Caribbean, and South America (Morrison & Hobson 2004). Red knots migrate between wintering and breeding grounds using stopover areas along the Atlantic coast of the USA, one of which is the shore of the Delaware Bay (Clark et al. 1993, Botton et al. 1994; Fig. 1A). Effective 12 January 2015, the red knot was designated under the US Endangered Species Act as 'Threatened' in the USA by the US Fish and Wildlife Service; Canada listed it as Endangered in 2007.

Concern has recently been raised about the ability of horseshoe crabs to traverse intertidal rack-and-bag oyster farms, and how farms may change shorebird foraging activity leading to the implementation of precautionary risk-averse conservation measures until additional information is obtained (Walsh et al. 2016). During the spawning season in 2016, we conducted a series of experiments to assess the ecological interactions among intertidal rack-and-bag oyster farms and horseshoe crabs and shorebird feeding opportunities. These included tests of whether horseshoe crabs are impinged by farm gear or alter their movement in, around, and among the oyster farms as they make their way to mate and spawn on nesting beaches, and a census of the spatial distribution of washed up eggs upon which red knots feed.

MATERIALS AND METHODS

All experiments were conducted in the Cape Shore region along the New Jersey side of the southern Delaware Bay (Fig. 1A) during late spring through early summer of 2016 (mid-April through late June). The study region is characterized by extensive mudflats that are exposed at low tide, and sloping sand beaches bordered by salt marsh at the upper intertidal. Tides in the Delaware Bay are semidiurnal, with a mean range of 1.6 m.

Testing rack heights for impairment of crab passage

During a daytime low tide, oyster racks were set at 4 heights above the surface of the sand: 7.5, 10.2, 15.2, and 30.5 cm. A range of 10 to 20 male horseshoe crabs ranging in prosoma width from 17.5 to 23.0 cm were collected from adjacent/nearby sloughs and

placed right side up approximately 1 m from and facing toward the rack, then observed as they walked beneath the rack. The prosoma width and success or failure of each crab to pass beneath the rack was recorded. Crabs that arrested movement as they physically came into contact with the racks were scored as having impaired movement.

Repeated transects to test if crabs avoid farms

Paired farm/control transects were established at each of 2 active oyster farms on the Cape Shore (n = 4 transects total). Each pair included 1 transect that intersected the farm, and 1 parallel control transect 20 m from the farm that passed through adjacent unfarmed intertidal habitat (Fig. 2). Each transect spanned a 1 m wide area perpendicular to the high tide line, and covered a zone inshore of the farm area (Zone 1) and through the farm (Zone 2) (Fig. 2). Transect lengths at each of the farms were unequal due to the nature of the mudflats on which they were located; they measured 190 and 138 m in total length (Farm A and B, respectively), with Zone 1 spanning the inshore 128 and 93 m of each (Farm A and B, respectively). The paired transects were surveyed a

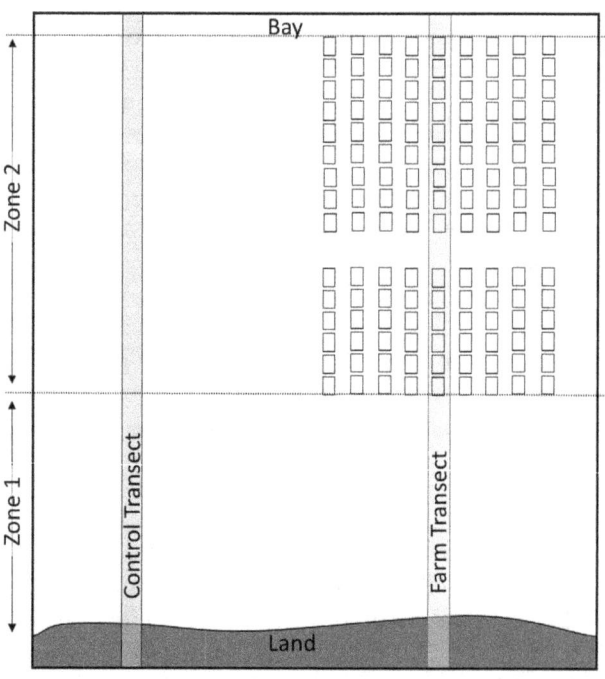

Fig. 2. Schematic layout of paired oyster farm and control transects. Transects at each of the farms were unequal; they measured 190 and 138 m in total length (Farms A and B, respectively), with Zone 1 spanning the inshore 128 and 93 m of each, respectively. Rectangles indicate individual oyster racks

total of 10 times during daytime low tides between 29 April and 15 June 2016. Location and activity of all crabs encountered (observed visually) along each transect were recorded, and all crabs whose mobility was impaired by a rack (e.g. crabs that were trapped, caught, or impinged) were noted.

Data were non-normally distributed, thus a non-parametric Wilcoxon signed rank test was performed to test for differences in the abundance of crabs observed on control versus farm transects. Absolute values of crab observations varied greatly over the course of the study due to seasonal spawning immigration and emigration; therefore, standardization of each survey was performed by summing paired transect counts for each survey and then calculating the fraction of the total that was observed on the farm. The fraction observed on the farm transect was then analyzed using a mixed model analysis of variance in which zone was nested within farm site to assess the influence of zone.

Nesting beach census

Three adjacent upper beach segments spanning approximately 91 m (150 yards) parallel to the shoreline were surveyed on 16 nights bracketing horseshoe crab spawning peaks around full and new moons between 22 April and 22 June 2016. The central segment was inshore of an active oyster farm, whereas the 2 segments on either side were not. Surveys were conducted at night 3 h after high tide to enumerate stranded crabs. All live crabs in the upper beach spawning habitat were counted and recorded along with their condition: flipped on their back (stranded) or right side up. The fractions of crabs flipped at farm and non-farm segments were compared using repeated measures analysis of variance.

Spatial survey of washed up egg distribution

Two photographic censuses were performed overnight on 8 and 30 May 2016, to document the distribution of horseshoe crab eggs washed up near the high tide line where red knot foraging is concentrated (Burger et al. 1977, 1997, Botton et al. 1994). The upper intertidal along the Cape Shore region (Fig. 1A) was divided evenly into 150 segments (approximately 91 m of shoreline per segment), and photographic samples documented a 0.5 m wide (alongshore) by 1 m (downbeach) quadrat perpendicular to the high tide line and crossing the wrack

line, located randomly alongshore near the center of each segment. Each photographic quadrat was assessed for the presence of horseshoe crab eggs that had been washed up at the surface, and each quadrat was coded as having no eggs, few eggs (defined as ≤5% of the quadrat containing eggs), and many eggs (defined as >5% of the quadrat containing eggs).

RESULTS

All crabs, regardless of prosoma width, were able to pass beneath racks ≥10 cm above the bottom (Fig. 3). Of the total 48 observations, 42 crabs passed beneath the racks, whilst 6 failed to transit the rack. Those 6 were among a total of 10 crabs tested at a rack height of 7.5 cm; the other 4 were able to pass the rack at a height of 7.5 cm. All of the 6 crabs that were unable to pass buried the front of their prosoma beneath the rack edge and stopped, but were not irreversibly stuck.

In total, we observed 853 crabs during the repeated transect surveys, with a relatively even distribution of crabs between farm and control transects (46 farm and 39 control transects at Site A, and 369 farm and 399 control transects at Site B, Table 1), both inshore

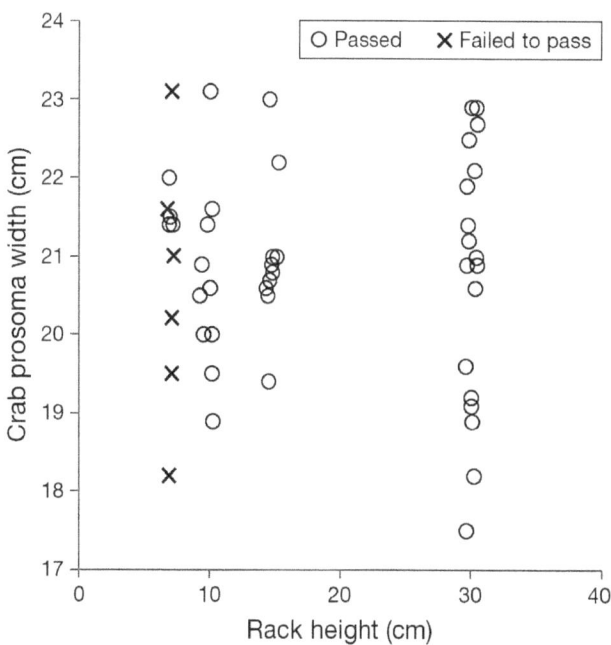

Fig. 3. Success (open circles) and failure (marked with 'x') of horseshoe crabs *Limulus polyphemus* to pass beneath oyster racks set to 7.5, 10.2, 15.2 and 30.5 cm above the sediment surface. Data points represent individual crabs for each rack height and have been jittered to help illustrate the distribution

Table 1. Total horseshoe crab *Limulus polyphemus* counts observed during 10 transect surveys performed periodically during daytime low tide between 29 April and 15 June 2016. See Fig. 2 for the layout of the transects

	Farm A	Farm B
Control transect		
Zone 1	36	188
Zone 2	3	211
Farm transect		
Zone 1	23	215
Zone 2	23	154

of farm activities (Zone 1) and within the farms (Zone 2). In total, 2 out of 853 (<0.5%) crabs were observed to be obstructed by racks.

The first 2 and final farm transect surveys were removed from statistical analyses because very few crabs were observed at the beginning and end of the study (i.e. many 0 counts). We found no significant difference in the number of crabs among farm and control transects ($V = 76.5$, p = 0.3, paired Wilcoxon signed rank test). Likewise, the fraction of crabs observed on the farm transects approximated 0.5, and zone ($F = 0.94$, p = 0.51) did not significantly affect the fraction of crabs observed on the farm transects.

In total, 3527 live crabs were counted on the 3 nesting beach segments during overnight censuses, with nearly half of those (48%) observed in the central segment inshore of the oyster farm. Segment 2 (inshore of the farm) tended to have more crabs observed during each census (Fig. 4), with the highest numbers of crabs on all 3 segments associated with full and new moons in the middle of the spawning season (Fig. 4). The fraction of flipped (stranded) crabs varied greatly among censuses; however, there was no significant difference in the fraction of flipped (stranded) crabs among farmed and non-farmed segments (Fig. 5; $F = 15.2$, p = 0.16).

The amount of washed up eggs observed in the survey was relatively lower in early May, before the majority of crabs returned to spawn. Densities of crab eggs washed up at the high tide line were distributed unevenly throughout the region surveyed (Fig. 1B,C). Washed up eggs were observed most frequently in the

central portion of the survey area, and were not highly concentrated in the area of farms.

DISCUSSION

Complementary experiments and surveys were conducted during the 2016 horseshoe crab spawning season to assess the ecological interactions among intertidal rack-and-bag oyster farms and horseshoe crabs and shorebird feeding opportunities. These included tests of whether horseshoe crabs become trapped by gear or alter their movement in, around, and among the oyster farms as they make their way to mate on nesting beaches, and a census of the spatial distribution of washed up eggs upon which red knots feed. All crabs, regardless of size, passed easily beneath racks positioned ≥10 cm above the bottom. This rack height corresponds with prosoma heights measured across a random selection of spawning male and female crabs (7.5 and 10.0 cm, respectively) from this same region (Kraeuter & Fegley 1994). Thus, the regulated rack height of 30.5 cm should be sufficiently precautious to allow crab movement beneath racks.

Repeated surveys of paired farm–control transects showed no evidence that crabs avoid or are trapped by farms as they come ashore to mate and lay eggs. The 2 farms used in this study showed a large difference in the number of crabs observed during transect surveys, with Site A being a longer transect with fewer crabs. The Cape Shore shoreline is somewhat

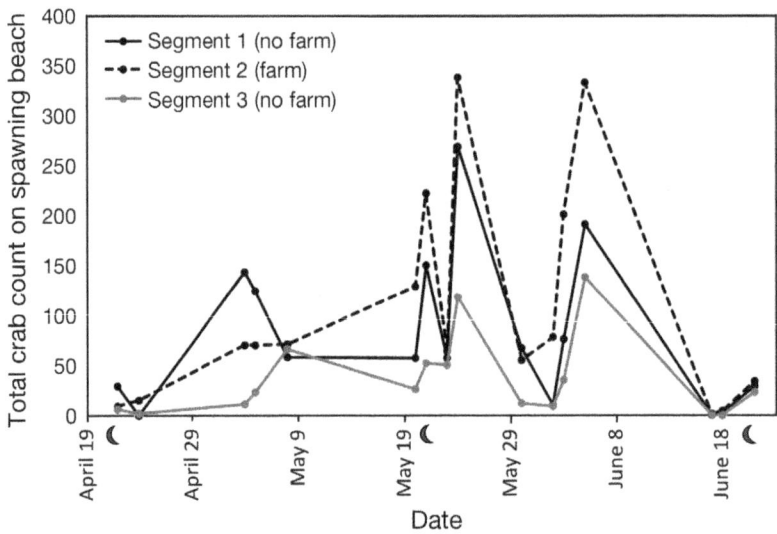

Fig. 4. Time series of total live horseshoe crabs *Limulus polyphemus* counted during each overnight census of nesting habitat conducted during the 2016 horseshoe crab spawning season. Solid black, dashed black, and solid grey lines show counts on Segments 1, 2, and 3, respectively. Moon symbols along the x-axis show the full moon events

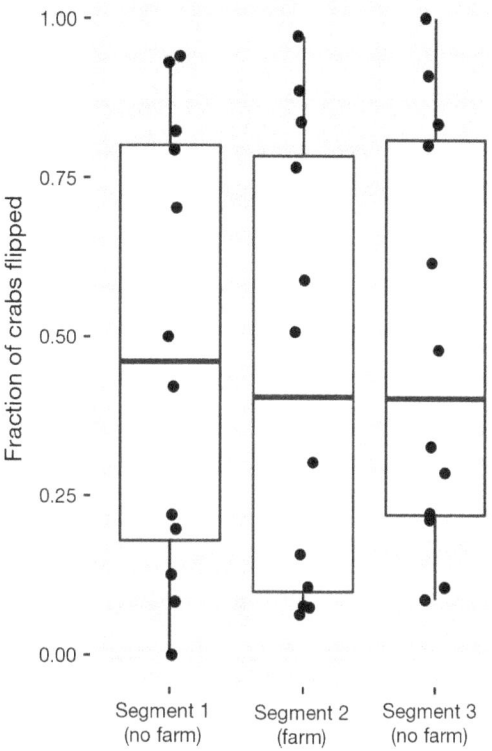

Fig. 5. Fraction of horseshoe crabs *Limulus polyphemus* flipped over (stranded) in the upper intertidal in each of 3 surveyed segments, with each census observation overlaid (black dots). Box plot—line: median, box: 25th and 75th percentile, whiskers: minimum and maximum observations

heterogeneous in the quality of crab habitat, with high-quality habitat, erosional areas, and some bulkheaded sections; the 2 farms also differed in the quality of the crab habitat near the upper intertidal. At Site A, the upper intertidal had been eroded with exposed peat, which has been shown to lead to fewer crabs returning to spawn (Botton et al. 1988). Site B was located within a section of higher quality, predominantly sandy crab habitat.

Farm gear did not alter the number of horseshoe crabs onshore on nesting beaches (Fig. 4). In fact, the trend was for more crabs to be present on nesting beaches inshore of where farming was occurring. The fraction of crabs on nesting beaches that were flipped over (stranded) varied greatly among the census walks, with a tendency for a greater fraction flipped on nights when wind and waves were greater (data not presented), which is consistent with previous studies (Botton & Loveland 1989). Importantly, the fraction of stranded crabs was not higher on nesting habitat inshore of farm gear (Fig. 5).

The density of crab eggs washed up in the wrack zone was unevenly distributed throughout the sur-

veyed region of the Cape Shore. Washed up eggs were observed most frequently in the central portion of the survey area, and were not highly concentrated in the area of farms. Once eggs are resuspended from nests by wave action or other means, they are passively dispersed on waves and wind-driven currents (Smith et al. 2002) and tend to accumulate at concentrating features such as sand spits and jetties (Botton et al. 1994). Thus, the spatial pattern of egg concentration in the wrack line observed here does not necessarily reflect where concentrations of spawning females are coming ashore. Considering that 20% of the surficial eggs on the entire beach are found in the uppermost wrack line (Nordstrom et al. 2006), the spatial patterns of eggs in the wrack line should be a good relative predictor of the best shorebird foraging habitat within the survey region.

In general, more eggs are available to birds in the swash zone on a rising tide than are available remaining on the beach; therefore, active foraging in the swash zone during high, rising tides may provide the best foraging opportunities, and counts of eggs remaining on the beach may underestimate the absolute foraging opportunities (Nordstrom et al. 2006). Likewise, the distribution of available eggs baywide varies, with the highest densities of resuspended eggs found upbay in the areas of Sea Breeze on the New Jersey side of the bay, and Kitts Hummock and North Bowers on the Delaware side (Smith et al. 2002). Our surveys showed no evidence that farms interfere with the ability of spawning crabs to reach nest sites, and that optimal horseshoe crab egg-foraging opportunities for shorebirds are not coincident with farm locations.

Acknowledgements. We thank the field crew that helped count crabs and eggs at all hours of day and night: M. Acquafredda, S. Borsetti, B. Campbell, B. Cerione, M. Danihel, B. Dixon, E. Gilardi, J. Gius, F. Klie, H. Lehmann, L. O'Neil, N. Pray, B. Schum, J. Shinn, and M. Whiteside. Beach segments used for egg counts were established as part of the New Jersey Sea Grant project 'Identifying the Impacts of Commercial Oyster Aquaculture on Foraging Behavior of Red Knots in Delaware Bay.' Some crab stranding data reported here were collected concurrent with participation in the Wetlands Institute citizen science program, Return The Favor program, and we appreciate the training provided and important work they continue to do. We are grateful for the continued cooperation of the oyster farmers of Delaware Bay who allow us access to their farms. Thoughtful comments from 3 anonymous reviewers improved an earlier draft of the manuscript. This work was partially supported by the USDA National Institute of Food and Agriculture Hatch project accession numbers 1002345 and 1009201 through the New Jersey Agricultural Experiment Station, Hatch projects NJ32115 and NJ32114.

LITERATURE CITED

Botton ML, Loveland RE (1989) Reproductive risk: high mortality associated with spawning horseshoe crabs (*Limulus polyphemus*) in Delaware Bay, USA. Mar Biol 101: 143–151

Botton ML, Loveland RE, Jacobsen TR (1988) Beach erosion and geochemical factors: influence on spawning success of horseshoe crabs (*Limulus polyphemus*) in Delaware Bay. Mar Biol 99:325–332

Botton ML, Loveland RE, Jacobsen TR (1994) Site selection by migratory shorebirds in Delaware Bay, and its relationship to beach characteristics and abundance of horseshoe crab (*Limulus polyphemus*) eggs. Auk 111:605–616

Burger J, Howe MA, Hahn C, Chase J (1977) Effects of tidal cycles on habitat selection and habitat partitioning by migrating shorebirds. Auk 94:743–758

Burger J, Niles L, Clark KE (1997) Importance of beach, mudflat and marsh habitats to migrant shorebirds on Delaware Bay. Biol Conserv 79:283–292

Calvo LM (2017) New Jersey shellfish aquaculture situation and outlook report 2015 production year. New Jersey Sea Grant Publication NJSG-17-912

Castro G, Myers JP (1993) Shorebird predation on eggs of horseshoe crabs during spring stopover on Delaware Bay. Auk 110:927–930

Clark KE, Niles LJ, Burger J (1993) Abundance and distribution of migrant shorebirds in Delaware Bay. Condor 95: 694–705

Dégremont L, Garcia C, Allen SK Jr (2015) Genetic improvement for disease resistance in oysters: a review. J Invertebr Pathol 131:226–241

Dumbauld BR, Ruesink JL, Rumrill SS (2009) The ecological role of bivalve shellfish aquaculture in the estuarine environment: a review with application to oyster and clam culture in the West Coast (USA) estuaries. Aquaculture 290:196–223

FAO (Food and Agriculture Organization of the United Nations) (2016) The state of world fisheries and aquaculture 2016. Contributing to food security and nutrition for all. FAO, Rome

Ford SE (1997) History and present status of molluscan shellfisheries from Barnegat Bay to Delaware Bay. In: MacKenzie CL, Burrell VG, Rosenfield A, Hobart WL (eds) The history, present condition, and future of the molluscan fisheries of North and Central America and Europe, Vol 1: North America. Tech Rep. U.S. Department of Commerce, NOAA, NMFS, Seattle, WA, p 119–140

Godet L, Toupoint N, Fournier J, Le Mao P, Retière C, Olivier F (2009) Clam farmers and oystercatchers: effects of the degradation of *Lanice conchilega* beds by shellfish farming on the spatial distribution of shorebirds. Mar Pollut Bull 58:589–595

Haramis GM, Link WA, Osenton PC, Carter DB and others (2007) Stable isotope and pen feeding trial studies confirm the value of horseshoe crab *Limulus polyphemus* eggs to spring migrant shorebirds in Delaware Bay.

J Avian Biol 38:367–376

Kraeuter JN, Fegley SR (1994) Vertical disturbance of sediments by horseshoe crabs (*Limulus polyphemus*) during their spawning season. Estuaries 17:288–294

Lathrop RGJ, Allen M, Love A (2006) Mapping and assessing critical horseshoe crab spawning habitats of Delaware Bay. Walton Center For Remote Sensing and Spatial Analysis, Rutgers University, New Brunswick, NJ

Loveland RE, Botton ML (2015) Sea level rise in Delaware Bay, USA: adaptations of spawning horseshoe crabs (*Limulus polyphemus*) to the glacial past, and the rapidly changing shoreline of the Bay. In: Carmichael RH, Botton ML, Shin PKS, Cheung SG (eds) Changing global perspectives on horseshoe crab biology, conservation and management. Springer, New York, NY, p 41–64

Morrison RIG, Hobson KA (2004) Use of body stores in shorebirds after arrival on high-Arctic breeding grounds. Auk 121:333–344

Newell RIE (2004) Ecosystem influences of natural and cultivated populations of suspension-feeding bivalve molluscs: a review. J Shellfish Res 23:51–61

Nordstrom KF, Jackson NL, Smith DR, Weber RG (2006) Transport of horseshoe crab eggs by waves and swash on an estuarine beach: implications for foraging shorebirds. Estuar Coast Shelf Sci 70:438–448

Rose JM, Bricker SB, Ferreira JG (2015) Comparative analysis of modeled nitrogen removal by shellfish farms. Mar Pollut Bull 91:185–190

Roycroft D, Kelly TC, Lewis LJ (2004) Birds, seals and the suspension culture of mussels in Bantry Bay, a nonseaduck area in Southwest Ireland. Estuar Coast Shelf Sci 61:703–712

Shumway SE, Davis C, Downey R, Karney R and others (2003) Shellfish aquaculture–in praise of sustainable economies and environments. World Aquacult 34:8–10

Smith DR (2007) Effect of horseshoe crab spawning density on nest disturbance and exhumation of eggs: a simulation study. Estuaries Coasts 30:287–295

Smith DR, Pooler PS, Loveland RE, Botton ML, Michels SF, Weber RG, Carter DB (2002) Horseshoe crab (*Limulus polyphemus*) reproductive activity on Delaware Bay beaches: interactions with beach characteristics. J Coast Res 18:730–740

Smith DR, Brockmann HJ, Beekey MA, King TL, Millard MJ, Zaldívar-Rae J (2017) Conservation status of the American horseshoe crab, (*Limulus polyphemus*): a regional assessment. Rev Fish Biol Fish 27:135–175

Walsh WL, Powposki R, Schrading E (2016) Biological opinion on the effects of existing and expanded structural aquaculture of native bivalves in Delaware Bay, Middle and Lower Townships, Cape May County, New Jersey on the federally listed red knot (*Calidris canutus rufa*). US Fish and Wildlife Service, New Jersey Field Office, Galloway, NJ

Żydelis R, Esler D, Kirk M, Boyd SW (2009) Effects of offbottom shellfish aquaculture on winter habitat use by molluscivorous sea ducks. Aquat Conserv 19:34–42

11

Differential effects of adult mussels on the retention and fine-scale distribution of juvenile seed mussels and biofouling organisms in long-line aquaculture

Paul M. South[1,2,*], Oliver Floerl[1], Andrew G. Jeffs[2]

[1]Cawthron Institute, 98 Halifax Street East, Nelson 7010, New Zealand
[2]Institute of Marine Science, University of Auckland, Private bag 92019, Auckland, New Zealand

ABSTRACT: The majority of juvenile seed mussels are lost in aquaculture production. Understanding the causes of the losses of seed mussels is critical to reducing uncertainties in mussel aquaculture production. One major cause of loss appears to be the secondary settlement behaviour of mussels, which is thought to be a behavioural process by which larger juveniles can safely recruit among adults. This implies that once a juvenile mussel has settled among adults, there is either some impetus to remain or other positive interactions that promote increased survival. In this study, 2 densities (5 and 20 per 45 cm experimental rope) of adult green-lipped mussels *Perna canaliculus* were deployed alongside juvenile seed mussels to test whether this enhanced the retention of the juveniles in a typical suspended culture. Adult shells were also deployed to ascertain whether any effects were due to the physical presence of the mussels or the influence of their biological functioning. The presence of adult mussels or shells did not increase the retention of juvenile *P. canaliculus*, but small-scale movements of juveniles were increased by the addition of 20 live adult mussels per experimental rope. However, the presence of adult mussels and mussel shells on experimental ropes greatly increased the abundance of biofouling organisms. While the addition of live adult mussels or shells failed to provide a simple tool for increasing retention of seed mussels on aquaculture lines, they offer new insights into the identity and ecology of key biofouling organisms that can be problematic in mussel aquaculture production.

KEY WORDS: Mussel retention · Secondary settlement · Green-lipped mussel · *Perna canaliculus* · *Mytilus galloprovincialis* · Recruitment

INTRODUCTION

Increases in the global human population and its pressure on wild stocks of fish and shellfish have resulted in the rapid emergence of aquaculture to meet our growing demand for seafood (Naylor et al. 2000, Pauly et al. 2002). Mussel aquaculture has grown to become a major global industry during the last 30 yr (Smaal 2002, Carrasco et al. 2014, FAO 2016). The majority of mussel aquaculture relies on wild sources of larval and juvenile mussels to seed aquaculture substrata (e.g. ropes, rafts and benthic mussel beds) in coastal production facilities. However, natural settlement of mussels is variable in space and time, and this variability causes considerable uncertainty in the continuity of aquaculture production (Carrasco et al. 2014). Access to wild seed sources for aquaculture is also increasingly being constrained by regulations such as quota allocation and reduced or managed access to seed-catching areas and seasons (de Vooys 1999, Smaal 2002). The early stages of mussel aquaculture production can be extremely inefficient, with massive quantities of juveniles frequently lost soon after capture, further

*Corresponding author: paul.south@cawthron.org.nz

compounding the vulnerability of aquaculture opera-tions to natural population fluctuations (Peteiro et al. 2007, Capelle et al. 2014). Retaining juvenile mussels within aquaculture production systems in the face of a natural tendency for losses is commonly referred to as 'retention'. Increasing the retention of juvenile seed mussels in the early production cycle would greatly increase the overall efficiency of the mussel aquaculture industry and lessen its susceptibility to natural variations in larval supply and settlement.

The New Zealand mussel industry is based on the aquaculture of the endemic green-lipped mussel *Perna canaliculus* (Gmelin, 1791) and has grown to an annual production of 101 311 t since its develop-ment in the 1970s (Aquaculture New Zealand 2016). The issue of retention has received particular atten-tion in New Zealand, where the mussel aquaculture industry is almost entirely reliant on 1 ephemeral wild source of seed mussels (Alfaro & Jeffs 2003, Alfaro et al. 2010). In recent years, overall production has been severely impacted by intermittent periods of limited availability of wild mussel seed. Further-more, the retention of juvenile *P. canaliculus* on aquaculture growing substrata, such as fibrous nurs-ery ropes, can be very poor, with losses that range from 50 to 100 % (Jeffs et al. 1999, Webb & Heasman 2006, Hayden & Woods 2011). Consequently, a few studies have specifically addressed the causes of low retention, focussing on methods of determining the quality of juveniles captured from the wild (Webb & Heasman 2006, Sim-Smith & Jeffs 2011), impacts on fitness and behaviour that can occur during transport or due to poor handling (Webb & Heasman 2006, Carton et al. 2007) and environmental conditions during early production (Alfaro 2006c, Carton et al. 2007, Hayden & Woods 2011). However, the issue of retention is complex and far from understood, largely due to the small number of studies that have addressed this problem.

One of the most important factors governing the retention of *P. canaliculus* is likely to be the second-ary settlement behaviour that is a pronounced fea-ture of this species. Secondary settlement has been observed in *P. canaliculus* juveniles of 0.3–6 mm in length (Buchanan & Babcock 1997, Jeffs et al. 1999, Hayden & Woods 2011). Primary settlement occurs when mussel larvae of 240–300 µm transition from pelagic to benthic modes of life and undergo meta-morphosis. Mussel larvae often initially settle on fila-mentous structures, such as algae, and this has been proposed as a mechanism to avoid consumption by, or competition with, conspecific adults, although there is supporting (Bayne 1964, Alfaro 2006a) and

refuting (Lasiak & Barnard 1995, Erlandsson et al. 2008) evidence for this being advantageous. Second-ary settlement is the process by which juvenile mus-sels detach and migrate away from their initial larval settlement sites to explore, select and re-settle in alternative locations (Bayne 1964). Movement can occur via the juvenile initiating mucus drifting or pedal crawling behaviours that operate over medium (10s to 100s m) and small (cm) scales, respectively (Bayne 1964, Buchanan & Babcock 1997, Carton et al. 2007, Le Corre et al. 2013, South 2016). The ability of large numbers of juvenile mussels to migrate away from aquaculture growing structures is enormously problematic not only because these mussels are lost from the production cycle, but also because it creates vacant space for colonisation by biofouling organ-isms which can compete with the remaining cultured mussels and create problems for subsequent aqua-culture processing (Fitridge et al. 2012). Further-more, small-scale movements of juvenile mussels are of particular interest in New Zealand because seed mussels are usually attached to degradable substrata (e.g. algae or, in the case of hatchery-reared juve-niles, coir ropes) when they are deployed at farm sites for on-growing. Therefore, it is essential that seed mussels migrate to permanent substrata (i.e. suspended growing ropes) if they are to remain in production.

The triggers of secondary settlement behaviour in juvenile mussels are not well understood and cur-rently cannot be managed. There appear to be many potential triggers of secondary settlement processes in mytilid mussels, including changes in the local environment (Carton et al. 2007, Hayden & Woods 2011) and changing habitat requirements or behav-iour (Alfaro & Jeffs 2002, Alfaro et al. 2004, von der Meden et al. 2010). Furthermore, juvenile seed mussels on growing structures will likely experience intense pressure from the ongoing settlement of a wide range of biofouling organisms competing for the same space (Holthuis et al. 2015). While biofoul-ing organisms are known to impact the later stages of mussel aquaculture production, their effect on the retention of juveniles has not been examined in any detail.

One potential cause of secondary settlement mi-gration is the proximity of the juvenile mussels to populations of conspecific adults. In natural habitats, mussels either settle as larvae directly among adults (Lasiak & Barnard 1995, Dobretsov & Wahl 2001) or arrive as secondary settlers following dispersal from primary settlement sites (Bayne 1964, Alfaro 2006b, Newell et al. 2010). Behavioural responses to physical and chemical properties of the substratum can be important determinants of secondary settlement location in mussels (Alfaro et al. 2004, von der Meden et al. 2010). For example, juvenile *Perna perna* actively seek out adult conspecifics over small (cm) scales (Cáceres-Martínez et al. 1994, Erlandsson et al. 2008, Porri et al. 2016). The mechanisms used by juveniles to remotely locate adult mussels is unknown, but is likely to be a waterborne chemical cue, as has been observed for the primary larval settlement in a range of invertebrates, including mussels (Anderson 1996, Steinberg et al. 2002, de Vooys 2003, Alfaro et al. 2006, Morello & Yund 2016). It is possible that as the juvenile mussels grow, their affinities for odours or surficial features shift to those associated with adults and therefore trigger secondary settlement (Bayne 1964, von der Meden et al. 2010). Determining whether this is the case could be of tremendous relevance for the development of approaches for reducing the losses of mussel juveniles in aquaculture operations.

If juvenile mussels undergo secondary settlement in order to recruit among adults, then it is possible that contact or close proximity to adult mussels might suppress secondary settlement behaviour. For example, juvenile *P. canaliculus* >2 mm have been shown to secondarily settle into adult beds and recruit to the adult population (Alfaro 2006b). In addition, adult mussels can provide a structural refuge from predators and abiotic stressors to reduce post-settlement mortality (Bertness & Grosholz 1985). If the presence of adult mussels suppresses secondary settlement of juveniles by providing chemical cues or physical refuge, then seeding adult individuals alongside conspecific juveniles for on-growing might offer an opportunity to reduce the losses of juveniles currently observed in many mussel farming operations.

This study used a field experiment to test whether deploying conspecific adults with juvenile seed mussels increases the retention of juveniles. In addition, small-scale movements of seed mussels were assessed to consider how the addition of adults might affect the distribution of seed mussels on nursery structures typically used in aquaculture production. The development of the biofouling assemblages was also described to gain a better understanding of which organisms might be problematic for mussel seed during the early stages of production.

MATERIALS AND METHODS

Study site and source of juvenile mussels

This study was undertaken on a long-line mussel farm operated by Sanford Ltd in outer Pelorus Sound in the Marlborough Sounds, New Zealand (40° 57′ 18″ S, 174° 3′ 39″ E, Fig. 1). The juvenile mussels used in this study were hatchery-reared by SpatNZ Ltd (Nelson, New Zealand). Hatchery-reared juveniles within a single cohort have consistently shared developmental histories, such as shared pa-

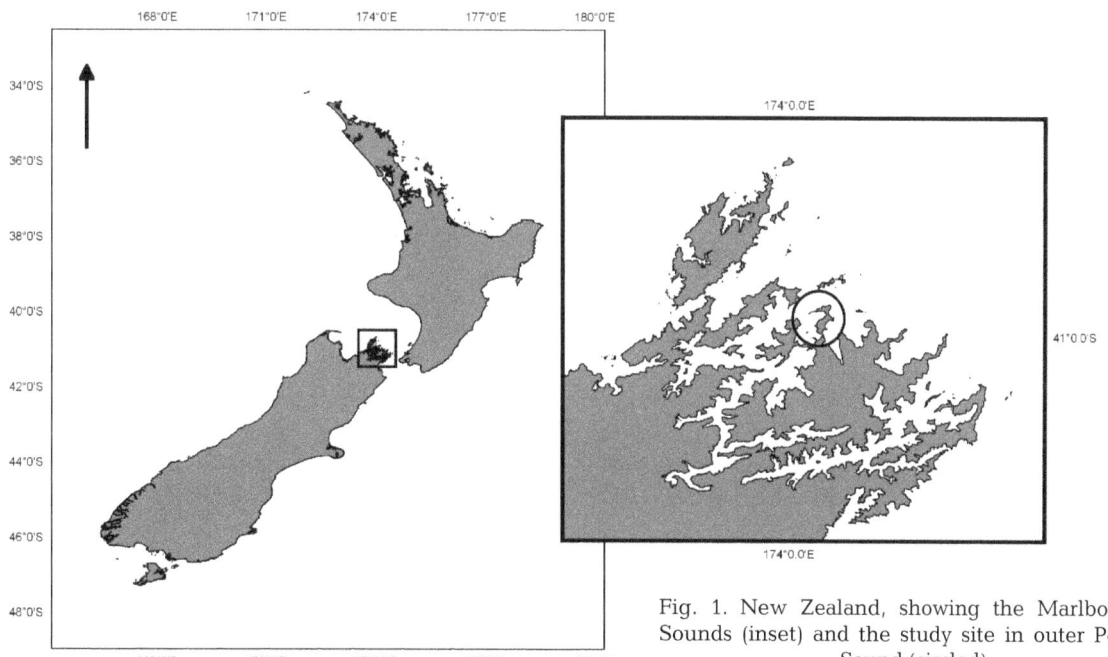

Fig. 1. New Zealand, showing the Marlborough Sounds (inset) and the study site in outer Pelorus Sound (circled)

rentage, ad libitum access to food and managed densities, and are therefore less likely to show wide variation in their response to experimental conditions than are wild juveniles. In the hatchery, larval mussels were settled onto fibrous coir (coconut fibre ropes ca. 10 mm in diameter) ropes that are transported to the nursery site and deployed alongside a typical fibrous polypropylene nursery rope which is suspended in the water column from a buoyed surface line to form a nursery rope. The polypropylene rope and coir rope containing the juveniles are held together with a mesh sock placed over the 2 strands which collectively form a nursery rope. However, only the polypropylene rope is a permanent structure and the coir and sock degrade and are subsequently lost during production.

Experimental design

A field experiment was initiated on 13 October 2015 to test whether the presence of adult *Perna canaliculus* on nursery ropes would increase the retention of conspecific juvenile seed mussels. This experiment was deployed in October (i.e. early austral spring) because this is typically a period of low primary settlement of *P. canaliculus* at the study site and therefore our estimates of retention were less likely to be confounded by over-settling conspecifics early in the experimental period. At other times of the year, new cohorts are easy to determine using size and morphological characteristics (Redfearn et al. 1986, Atalah et al. 2016). Furthermore, the mussel farm used in this study was in 30 m deep water, with the experiment deployed at 4 m depth, while the secondary settlement by wild mussels typically occurs deeper in the water column (Alfaro & Jeffs 2003).

Fifty experimental nursery ropes (45 cm in length, hereafter 'experimental ropes') were assigned to 1 of 5 treatments that included 2 densities of live adult *P. canaliculus* (5 and 20 adults per experimental rope, hereafter PL [*P. canaliculus* Low] and PH [*P. canaliculus* High], respectively), 2 densities of empty adult shells that were included to examine the structural effects of adult mussels without their biological functioning such as defecation, byssus production and filter feeding (5 and 20 shells per experimental rope, hereafter SL [Shell Low] and SH [Shell High], respectively), and a control to which no adults or shells were added (hereafter, C). The high adult mussel and shell density of 20 rope^{-1} (i.e. 44 m^{-1}) was chosen to reflect typical densities of mussels of the size used in this experiment at cropping, in a com-

mercial aquaculture setting. An additional 10 ropes were seeded and transported to the study site, but were not deployed, being retained for analyses to provide an estimate of the starting density on experimental ropes. The live adult mussels used in this experiment were obtained from a nearby mussel farm and were 93.1 ± 2.1 SE mm (n = 20) in shell length. Biofouling organisms were removed from the live adults, which were then deployed at even intervals along the experimental ropes to replicate a typical arrangement of adult mussels on a growing line. The rope and the mussels were then held together by covering with a mesh stocking. The adult mussel shells were 91.7 ± 1.0 mm (n = 20) in shell length and were glued together to represent the physical form of adult mussels, and glued to the polypropylene rope at regular intervals. Similar amounts of glue were added to all of the other polypropylene ropes used in this experiment as a procedural control.

P. canaliculus are grown in a longline culture system that comprises continuous looped ropes suspended from 2 buoyed longlines (Jeffs et al. 1999, Woods et al. 2012). Each buoy is 6–8 m apart, forming 'bays' between buoys from which the nursery and the later grow-out ropes are hung. The present experiment occupied 1 such bay made available on a commercial mussel farm and involved the deployment of 5 rectangular frames (100 × 90 cm) consisting of 2 wooden vertical rods (20 × 10 × 900 mm) intersected horizontally at the top, middle and bottom by 3 cylindrical nylon rods (10 × 1000 mm; Fig. 2). The frames were lashed at 1 m intervals to the backbone lines and hung at a depth of 4 m. The frames were weighted in order to align them vertically in the water column. Two replicates of each experimental treatment were cable-tied to each of the frames (Fig. 2). These experimental ropes were positioned in a random order, ~15 cm apart on the frame and sat vertically in the water column to reflect the typical position of a commercial mussel aquaculture nursery rope used for rearing juvenile mussels of this species. Each experimental rope was seeded with 45 cm length of coir rope coated in hatchery-raised juveniles that were seeded according to standard industry practices. The hatchery-raised juveniles were approximately 6 wk post-settlement, and their shell length ranged from 0.29 to 1.76 mm with a mean length of 1.01 mm ± 0.01 (n = 498). All of the experimental ropes were transported to the study site in a cool and damp environment to reduce the likelihood of mortality or modified secondary settlement behaviour as artefacts of factors occurring during transport (Carton et al. 2007).

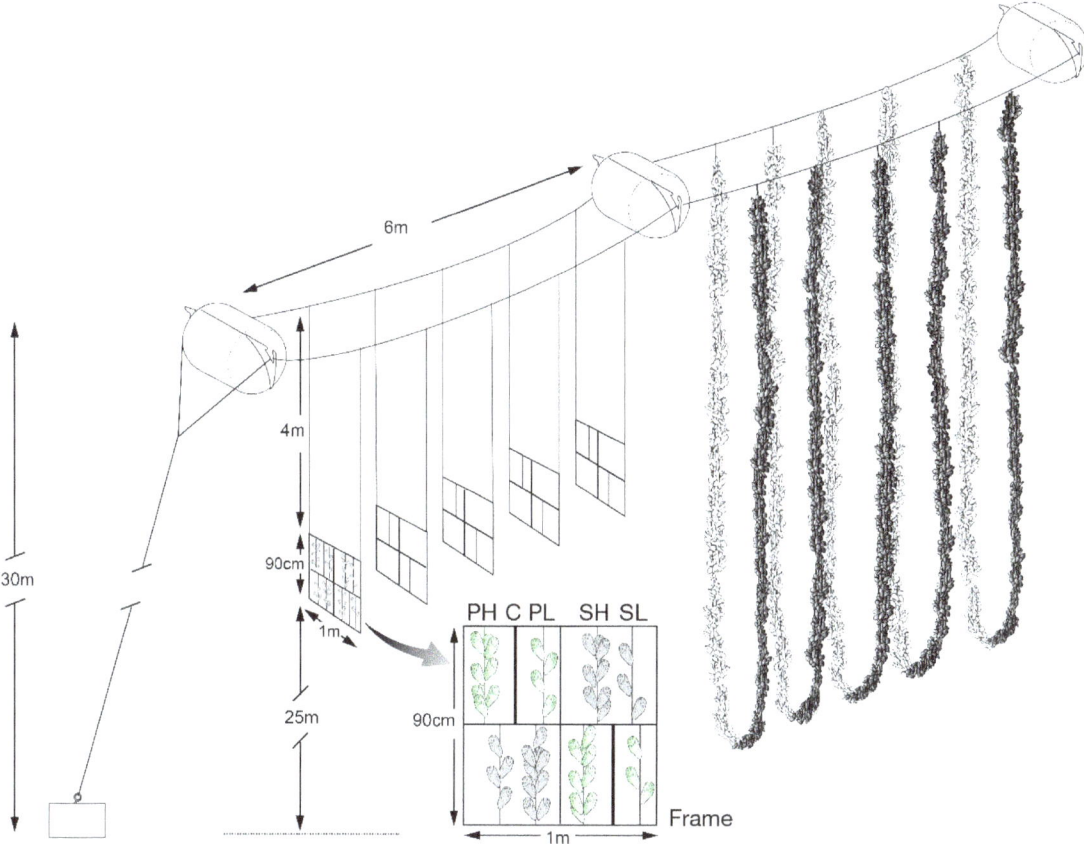

Fig. 2. Experimental set-up on a long-line culture system at the study site in outer Pelorus Sound. Frames were 1 m apart. The inset depicts the spatial arrangement of replicates on the frames. The treatments were a control with no mussels added to experimental ropes (C; thick black line) and experimental ropes with low or high densities of either live adult *Perna canaliculus* (*Perna* Low [PL] and *Perna* High [PH], green mussels) or their shells (Shells Low [SL] and Shells High [SH], grey mussels). Experimental ropes were ~15 cm apart on the frames with 1 replicate treatment^{-1} frame^{-1} sampled after 2 deployment durations (1 and 5 mo)

After 30 d (hereafter, 1 mo), 1 experimental rope from each treatment from each frame was randomly collected by removing the frames from the water and cutting the cable ties. Sampling was repeated at the conclusion of the experiment after 145 d (hereafter, 5 mo). At each sampling, 5 experimental ropes per treatment were collected. Experimental ropes were returned to the laboratory and separated into the 3 substrata (coir, polypropylene rope and socks). Large biofouling organisms were removed with tweezers, and each substratum was then washed over a 250 μm sieve to ensure any wild primary mussel settlers were also captured. Experimental substrata were checked after washing to ensure that all juvenile mussels were sampled. The numbers of juvenile *P. canaliculus* within each substratum were counted, and 50 individuals were measured from each experimental rope using image analysis (to 0.01 mm precision) and Vernier callipers (to 0.5 mm precision) at 1 and 5 mo, respectively. The shell length of 20 randomly selected *Myti-*

lus galloprovincialis that arrived on each experimental rope were measured for the experimental ropes that were sampled at 1 mo. All biofouling organisms (>1 mm) were identified to the highest possible taxonomic resolution (usually to family level or higher) and counted. The dry weights (after 48 h at 50°C) of biofouling algae and sessile invertebrates were quantified. All settlers of *P. canaliculus* (assessed by smaller size than hatchery juveniles), *M. galloprovincialis* and *Modiolarca impacta* >250 μm on the experimental ropes were identified and counted.

Statistical analyses

Variation among experimental factors for all analyses in this study was tested using permutational analysis of variance (PERMANOVA), which has no assumption of data normality, but assumes homogeneity of variances (Anderson et al. 2008). Cochran's C-test was used to assess the variance structure of data used for each univariate analysis. Where there was significant heterogeneity, data were transformed (square root, log[x + 1] or arcsine-square root) to stabilise variances. When transformation failed to homogenise the data, the results from analysed data were considered significant only at p < 0.01 to decrease the risk of Type 1 error. For multivariate tests, the PERMDISP routine was used to assess data dispersion. PERMANOVA was used to analyse univariate and multivariate data using distance matrices based on Euclidean distances and Bray Curtis similarities, respectively. Non-metric multidimensional scaling (nMDS) was used to visualise differences in biofouling assemblage composition prior to multivariate tests. Pairwise comparisons were used to determine between-treatment differences for significant effects in the full models. Tests were based on 999 (size data) or 9999 (all other data) permutations.

Retention of juvenile mussels

The effect of Duration (0, 1 and 5 mo) on the number of juvenile mussels was tested with a PERMANOVA using only the untreated control experimental ropes in order to determine patterns of retention on typical aquaculture substrata.

Effects of adult mussels on juvenile retention, distribution and size

The planned analysis included the factors Treatment (C, SL, SH, PL and PH), Substratum (coir, polypropylene rope and sock) and Duration (1 mo, 5 mo). However, separating the substrata turned out to be impossible after 5 mo due to the interwoven biofouling and mussel byssus. Therefore, the number of juveniles retained was summed across substrata for each experimental rope to test for the effects of Treatment and Duration on juvenile retention using a PERMANOVA model with the factors Treatment (C, SL, SH, PL and PH, fixed effect), Duration (1 and 5 mo, fixed) and Frame (1–5, random). Where Frame was used as a factor in the analysis, variation due to the highest-order interaction could not be calculated due to insufficient replication and was assumed to be a component of the residual variation (Anderson et al. 2008). Therefore, factors are tested for the presence of main effects over and above interactions involving frames (Quinn & Keough 2002), with the caveat that the significance of main effects might not be spatially consistent. A separate PERMANOVA was undertaken using only the 1 mo data to examine the distribution of juveniles on the 3 substrata, thereby assessing small-scale migrations (i.e. the number of juvniles that had moved from coir to the other substrata). The factors Treatment (C, SL, SH, PL and PH) and Substratum (coir, polypropylene rope and sock) were fixed while Frame (1–5) was a random factor. Additional separate analyses were done for each individual substratum (coir, polypropylene rope and sock) after 1 mo to test the effects of Treatment (fixed) and Frame (random) on the percentage of the total number of juveniles on each experimental rope. Finally, the size of the juvenile P. canaliculus was analysed separately at 1 and 5 mo for the effects of Treatment (fixed) and Frame (random). Size frequency distributions for each of the durations (0, 1 and 5 mo) were inspected to determine whether any natural settlement of P. canaliculus was present. Data are presented as means ± SE, unless stated otherwise.

Effects of adult mussels on the recruitment of biofouling organisms

Data for the 6 most abundant taxa, the total biomass of sessile invertebrates and algae and the entire biofouling assemblage (multivariate data were fourth-root transformed) were pooled across substrata to test for the effects of Treatment (C, SL, SH, PL and PH, fixed), Duration (1 and 5 mo, fixed) and Frame (1–5, random). The similarity of percentages (SIMPER) routine was used to determine the proportional contribution of individual biofouling taxa to variation among treatments for multivariate data. To test whether the most abundant biofoulers after 1 mo were distributed differentially among substrata, the factors Treatment (C, SL, SH, PL and PH, fixed), Substrata (coir, polypropylene rope and sock, fixed) and Frame (1–5, random) were analysed with PERMANOVA.

After 5 mo, many of the adult mussels had been lost from the experimental ropes. Therefore, correlation analyses (Pearson's product moment) were used across live mussel treatments (i.e. PL and PH) to determine whether the number of remaining adults might be associated with the magnitude of biofouling development. The number or biomass of common biofouling taxa was correlated (Pearson's product moment on log[x + 1] transformed data) to the number of juvenile P. canaliculus after 1 mo and 5 mo separately to determine whether biofouling may be implicated in juvenile losses. Correlation analyses were done across experimental mussel treatments.

Analyses were done in PRIMER 6 & PERM-ANOVA+ (PRIMER-E) and STATISTICA 12 (Stat-Soft).

RESULTS

Retention of juvenile mussels

At the beginning of the experiment, 787.5 ± 20.4 (mean ± SE, n = 10) juvenile *Perna canaliculus* were attached to the experimental ropes. Considerable reductions in the abundance of juvenile mussels were observed on the untreated control ropes after 1 mo (decreasing to 422.4 ± 61.1 mussels, i.e. 46.4 % less) and then again after 5 mo (145.6 ± 10.2 mussels), resulting in a significant effect of Duration ($F_{2,12}$ = 65.7, p < 0.001, Table 1) in the analysis. On average, retention of juveniles after 5 mo was 18.5 % (i.e. an 81.5 % loss relative to initial abundance).

Effects of adult mussels on juvenile retention, distribution and size

After 1 mo, all of the live adult mussels added to the experimental treatments (i.e. PL, PH) had attached to the polypropylene rope, survived and remained. Despite being spread out along the polypropylene rope on deployment, the adult mussels moved along the ropes to form clumps, and generally all adults on individual experimental ropes were attached to one another. Several vacant patches of adult mussel byssus on the ropes suggested that adults had attached and subsequently moved on multiple occasions during this first month. After 5 mo, the number of adult mussels remaining ranged from 0 to 18 (0, 0, 8, 16, 18) in the PH and 2 to 5 (2, 3, 4, 5, 5) in PL treat-

ments. However, there was no correlation between the number of remaining adult mussels and the number of juveniles on the experimental ropes (Pearson's r = 0.31, p = 0.38). Additionally, 2 live adult mussels were found on 1 of both the C and SH experimental ropes, after presumably having byssus-walked around the frame. All mussel shells (SL and SH treatments) remained at the end of the experiment. It was not possible to separate the different substrata from the experimental ropes after 5 mo in any of the treatments because the remaining coir and sock were tightly bound to the polypropylene rope with byssus from the juveniles and fouling by blue mussels *Mytilus galloprovincialis*.

The presence of live adult mussels or mussels shells, whether at high or low density on experimental ropes, had no effect on the retention of juvenile *P. canaliculus* (Treatment $F_{4,16}$ = 1.8, p = 0.169; Table 2, Fig. 3). Among the 5 treatments, the number of juvenile mussels declined between 1 and 5 mo (Duration $F_{4,16}$ = 98.83, p < 0.01; Fig. 3, Table 2). On average, all treatments had less than 25 % of the original starting abundance of juvenile mussels at the end of the experiment.

One month following deployment of the experimental ropes, there were differences in small-scale migrations (i.e. juveniles that had moved from coir to ropes or outer socks) of juvenile mussels among the 3 substrata for the 5 treatments (Treatment × Substratum, $F_{8,56}$ = 3.1, p < 0.01, Table 3), with greater numbers of juveniles found on the polypropylene rope in PH (64.6 ± 8.7) than in SH (25.2 ± 7.4), SL (25.0 ± 5.7) and PL (43.6 ± 6.8) treatments (pairwise *t*-tests p < 0.05).

The majority (>66 % in all treatments) of the juvenile seed mussels remaining at 1 mo were still attached to the coir, and there were no differences among treatments (Treatment $F_{4,16}$ = 1.82, p = 0.17, Fig. 4a, Table 4). A greater percentage (>10 %) of

Table 1. PERMANOVA testing for the effects of deployment duration (0, 1 and 5 mo) on the number of juvenile *Perna canaliculus* retained on experimental ropes. f/r: fixed/random. **Bold** text indicates significance at p < 0.05. Perms: number of unique permutations for each factor in the analysis

Source	f/r	df	MS	*F/t*	p	Perms
Main test						
Duration	f	2	**4.608**	**65.721**	**0.0001**	9952
Residual		17	7.011			
Pairwise *t*-tests						
0 > 1				3.567	0.002	2855
0 > 5				6.844	0.0005	2228
1 > 5				4.467	0.0072	91

Table 2. PERMANOVA testing for the effects of seeding live adult mussels or their shells and deployment duration (1 and 5 mo) on the number of juvenile *Perna canaliculus* remaining on experimental ropes. f/r: fixed/random. **Bold** text indicates significance at p < 0.05. Refer to Table 1 for 'Perms'

Source	f/r	df	MS	*F*	p	Perms
Treatment (T)	f	4	0.22	1.85	0.169	9948
Duration (D)	f	1	**10.11**	**98.835**	**0.008**	**5264**
Frame (F)	r	4	0.05	0.70	0.602	9946
T × D	f	4	0.05	0.66	0.636	9956
T × F	r	16	0.12	1.68	0.151	9943
D × F	r	4	0.1	1.41	0.269	9953
Residual		16	0.07			

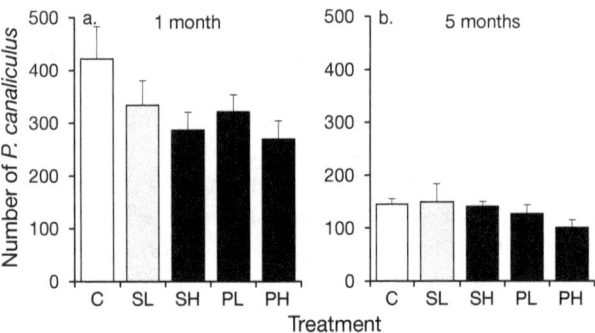

Fig. 3. Effects of the factors Treatment and Duration of deployment on the mean (± SE) number of juvenile *Perna canaliculus* remaining on 45 cm experimental ropes at (a) 1 mo and (b) 5 mo. C: control (0 adult mussels or shells), SL: shells at low density (5 empty *P. canaliculus* shells), SH: shells at high density (20 empty shells), PL: *P. canaliculus* at low density (5 live adults), and PH: *P. canaliculus* at high density (20 live adults)

age of mussels attached to the socks among treatments (Treatment $F_{4,16}$ = 0.49, p < 0.743, Fig. 4c, Table 4). The number of *P. canaliculus* did not vary among experimental frames in any of the analyses undertaken.

One month following deployment, the juvenile *P. canaliculus* had attained an average size of 1.66 ± 0.01 mm in length (range = 0.69–2.92 mm), representing an average growth increment of 0.65 mm. By 5 mo, the juveniles had grown to 20 ± 0.004 mm (range = 6–36 mm). However, at both deployment periods, the size of juveniles varied among frames (1 mo; Treatment × Frame $F_{16,1225}$ = 2.09, p < 0.01, 5 mo $F_{16,1225}$ = 2.2, p < 0.01, Fig. 5, Table 5), but no clear patterns among treatments emerged from post hoc analyses (Fig. 5). Mussels smaller than experimental cohorts were not observed throughout the experiment, indicating that there was an absence of primary settlement of *P. canaliculus*.

juvenile mussels was found on the polypropylene rope in PH (24.1 ± 1.6 SE) compared to all other treatments (Treatment $F_{4,16}$ = 23.01, p < 0.0001, Fig. 4b, Table 4). There were no differences in the percent-

Table 3. PERMANOVA testing for the effects of seeding live adult mussels or their shells on the distribution of juvenile *Perna canaliculus* on aquaculture substrata after 1 mo. f/r: fixed/random. **Bold** text indicates significance at p < 0.05. Refer to Table 1 for 'Perms'

Source	f/r	df	MS	F	p	Perms
Treatment (T)	f	4	0.32	0.72	0.5947	9962
Substratum (S)	f	2	**31.54**	**415.41**	**0.0009**	**9914**
Frame (F)	r	4	0.18	1.66	0.1784	9941
T × S	f	8	**0.62**	**5.81**	**0.0004**	**9955**
T × F	r	16	**0.44**	**4.11**	**0.0002**	**9924**
S × F	r	8	0.08	0.71	0.6828	9941
Residual		32	0.11			

Table 4. PERMANOVA testing for the effects of seeding live adult mussels or their shells on the percentage of the remaining juvenile *Perna canaliculus* on aquaculture substrata after 1 mo. f/r: fixed/random. **Bold** text indicates significance at p < 0.05. Refer to Table 1 for 'Perms'

Source	f/r	df	MS	F	p	Perms
			Coir			
Treatment	f	4	12.74	0.49	0.7425	9952
Frame	r	4	16.68	0.65	0.6453	9967
Residual		16	25.51			
			Rope			
Treatment	f	4	**254.87**	**23.01**	**0.0002**	**9956**
Frame	r	4	7.95	0.71	0.5952	9959
Residual		16	11.07			
			Outer sock			
Treatment	f	4	12.74	0.49	0.7425	9964
Frame	r	4	16.68	0.65	0.6453	9955
Residual		16	25.51			

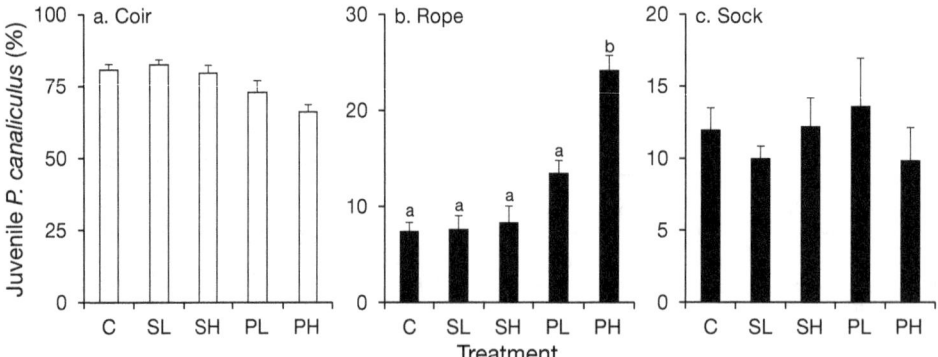

Fig. 4. Mean percentage (± SE) of remaining juvenile *Perna canaliculus* on (a) coir, (b) polypropylene rope and (c) sock substrata among the 5 experimental treatments 1 mo after placement in the field. Treatment abbreviations as in Fig. 3. Lower case letters indicate significant differences between treatments. Note different scales on the *y*-axes

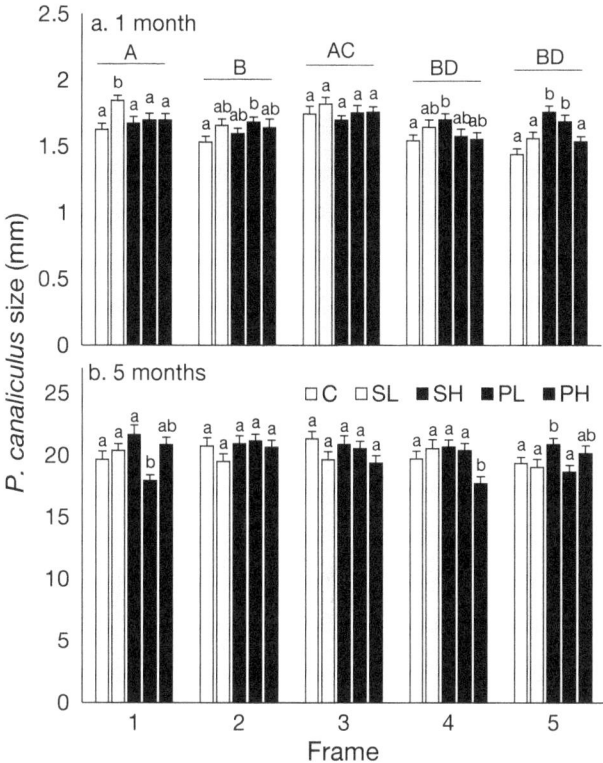

Fig. 5. Effects of the factors Treatment and Frame on the size (mean length + SE in mm) of *Perna canaliculus* at (a) 1 and (b) 5 mo. Treatment abbreviations as in Fig. 3. Lower case letters indicate significant differences between treatments following a significant Treatment × Frame interaction. Capitals indicate significant differences among frames where a main effect of Frame was observed. Note different scales on the y-axes

Effects of adult mussels on the recruitment of biofouling organisms

In this experiment, 68 biofouling taxa recruited to the experimental ropes, including amphipods, bivalves, ascidians, macroalgae and bryozoans. The 6 taxa that numerically dominated the biofouling assemblage were the mytilid mussels *M. galloprovincialis* and *Modiolarca impacta* and the amphipods

Ischyroceridae, *Paradexamine* spp., *Caprella* spp. and *Parawaldeckia* sp. There were significant negative correlations among the number of *Caprella* spp. (r = −0.5000, p < 0.05), *M. galloprovincialis* (r = −0.501, p < 0.05) and *P. canaliculus* juveniles at 1 mo, but not at 5 mo.

The factor Duration (1 and 5 mo) explained the greatest amount of variation in the abundance of biofouling taxa (Fig. 6, Table 6). The abundance of Ischyroceridae, *Paradexamine* spp. and *Caprella* spp. decreased between 1 and 5 mo, while *M. galloprovincialis*, *Parawaldeckia* sp. and *M. impacta* increased in abundance. In addition, the biomass of sessile invertebrates and algae increased significantly between 1 and 5 mo.

Variation in the abundance of biofouling taxa among the treatments was more complex. The addition of shells (SH) or live adult mussels (PH) had significant positive effects on the number of *M. galloprovincialis* at 1 mo (Treatment × Duration, $F_{4,16} = 6.41$, p < 0.01, Fig. 6a, Table 6) with the abundance of *M. galloprovincialis* in the PH (371.8 ± 22.0 SE) treatment being greater (30.4–58.8%) than in any other treatment (Fig. 6a). The SH (258.0 ± 14.8) treatment had significantly greater numbers of *M. galloprovincialis* than SL (199.8 ± 21.5) and C (153.2 ± 18.1, Fig. 6a) at 1 mo. These *M. galloprovincialis* were between 0.25 and 1.5 mm in shell length, with a mean length of 0.81 ± 0.01 mm. There were no differences in the number of *M. galloprovincialis* among treatments at 5 mo (Fig. 6b), although there was a significant positive correlation between the number of *M. galloprovincialis* and the number of adult *P. canaliculus* remaining on the experimental ropes (Pearson's r = 0.78, p = 0.008).

There were greater numbers of the amphipods Ischyroceridae, *Paradexamine* spp. and *Caprella* spp. on PH compared to C treatments indicated by either significant main effects of Treatment (Table 6) or Treatment × Duration interactions (Fig. 6c–h, Table 6). For example, the tube-building amphipods Ischyro-

Table 5. PERMANOVAs testing for the effects of adding live mussels or their shells to aquaculture substrata on the size (length in mm) of juvenile *Perna canaliculus* after 1 and 5 mo. f/r: fixed/random. **Bold** text indicates significance at p < 0.05. Refer to Table 1 for 'Perms'

Source	f/r	df	1 mo				5 mo			
			MS	F	p	Perms	MS	F	p	Perms
Treatment	f	4	0.018	2.738	0.074	999	0.029	1.277	0.315	999
Frame	r	4	**0.035**	**10.944**	**0.001**	**999**	0.016	1.783	0.129	999
T × F	r	16	**0.007**	**2.090**	**0.007**	**998**	**0.023**	**2.492**	**0.002**	**999**
Residual		1225	0.003				0.009			

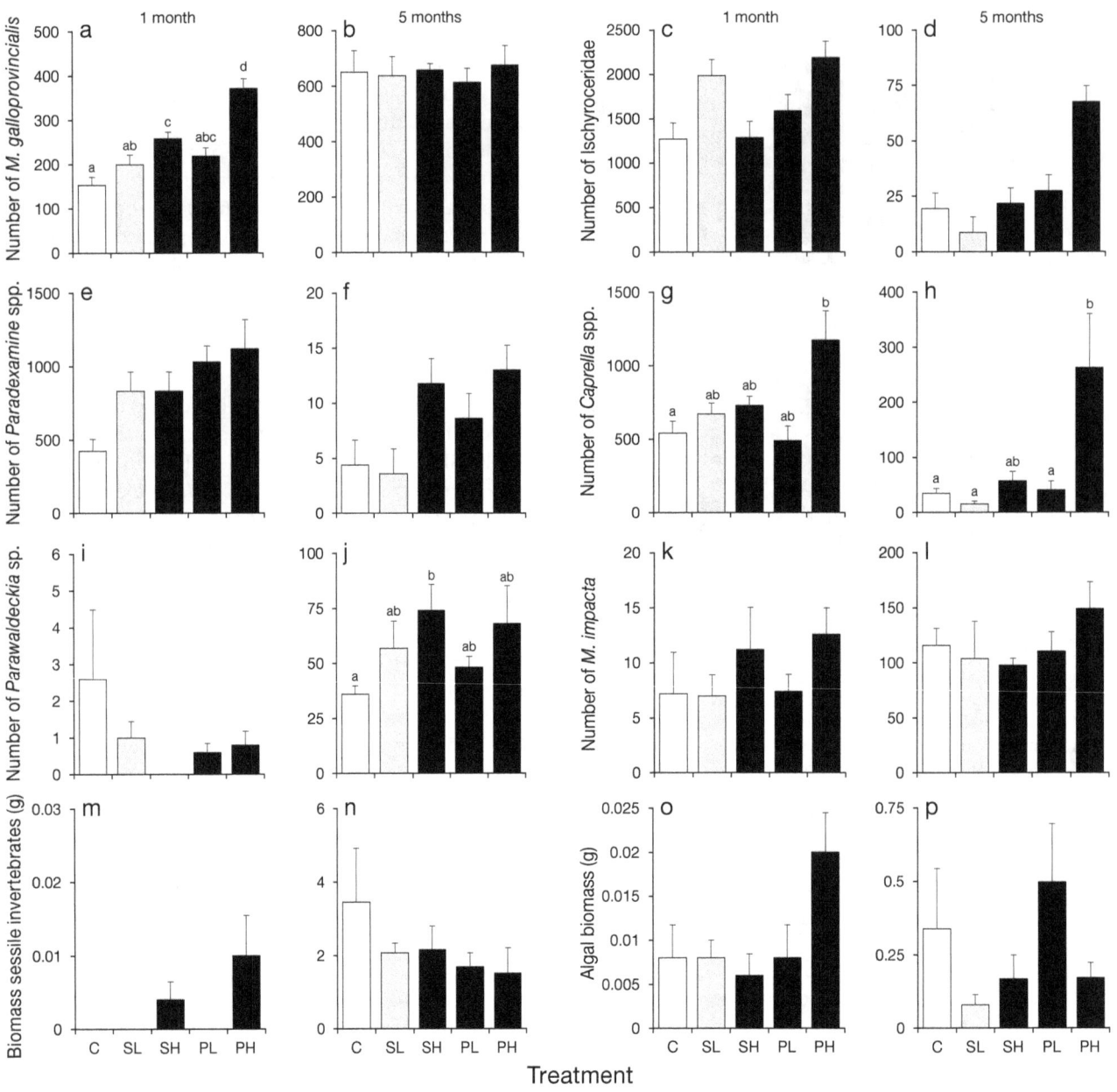

Fig. 6. Effects of the factors Treatment and Duration (1 and 5 mo) on the mean (+ SE) abundance of the main biofouling organisms on the 45 cm experimental ropes. Treatment abbreviations as in Fig. 3. Different letters above bars indicate pairwise differences at p < 0.05 for significant Treatment × Duration interactions. Pairwise results for significant main effects of Treatment are presented in Table 6. Duration was significant for all taxa (p < 0.01). Note different scales on the y-axes

ceridae were significantly more abundant on PH compared to C, SL and SH treatments (Treatment, $F_{4,16}$ 8.21, p < 0.001, Fig. 6c,d, Table 6) overall. By contrast, the amphipod *Parawaldeckia* sp. was more abundant on SH (74 ± 11.24) compared to C (36 ± 3.8) experimental ropes at 5 mo (Treatment × Duration, $F_{4,16}$ =

4.19, p < 0.05; Fig. 6i,j, Table 6). There were no main or interactive effects of Treatment on the numbers of the bivalve *M. impacta* (Fig. 6k,l, Table 6) or the biomass of sessile invertebrates or algae (Fig. 6m–p, Table 6). There were no effects of Frame in the analysis of count or biomass data for biofouling taxa.

Table 6. PERMANOVAs testing for the effects of seeding live adult *Perna canaliculus* or their shells and deployment duration on the abundance and assemblage of biofouling organisms after 1 and 5 mo. f/r: fixed/random. **Bold** text indicates significance at p < 0.05 or 0.01 where data were heterogeneous. Perms: number of unique permutations for each factor in the analysis. See Fig. 6 for pairwise tests following significant Treatment × Duration interactions (note these were inconclusive for Ischyroceridae). Pairwise t-tests for main effects (see Fig. 3 for treatment abbreviations); Ischyroceridae: PH > C, SL, SH (t = 3–4.9, p < 0.05); *Paradexamine* spp.: C < SH (t = 3.9, p < 0.01), C < PL (t = 3.8, p < 0.01), C < PH (t = 3.9, p < 0.05), SL < SH (t = 2.9, p < 0.05); *Parawaldeckia* sp.: C < SH (t = 5.3, p < 0.01)

Source	f/r	df	*Mytilus galloprovincialis* MS	F	p	Perms	Ischyroceridae[a] MS	F	p	Perms	*Paradexamine* spp. MS	F	p	Perms
Treatment (T)	f	4	**0.32**	**12.67**	**0.0002**	**9939**	**1.59**	**8.2**	**0.001**	**9959**	**1.99**	**5.87**	**0.004**	**9961**
Duration (D)	f	1	**13.18**	**123.97**	**0.0099**	**5262**	**234.9**	**4055**	**0.009**	**5194**	**279.67**	**685.9**	**0.009**	**5304**
Frame (F)	r	4	0.06	1.78	0.1846	9951	0.04	0.1	0.978	9959	0.42	1.02	0.43	9958
T × D	f	4	**0.24**	**6.41**	**0.0024**	**9940**	**1.23**	**3.25**	**0.039**	**9949**	0.49	1.16	0.368	9957
T × F	r	16	0.02	0.66	0.7865	9932	0.19	0.51	0.907	9950	0.34	0.81	0.665	9944
D × F	r	4	0.1	2.77	0.0626	9948	0.05	0.15	0.961	9949	0.4	0.97	0.454	9962
Residual		16	0.03				0.37				0.41			

Source	f/r	df	*Caprella* spp.[b] MS	F	p	Perms	*Parawaldeckia* sp. MS	F	p	Perms	*Modiolarca impacta* MS	F	p	Perms
Treatment	f	4	**3.85**	**9.74**	**0.0004**	**9954**	0.05	0.17	0.945	9957	1.03	1.66	0.188	9938
Duration	f	1	**97.99**	**225.76**	**0.0082**	**5311**	**152.3**	**683.4**	**0.008**	**5261**	**75.21**	**92.79**	**0.011**	**5276**
Frame	r	4	0.73	2.01	0.1449	9950	0.39	2.22	0.117	9956	0.83	0.97	0.453	9944
T × D	f	4	**1.44**	**3.96**	**0.0223**	**9951**	**0.73**	**4.19**	**0.017**	**9959**	0.47	0.55	0.717	9957
T × F	r	16	0.39	1.08	0.4314	9937	0.3	1.74	0.142	9940	0.61	0.71	0.762	9931
D × F	r	4	0.43	1.19	0.3602	9957	0.22	1.26	0.325	9957	0.81	0.94	0.476	9970
Residual		16	0.36				0.17				0.86			

Source	f/r	df	Sessile invertebrates MS	F	p	Perms	Algae MS	F	p	Perms	Assemblage MS	F	p	Perms
Treatment	f	4	0.06	1.29	0.3047	5263	0.06	1.29	0.305	5269	368.31	1.43	0.058	9876
Duration	f	1	**0.72**	**47.22**	**0.0072**	**9958**	**0.72**	**47.22**	**0.007**	**9941**	**23637**	**90.9**	**0.008**	**5303**
Frame	r	4	0.01	0.27	0.8962	9944	0.01	0.27	0.896	9941	206.14	1.36	0.162	9925
T × D	f	4	0.07	1.28	0.3147	9952	0.07	1.28	0.315	9939	**369.7**	**2.45**	**0.001**	**9900**
T × F	r	16	0.05	0.96	0.5415	9950	0.05	0.96	0.542	9955	**256.96**	**1.7**	**0.004**	**9903**
D × F	r	4	0.01	0.28	0.8934	9937	0.01	0.28	0.893	9935	**260.01**	**1.72**	**0.038**	**9842**
Residual		16	1.89				0.05				150.56			

[a]Cochran's C = 0.23, p < 0.05; [b]Cochran's C = 0.24, p < 0.05

[a]Cochran's C = 0.24, p < 0.05

There were distinct differences in the composition of the biofouling assemblages on the experimental ropes between 1 and 5 mo (Duration, $F_{1,16}$ = 90.91, p < 0.01, Fig. 7a, Table 6). The biofouling assemblage on the PH treatment at 1 mo varied from all other treatments except PL (Treatment × Duration, $F_{4,16}$ = 2.45, p < 0.01, pairwise t-tests: PH different from C, SL and SH, T = 1.94 – 2.35, p < 0.05; Fig. 7b). There were no differences in biofouling composition among treatments after 5 mo of immersion (pairwise t-tests: t = 0.724 – 1.46, p > 0.12; Fig. 7c). Differences in assemblage composition among treatment or Duration levels were not always consistent among replicate frames (Treatment × Frame, $F_{16,16}$ = 1.70, p < 0.01; Duration × Frame, $F_{4,16}$ = 1.72, p < 0.01, Table 6). Following the significant effect of Duration, SIMPER analysis indicated that there were greater numbers of Ischyroceridae, *Paradexamine* spp. and *Caprella* spp., and fewer *M. impacta*, Tanaidacea data not presented due to low overall relative abundance and *M. galloprovincialis* (accounting for >50 % of the dissimilarity) after 1 mo compared to 5 mo. After 1 mo, the taxa that contributed >50 % of the dissimilarity to among-treatment differences (Ischyroceridae, *Paradexamine* spp., *Caprella* spp. and the masking crab *Notomithrax minor*, which was in low relative abundance) were in greater abundance in the PH treatment.

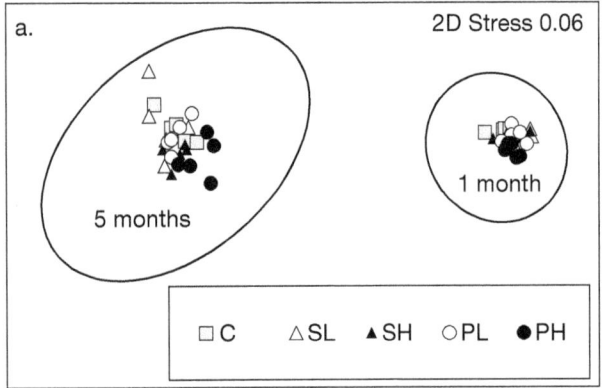

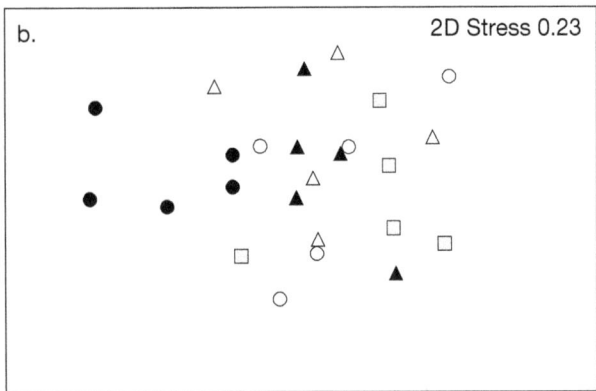

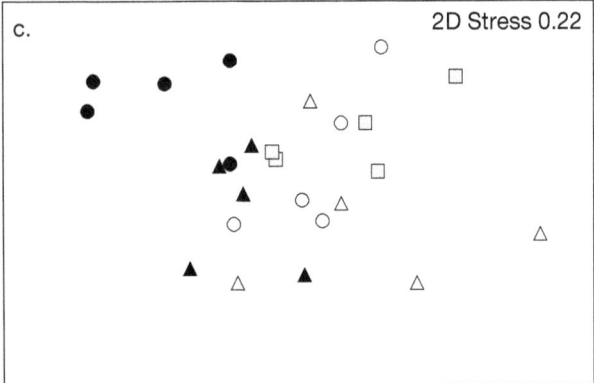

Fig. 7. Non-metric multidimensional scaling plot showing data dispersion in 5 treatments (a) across, (b) within 1 mo and (c) within 5 mo deployment durations. Replicates for each of 2 deployment durations (1 and 5 mo) are encircled in (a). Treatment abbreviations as in Fig. 3

Distribution of biofouling organisms on aquaculture substrata

At 1 mo, the most abundant biofouling organisms predominantly recruited to the sock and polypropylene rope (Fig. 8, Table 7). Greater numbers of *M. galloprovincialis* were found on the polypropylene rope in PH compared to all other treatments, and on the

sock in PH compared to SL and C (Treatment × Substratum, $F_{8,32} = 5.3$, p < 0.001, Fig. 8a, Table 7). Additionally, the C treatment had fewer *M. galloprovincialis* individuals than SH on the polypropylene rope, and SH and PL on the sock; there were no differences among treatments on the coir. There was also some spatial variability in the abundance of *M. galloprovincialis* (Treatment × Frame, $F_{16,32} = 2.6$, p < 0.05, Table 7). At 1 mo, the number of Ischyroceridae was greater on the polypropylene rope substratum in SL compared to C, and greater on the sock in PH compared to C treatments (Treatment × Substratum, $F_{8,16} = 5.01$, p < 0.001, Fig. 8b, Table 7). On the coir, there were significantly greater numbers of Ischyroceridae in C compared to SL, SH and PL treatments. *Paradexamine* spp. were more abundant on the polypropylene rope and sock than on the coir (Substratum, $F_{2,16} = 33.0$, p < 0.001, Fig. 8c, Table 7), but did not vary in abundance among treatment levels. The number of *Caprella* spp. varied among the treatments on each of the substrata (Treatment × Substratum, $F_{8,16} = 6.7$, p < 0.001, Fig. 8d, Table 7), with more on the coir in C than all others, except PH, and fewer on the sock in C compared to PH treatments. There were greater numbers of *Caprella* spp. on the rope in PH compared to SL treatments.

DISCUSSION

The results of this study demonstrate that high losses of juvenile *Perna canaliculus* from nursery ropes can be experienced early in commercial aquaculture production. On average 46% of the *P. canaliculus* juveniles were lost from experimental ropes after only 1 mo (average of 85.2 lost mussels rope^{-1} wk^{-1}) and fewer than 19% of juvenile mussels remained after 5 mo (average of 13.4 lost mussels rope^{-1} wk^{-1}). These data support earlier estimates of loss and suggest that understanding the causes of losses and mitigating against them would be beneficial to mussel production in New Zealand (Jeffs et al. 1999, Hayden & Woods 2011). Including live adult mussels, or their shells, at 2 different densities had no effect on the retention of juvenile mussels over 1 and 5 mo under conditions typical of the early production cycle. Therefore, adult *P. canaliculus* did not mitigate stressors or suppress secondary settlement behaviour in juvenile mussels. Furthermore, many of the adult mussels were themselves lost when they were deployed at high density, providing additional evidence against the usefulness of simultaneous deployments of adults and juveniles. However, the addition of live

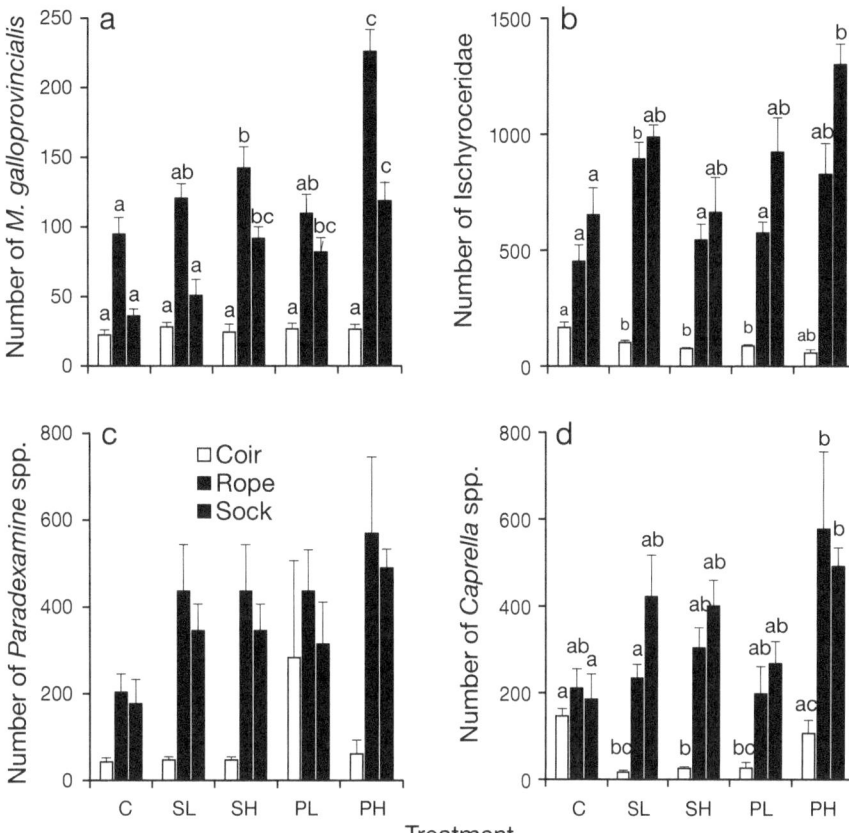

Fig. 8. Mean (+ SE) number of (a) *Mytilus galloprovincialis*, (b) Ischyroceridae, (c) *Paradexamine* spp. and (d) *Caprella* spp. on the coir (white bars), polypropylene rope (grey bars) and sock (black bars) among treatment after 1 mo in the field. Treatment abbreviations as in Fig. 3; n = 5 ropes treatment[-1], all experimental ropes were 45 cm in length. Different letters above bars indicate significant differences among treatments within substrata following a significant Treatment × Substratum interaction. Note different scales on the *y*-axes

adult mussels or adult mussel shells did increase the abundance of some common biofouling organisms, in particular *Mytilus galloprovincialis*, which may have detrimental effects on mussel production (Fitridge et al. 2012, Sievers et al. 2013, Lacoste & Gaertner-Mazouni 2015, Forrest & Atalah 2017).

Secondary settlement behaviour is one likely cause of the observed losses of juvenile *P. canaliculus* (Jeffs et al. 1999, Hayden & Woods 2011). Juvenile *P. canaliculus* of the same size as the juveniles used in our study have been observed to settle into natural habitats, indicating that juveniles of this size range readily use secondary settlement behaviour to change location (Buchanan & Babcock 1997, Alfaro & Jeffs 2003, Alfaro 2006b, South 2016). For example, pulses of settlement of *P. canaliculus* sized >500 μm were observed on collectors deployed for periods of only 1 d at a time on a rocky shore in southern New Zealand (South 2016). In northern New Zealand, settlers arriving at an intertidal mussel bed were >2 mm in length (Alfaro 2006b). Therefore, it appears that *P. canaliculus* juveniles of the size deployed in this study (mean = 1.01 ± 0.01 mm) are highly mobile and could have migrated away from the experimental ropes. While it might be possible that juveniles could have migrated among experimental ropes in this

study, it is highly unlikely, as there was no significant variation in juvenile numbers among treatments that would be consistent with such movements (Fig. 3).

Secondary settlement migration from the coir or algae on which mussels are initially seeded is essential to mussel production in New Zealand, because the coir and algae degrade over time, while the polypropylene rope offers a stable, permanent substratum. After 1 mo, a greater percentage (~10%) of juvenile mussels remaining on the experimental ropes had migrated from the coir to the polypropylene ropes when live adult mussels were added at the higher density. The underlying triggers of secondary settlement are not clear and could occur due to external factors such as chemical cues, changes in abiotic conditions or developmental changes in the juveniles. Alternatively, secondary settlement might be a response to negative changes in the immediate local environment, such as the recruitment of predatory species or organisms that modify food availability to the juveniles. It is possible that movements by the live adults along the nursery ropes could have displaced juveniles, causing them to re-locate onto the polypropylene rope. Juveniles might also have actively migrated away from adults, as a response to their biological functioning (defecation, byssus pro-

Table 7. PERMANOVAs testing for the effects of adding live adult *Perna canaliculus* or their shells on the distribution of key biofouling organisms within experimental ropes after 1 mo in the field. f/r: fixed/random. **Bold** text indicates significance at p < 0.05 or 0.01 where data were heterogeneous. Refer to Table 1 for 'Perms'

Source	f/r	df	MS	F	p	Perms	MS	F	p	Perms
			Mytilus galloprovincialis				Ischyroceridae[a]			
Treatment (T)	f	4	**1.167**	**6.546**	**0.0028**	**9956**	0.397	0.733	0.6043	9955
Substratum (S)	f	2	**17.127**	**68.953**	**0.0001**	**9953**	**39.887**	**106.31**	**0.0003**	**9939**
Frame (F)	r	4	0.029	0.414	0.7986	9953	0.228	0.912	0.4788	9949
T × S	f	8	**0.365**	**5.3**	**0.0004**	**9956**	**1.255**	**5.013**	**0.0003**	**9940**
T × F	r	16	**0.178**	**2.591**	**0.0106**	**9941**	0.542	2.164	0.0192	9918
S × F	r	8	**0.248**	**3.611**	**0.0039**	**9937**	0.375	1.499	0.1826	9956
Residual		32	0.069				0.25			
			Paradexamine spp.				*Caprella* spp.[b]			
Treatment	f	4	1.116	1.818	0.1712	9958	**3.076**	**6.929**	**0.0017**	**9973**
Substratum	f	2	**28.921**	**33.009**	**0.0009**	**9950**	**34.246**	**90.572**	**0.0006**	**9951**
Frame	r	4	0.991	1.41	0.245	9954	0.21	0.609	0.6642	9949
T × S	f	8	0.889	1.264	0.2909	9948	**2.315**	**6.701**	**0.0001**	**9947**
T × F	r	16	0.614	0.873	0.6102	9930	0.444	1.285	0.2602	9915
S × F	r	8	0.876	1.246	0.3033	9944	0.378	1.095	0.3909	9938
Residual		32	0.703				0.345			

[a]Cochran's C = 0.71, p < 0.001; [b]Cochran's C = 0.35, p < 0.001

duction) or competition for food. Alternatively, juvenile mussels might have moved towards the adults that were mostly attached to the polypropylene rope (Porri et al. 2016). Movement of juveniles towards adults would provide support for the original hypothesis that adults might have some influence on the retention of juveniles, but the number of mussels undertaking such movements was insufficient to have any effect on the number of mussels retained during the course of this experiment. The addition of live adult mussels or mussel shells to the experimental ropes led to an increase in the abundance of biofouling organisms, and this, or the presence of the adults, may have prompted the relocation of the juvenile mussels.

The ability of *P. canaliculus* juveniles to mucus-drift to facilitate secondary settlement ends at around 6 mm in shell length, and this might explain the reduced rate of loss of juvenile mussels from the experimental ropes observed after 1 mo (Fig. 3 in the present study, Buchanan & Babcock 1997). Losses of juvenile mussels from aquaculture substrata could also be due to other factors that include variations in genetics and fitness among individuals (Phillips 2002, Alcapán et al. 2007, Sim-Smith & Jeffs 2011), disease (Jones et al. 1996), predation pressure (Hayden 1995, Peteiro et al. 2010), biofouling (Fitridge et al. 2014) and stressors associated with the relay of juveniles from the hatchery or wild collection sites to the nursery farm location that might increase mortality or trigger secondary settlement (Webb & Heasman 2006, Carton et al. 2007). Quantifying the rela-

tive importance of the causes of loss has not been satisfactorily achieved in this and other studies and remains an important research priority.

The addition of live adult mussels in high density to experimental ropes generally increased the abundance and composition of biofouling organisms after 1 mo in the field. Adding low densities of live adults, or only the shells of adults, had weaker effects on biofouling development (Fig. 6). Together this indicates that factors such as increased habitat area, refuge from predation and abiotic stress, as well as modification of food availability that are associated with adult mussels in high density, might have been important for many of the biofouling taxa recorded in this study (Commito & Rusignuolo 2000). Biofouling organisms can compete for food and space with cultured organisms, leading to reduced biomass and crop losses (Ramsay et al. 2008, De Nys & Guenther 2009, Sievers et al. 2013). The increased abundance of biofouling organisms associated with the presence of adult mussels (alive or shells) did not appear to affect the retention of juveniles. For example, the number of *P. canaliculus* juveniles was not significantly smaller where live adults were added in high density and where biofoulers such as *M. galloprovincialis* and *Caprella* spp. were in greater abundance. However, there were significant negative correlations between the number of remaining juveniles and the number of *Caprella* spp. and *M. galloprovincialis*, at 1 mo, a trend that warrants further targeted research to determine possible effects of these species.

M. galloprovincialis responded differentially to the presence of live adults versus shells after 1 mo (Fig. 6a). The number of *M. galloprovincialis* was more than double in the high live adult density treatment after 1 mo compared to the controls, clearly showing that experimental ropes bearing live mussels promote settlement or survival of *M. galloprovincialis*. Similar positive effects of adults on settlement have been reported in other studies of mytilid mussels (Nielsen & Franz 1995, Dobretsov & Wahl 2001, Sardiña et al. 2009, Dolmer & Stenalt 2010). Adding mussel shells in high density also had a positive, albeit a smaller, effect on the number of *M. galloprovincialis*, indicating that this species may be benefitting from the structural properties of *P. canaliculus* shells (Dolmer & Stenalt 2010). Given the small size (0.25–1.5 mm) of *M. galloprovincialis* settlers at 1 mo, it is likely that these arrived as primary settlers, although this experiment was not structured to test this. The greater number of *M. galloprovincialis* in treatments with live adult *P. canaliculus* could be the result of increased primary settlement due to chemical cues from the live adults (Alfaro et al. 2006) or the result of modified small-scale hydrodynamic patterns (Grizzle et al. 1996, Miron et al. 2000) and increased surface complexity associated with the adults (e.g. byssal threads) and shells (e.g. increased settlement area) (Cáceres-Martínez et al. 1994, Gribben et al. 2011). Alternatively, differences in the abundance of *M. galloprovincialis* among the treatments might have been due to variations in post-settlement processes such as mortality (Hunt & Scheibling 1997, von der Meden et al. 2012), secondary settlement (von der Meden et al. 2010, South 2016) or predation (Hayden 1995, Peteiro et al. 2010). The positive effects of live adults and shells of *P. canaliculus* on the abundance of its congener *M. galloprovincialis* contrasts with the absence of any effect on the conspecific juveniles deployed on the experimental ropes. It is possible that adult–juvenile interactions are only important early in the settlement process or that *M. galloprovincialis* and *P. canaliculus* have different underlying settlement-recruitment strategies. The lack of primary settlement of *P. canaliculus* during this study did not allow for any comparison among treatments or with the settlement of *M. galloprovincialis*.

The strong increase in the number of *M. galloprovincialis* on experimental ropes bearing high densities of adult *P. canaliculus* has important implications for the New Zealand mussel industry and the surrounding natural environment (Rius et al. 2011). The blue mussel *M. galloprovincialis* has a wide global range with southern and northern lineages and is a successful invader of native ecosystems in many countries (McQuaid & Phillips 2000, Braby & Somero 2006, Branch et al. 2008, Gardner et al. 2016). In the Marlborough Sounds, *M. galloprovincialis* has increased in abundance over the last 25 yr and has become a problematic biofouling organism that competes for food and displaces crops of *P. canaliculus* on growing lines (Woods et al. 2012, Atalah et al. 2017).

The most abundant biofouling taxa recruited differentially to the experimental substrata used to deploy seed mussels. For example, *M. galloprovincialis* settled heavily onto the polypropylene rope substratum (versus the socking or the coir), thereby occupying the space onto which *P. canaliculus* juveniles must migrate to remain in production. This result is in accordance with other studies that have shown *M. galloprovincialis* to have strong affinities for fibrous substrata such as the polypropylene ropes used in this study (Cáceres-Martínez et al. 1994, Carl et al. 2012). The amphipods Ischyroceridae built tubes constructed from fine sediment on the sock and polypropylene rope, potentially modifying small-scale hydrodynamic processes and food delivery to juvenile mussels (Fitridge et al. 2012), for which feeding is known to be limited due to rudimentary feeding structures (Gui et al. 2016). Critically, this suggests that the substrata used for mussel culture in New Zealand can promote the settlement of nuisance species and that more research is warranted to understand their impact and methods to deter them.

CONCLUSIONS

This study shows how high losses of seed mussels can greatly affect the efficiency of mussel production in a typical aquaculture setting in New Zealand. The majority of juvenile losses occurred within the first month, but continued over the following 5 mo. The addition of adult mussels to experimental ropes, while unsuccessful at increasing retention as originally hypothesised, provides considerable insight into how crops of mussels interact with their environment and the spatial and temporal dynamics of biofouling organisms in the early stages of aquaculture production. Overall, the presence of adult *Perna canaliculus* on experimental ropes increased the numbers of biofouling organisms subsequently arriving on these ropes, especially *Mytilus galloprovincialis*, most likely as a result of their provision of habitat and biological functioning. The polypropylene rope and outer sock substrata used by industry in nursery systems appear to promote the settlement of some biofouling organisms. Research into alternative systems is required to reduce the impacts of unwanted biofouling species on juvenile mussels in aquaculture.

Acknowledgements. We gratefully thank SpatNZ Ltd and in particular Dan McCall for providing juvenile *Perna canaliculus*, on-water support and aquaculture infrastructure for this project. Thanks to Fiona Gower and the taxonomy team at Cawthron Institute for their technical support. This research was funded by New Zealand Ministry of Business, Innovation and Employment under Contract CAWX1315 (The Cultured Shellfish Programme: Enabling, 447 Growing, and Securing NZ's Shellfish Aquaculture Sector). Vivien Ward kindly crafted Fig. 2. We thank 3 anonymous referees whose thoughtful and detailed critique greatly improved this work.

LITERATURE CITED

Alcapán AC, Nespolo RF, Toro JE (2007) Heritability of body size in the Chilean blue mussel (*Mytilus chilensis* Hupé 1854): effects of environment and ageing. Aquacult Res 38:313–320

Alfaro AC (2006a) Evidence of cannibalism and bentho-pelagic coupling within the life cycle of the mussel, *Perna canaliculus*. J Exp Mar Biol Ecol 329:206–217

Alfaro AC (2006b) Population dynamics of the green-lipped mussel, *Perna canaliculus*, at various spatial and temporal scales in northern New Zealand. J Exp Mar Biol Ecol 334:294–315

Alfaro AC (2006c) Byssal attachment of juvenile mussels, *Perna canaliculus*, affected by water motion and air bubbles. Aquaculture 255:357–361

Alfaro AC, Jeffs AG (2002) Small-scale mussel settlement patterns within morphologically distinct substrata at Ninety Mile Beach, northern New Zealand. Malacologia 44:1–15

Alfaro AC, Jeffs AG (2003) Variability in mussel settlement on suspended ropes placed at Ahipara Bay, Northland, New Zealand. Aquaculture 216:115–126

Alfaro AC, Jeffs AG, Creese RG (2004) Bottom-drifting algal/mussel spat associations along a sandy coastal region in northern New Zealand. Aquaculture 241:269–290

Alfaro AC, Copp BR, Appleton DR, Kelly S, Jeffs AG (2006) Chemical cues promote settlement in larvae of the green-lipped mussel, *Perna canaliculus*. Aquacult Int 14: 405–412

Alfaro AC, McArdle B, Jeffs AG (2010) Temporal patterns of arrival of beachcast green-lipped mussel (*Perna canaliculus*) spat harvested for aquaculture in New Zealand and its relationship with hydrodynamic and meteorological conditions. Aquaculture 302:208–218

Anderson MJ (1996) A chemical cue induces settlement of Sydney rock oysters, *Saccostrea commercialis*, in the laboratory and in the field. Biol Bull (Woods Hole) 190: 350–358

Anderson MJ, Gorley RN, Clarke KR (2008) PERMANOVA for PRIMER: guide to software and statistical methods. PRIMER-E, Plymouth

Aquaculture New Zealand (2016) Overview of New Zealand aquaculture. www.aquaculture.org.nz/industry/overview/ (accessed on 7 April 2016)

Atalah J, Rabel H, Forrest BM (2016) Blue mussel over-settlement predictive model and web application. Prepared for Marine Farming Association. Cawthron Report No. 2801. Cawthron Institute, Nelson

Atalah J, Rabel H, Forrest BM (2017) Modelling long-term recruitment patterns of blue mussels *Mytilus galloprovincialis*: a biofouling pest of green-lipped mussel aquaculture in New Zealand. Aquacult Environ Interact 9: 103–114

Bayne BL (1964) Primary and secondary settlement in *Mytilus edulis* L. (Mollusca). J Anim Ecol 33:513–523

Bertness MD, Grosholz E (1985) Population dynamics of the ribbed mussel, *Geukensia demissa*: the costs and benefits of an aggregated distribution. Oecologia 67:192–204

Braby CE, Somero GN (2006) Ecological gradients and relative abundance of native (*Mytilus trossulus*) and invasive (*Mytilus galloprovincialis*) blue mussels in the California hybrid zone. Mar Biol 148:1249–1262

Branch GM, Odendaal F, Robinson TB (2008) Long-term monitoring of the arrival, expansion and effects of the alien mussel *Mytilus galloprovincialis* relative to wave action. Mar Ecol Prog Ser 370:171–183

Buchanan S, Babcock R (1997) Primary and secondary settlement by the greenshell mussel *Perna canaliculus*. J Shellfish Res 16:71–76

Cáceres-Martínez J, Robledo JAF, Figueras A (1994) Settlement and post-larvae behavior of *Mytilus galloprovincialis*: field and laboratory experiments. Mar Ecol Prog Ser 112:107–117

Capelle JJ, Wijsman JWM, Schellekens T, van Stralen MR, Herman PMJ, Smaal AC (2014) Spatial organisation and biomass development after relaying of mussel seed. J Sea Res 85:395–403

Carl C, Poole AJ, Williams MR, de Nys R (2012) Where to settle—settlement preferences of *Mytilus galloprovincialis* and choice of habitat at a micro spatial scale. PLOS ONE 7:e52358

Carrasco AV, Astorga M, Cisterna A, Farias A, Espinoza V, Uriarte I (2014) Pre-feasibility study for the installation of a Chilean mussel *Mytilus chilensis* (Hupe 1854) seed hatchery in the Lakes region, Chile. Fish Aquacult J 5: 102, doi:10.4172/2150-3508.1000102

Carton AG, Jeffs AG, Foote G, Palmer H, Bilton J (2007) Evaluation of methods for assessing the retention of seed mussels (*Perna canaliculus*) prior to seeding for grow-out. Aquaculture 262:521–527

Commito JA, Rusignuolo BR (2000) Structural complexity in mussel beds: the fractal geometry of surface topography. J Exp Mar Biol Ecol 255:133–152

De Nys R, Guenther J (2009) The impact and control of biofouling in marine finfish aquaculture. Woodshead Publishing, Cambridge

de Vooys CGN (1999) Numbers of larvae and primary plantigrades of the mussel *Mytilus edulis* in the western Dutch Wadden Sea. J Sea Res 41:189–201

de Vooys CGN (2003) Effect of a tripeptide on the aggregational behaviour of the blue mussel *Mytilus edulis*. Mar Biol 142:1119–1123

Dobretsov S, Wahl M (2001) Recruitment preferences of blue mussel spat (*Mytilus edulis*) for different substrata and microhabitats in the White Sea (Russia). Hydrobiologia 445:27–35

Dolmer P, Stenalt E (2010) The impact of the adult blue mussel (*Mytilus edulis*) population on settling of conspecific larvae. Aquacult Int 18:3–17

Erlandsson J, Porri F, McQuaid CD (2008) Ontogenetic changes in small-scale movement by recruits of an exploited mussel: implications for the fate of larvae settling on algae. Mar Biol 153:365–373

FAO (Food and Agriculture Organization of the United Nations) (2016) The state of world fisheries and aquaculture 2016. Contributing to food security and nutrition for all. FAO, Rome

Fitridge I, Dempster T, Guenther J, de Nys R (2012) The impact and control of biofouling in marine aquaculture: a review. Biofouling 28:649–669

Fitridge I, Sievers M, Dempster T, Keough MJ (2014) Tackling a critical industry bottleneck: developing methods to avoid, prevent & treat biofouling in mussel farms. Report No. CC BY 3.0. University of Melbourne

Forrest BM, Atalah J (2017) Significant impact from blue mussel *Mytilus galloprovincialis* biofouling on aquaculture production of green-lipped mussels in New Zealand. Aquacult Environ Interact 9:115–126

Gardner JPA, Zbawicka M, Westfall KM, Wenne R (2016) Invasive blue mussels threaten regional scale genetic diversity in mainland and remote offshore locations: the need for baseline data and enhanced protection in the Southern Ocean. Glob Change Biol 22:3182–3195

Gribben PE, Jeffs AG, de Nys R, Steinberg PD (2011) Relative importance of natural cues and substrate morphology for settlement of the New Zealand Greenshell™ mussel, *Perna canaliculus*. Aquaculture 319:240–246

Grizzle RE, Short FT, Newell CR, Hoven H, Kindblom L (1996) Hydrodynamically induced synchronous waving of seagrasses: 'monami' and its possible effects on larval mussel settlement. J Exp Mar Biol Ecol 206:165–177

Gui Y, Zamora L, Dunphy B, Jeffs A (2016) Understanding the ontogenetic changes in particle processing of the greenshell™ mussel, *Perna canaliculus*, in order to improve hatchery feeding practices. Aquaculture 452:120–127

Hayden BJ (1995) Factors affecting recruitment of farmed greenshell mussels, *Perna canaliculus* (Gmelin) 1791, in Marlborough Sounds. PhD thesis, University of Otago

Hayden BJ, Woods CMC (2011) Effect of water velocity on growth and retention of cultured Greenshell™ mussel spat, *Perna canaliculus* (Gmelin, 1791). Aquacult Int 19:957–971

Holthuis TD, Bergström P, Lindegarth M, Lindegarth S (2015) Monitoring recruitment patterns of mussels and fouling tunicates in mariculture. J Shellfish Res 34:1007–1018

Hunt HL, Scheibling RE (1997) Role of early post-settlement mortality in recruitment of benthic marine invertebrates. Mar Ecol Prog Ser 155:269–301

Jeffs A, Holland R, Hooker S, Hayden B (1999) Overview and bibliography of research on the greenshell mussel, *Perna canaliculus*, from New Zealand waters. J Shellfish Res 18:347–360

Jones JB, Scotti PD, Dearing SC, Wesney B (1996) Virus-like particles associated with marine mussel mortalities in New Zealand. Dis Aquat Org 25:143–149

Lacoste E, Gaertner-Mazouni N (2015) Biofouling impact on production and ecosystem functioning: a review for bivalve aquaculture. Rev Aquacult 7:187–196

Lasiak TA, Barnard TCE (1995) Recruitment of the brown mussel *Perna perna* onto natural substrata: a refutation of the primary/secondary settlement hypothesis. Mar Ecol Prog Ser 120:147–153

Le Corre N, Martel AL, Guichard F, Johnson LE (2013) Variation in recruitment: differentiating the roles of primary and secondary settlement of blue mussels *Mytilus* spp. Mar Ecol Prog Ser 481:133–146

McQuaid CD, Phillips TE (2000) Limited wind-driven dispersal of intertidal mussel larvae: *in situ* evidence from the plankton and the spread of the invasive species *Mytilus galloprovincialis* in South Africa. Mar Ecol Prog Ser 201:211–220

Miron G, Walters LJ, Tremblay R, Bourget E (2000) Physiological condition and barnacle larval behavior: a preliminary look at the relationship between TAG/DNA ratio and larval substratum exploration in *Balanus amphitrite*. Mar Ecol Prog Ser 198:303–310

Morello SL, Yund PO (2016) Response of competent blue mussel (*Mytilus edulis*) larvae to positive and negative settlement cues. J Exp Mar Biol Ecol 480:8–16

Naylor RL, Goldburg RJ, Primavera JH, Kautsky N and others (2000) Effect of aquaculture on world fish supplies. Nature 405:1017–1024

Newell CR, Short F, Hoven H, Healey L, Panchang V, Cheng G (2010) The dispersal dynamics of juvenile plantigrade mussels (*Mytilus edulis* L.) from eelgrass (*Zostera marina*) meadows in Maine, U.S.A. J Exp Mar Biol Ecol 394:45–52

Nielsen KJ, Franz DR (1995) The influence of adult conspecifics and shore level on recruitment of the ribbed mussel *Geukensia demissa* (Dillwyn). J Exp Mar Biol Ecol 188:89–98

Pauly D, Christensen V, Guénette S, Pitcher TJ and others (2002) Towards sustainability in world fisheries. Nature 418:689–695

Peteiro LG, Filgueira R, Labarta U, Fernández-Reiriz MJ (2007) Settlement and recruitment patterns of *Mytilus galloprovincialis* L. in the Ría de Ares-Betanzos (NW Spain) in the years 2004/2005. Aquacult Res 38:957–964

Peteiro LG, Filgueira R, Labarta U, Fernández-Reiriz MJ (2010) The role of fish predation on recruitment of *Mytilus galloprovincialis* on different artificial mussel collectors. Aquacult Eng 42:25–30

Phillips NE (2002) Effects of nutrition-mediated larval condition on juvenile performance in a marine mussel. Ecology 83:2562–2574

Porri F, McQuaid CD, Erlandsson J (2016) The role of recruitment and behaviour in the formation of mussel-dominated assemblages: an ontogenetic and taxonomic perspective. Mar Biol 163:1–10

Quinn GP, Keough MJ (2002) Experimental design and data analysis for biologists. Cambridge University Press, Cambridge

Ramsay A, Davidson J, Landry T, Stryhn H (2008) The effect of mussel seed density on tunicate settlement and growth for the cultured mussel, *Mytilus edulis*. Aquaculture 275:194–200

Redfearn P, Chanley P, Chanley M (1986) Larval shell development of four species of New Zealand mussels: (Bivalvia, Mytilacea). N Z J Mar Freshw Res 20:157–172

Rius M, Heasman KG, McQuaid CD (2011) Long-term coexistence of non-indigenous species in aquaculture facilities. Mar Pollut Bull 62:2395–2403

Sardiña P, Cataldo DH, Boltovskoy D (2009) Effects of conspecifics on settling juveniles of the invasive golden mussel, *Limnoperna fortunei*. Aquat Sci 71:479–486

Sievers M, Fitridge I, Dempster T, Keough MJ (2013) Bio-
    fouling leads to reduced shell growth and flesh weight in
    the cultured mussel *Mytilus galloprovincialis*. Biofouling
    29:97–107

Sim-Smith CJ, Jeffs AG (2011) A novel method for deter-
    mining the nutritional condition of seed green-lipped
    mussels, *Perna canaliculus*. J Shellfish Res 30:7–11

Smaal A (2002) European mussel cultivation along the
    Atlantic coast: production status, problems and perspec-
    tives. Hydrobiologia 484:89–98

South PM (2016) An experimental assessment of measures
    of mussel settlement: effects of temporal, procedural and
    spatial variations. J Exp Mar Biol Ecol 482:64–74

Steinberg PD, de Nys R, Kjelleberg S (2002) Chemical cues
    for surface colonization. J Chem Ecol 28:1935–1951

von der Meden CEO, Porri F, McQuaid CD, Faulkner K,
    Robey J (2010) Fine-scale ontogenetic shifts in settle-
    ment behaviour of mussels: changing responses to bio-
    film and conspecific settler presence in *Mytilus gallo-
    provincialis* and *Perna perna*. Mar Ecol Prog Ser 411:
    161–171

von der Meden CEO, Porri F, McQuaid CD (2012) New esti-
    mates of early post-settlement mortality for intertidal
    mussels show no relationship with meso-scale coastline
    topographic features. Mar Ecol Prog Ser 463:193–204

Webb SC, Heasman KG (2006) Evaluation of fast green
    uptake as a simple fitness test for spat of *Perna canalicu-
    lus* (Gmelin, 1791). Aquaculture 252:305–316

Woods CMC, Floerl O, Hayden BJ (2012) Biofouling on
    Greenshell™ mussel (*Perna canaliculus*) farms: a prelim-
    inary assessment and potential implications for sustain-
    able aquaculture practices. Aquacult Int 20:537–557

12

Beneficial effects of fish stocking on performance and pest control in the lotus field system

Li Ma[1], Jie Zhu[1], Qi Chen[1], Wei Li[2,*], Guo-Hua Huang[1,*]

[1]Hunan Provincial Key Laboratory for Biology and Control of Plant Diseases and Insect Pests, Hunan Agricultural University, Changsha 410128, Hunan, PR China

[2]State Key Laboratory of Freshwater Ecology and Biotechnology, Institute of Hydrobiology, Chinese Academy of Sciences, Wuhan 430072, Hubei, PR China

ABSTRACT: The present study was conducted to evaluate the economic effects of fish stocking on plant and fish performance in lotus fields, as well as to determine its influence on controlling common pest species associated with lotus fields. Lotus yield, fish growth performance, economic returns and pest abundance at 5 fish stocking density treatments (0, 1500, 3000, 4500 and 6000 ind. ha^{-1}) were determined in 15 separate lotus fields at a lotus farming facility in Wugang City, Hunan Province, China. The results showed a tendency for the lotus yield to increase with fish stocking density, but significant differences were not found among the treatments. Survival rates were not significantly different among the fish stocking treatments for the 3 species of carp tested. Absolute growth rate and specific growth rate showed similar trends, with their values significantly decreasing as stocking density increased for each carp species. The total yield for the 3 fish species did show significant differences, with the highest yield in the 6000 treatment and the lowest in the 1500 treatment. Total net income was highest in the 4500 treatment and lowest in the control (0) treatment. No significant differences were observed between the 4500 and 3000 treatments. The abundance levels of 3 types of pests were lower in the fish stocking treatments than in the control treatment, and were shown to decrease with increasing stocking density. Based on these observations, stocking densities of 3000 to 4200 total ind. ha^{-1} were considered optimal levels in lotus–fish culture systems.

KEY WORDS: Growth efficiency · Lotus–fish culture · *Nelumbo nucifera* · Carps · Stocking density · Performance · Economic benefit · Pest control

INTRODUCTION

Nelumbo nucifera Gaertn., also known as Indian lotus or sacred lotus, is an aquatic herbaceous perennial plant that has an extremely long history in cultivation as a popular and economically important cash crop in many Asian countries (Yi et al. 2002). All parts of the lotus plant are highly valued, whether for their use as a vegetable food source, as ornamental plants or for medicinal applications. For example, lotus rhizomes are considered a major vegetable crop in Asia (Tian et al. 2009). Lotus has also become a potentially important crop in Australia (Nguyen 2001), New Zealand (Follett & Douglas 2003) and the United States (Tian et al. 2006). In China, lotus has a wide distribution, extending from southern Hainan Province to northern Heilongjiang Province, and from eastern Taiwan to the western Tian Mountains in Xinjiang Province. The cultivated area of lotus planted for vegetables in China currently exceeds 5000 km^2, which is now the largest crop area devoted to growing aquatic vegetable in the country (Ke et al. 2015).

Historically, lotus has been planted in ponds for vegetable production. Recently, however, planting in paddy fields has become increasingly popular. Previ-

*Corresponding authors: liwei@ihb.ac.cn; ghhuang@hunau.edu.cn

ous studies have shown that planting lotus in paddy fields could be more profitable than planting rice (e.g. Liu & Fu 2004). However, the widespread proliferation of pests and weeds in lotus fields has often led to a serious decline in output and quality in the harvest of lotus (Xiong et al. 2010). Traditional chemical control has been an effective means for pest and weed suppression in lotus fields, but it obviously presents a number of challenges; for example, the pesticides used may lead to water pollution (Tan et al. 2003). Biological control methods, however, may eliminate many of the drawbacks of chemical control. The introduction of fish culture has often been suggested as a feasible method for pest and weed control in lotus fields.

In China, the co-culture or rotated culture of lotus and fish in ponds has been practiced for many years (Yi et al. 2002). The practice of co-culturing lotus and fish in paddies could yield a higher net return than planting lotus alone (Yi et al. 2002). It also is effective in removing nitrogen and phosphorus contaminants from old pond sediment, while the lotus improves the aquacultural environment compared to the model of culturing fish alone (Chen & Li 2007). The model of rotating fish and aquatic vegetables would allow producers to bring 2 crops to market rather than 1, and would act as a buffer if a loss occurred in one of the two ventures (Edwards 1987). However, little is currently known about stocking parameters of fish in lotus fields and the effects they have on pest populations, although a limited number of papers are available on the production and economic benefits of lotus–fish co-culture techniques (Wu 2005). Because so little research has been conducted on identifying the optimal stocking parameters, including fish species to be stocked and stocking density, it is essential to determine the optimal parameters to maximize the performance of the fish species selected and their success in controlling pests in lotus fields.

Accordingly, we hypothesized that fish co-cultured in the lotus field systems at an appropriate stocking density could aid in pest control and result in increased economic benefits. In this study, 3 carp species with different trophic or spatial niches were selected and stocked in lotus fields at 5 stocking densities. We compared the production of lotus, the growth, survival and production of the stocked fish, the integrated economic benefit, and the abundance of pests with the 5 treatments. The aim of this study was to evaluate the economic effects of fish stocking on plant and fish performance in lotus fields, as well as to determine the influence of fish stocking on controlling common pest species associated with lotus

fields, and to identify the optimal co-culture fish species and density in the fields. Results from this study will enable efficient management and maximize profitability of the co-culture model of lotus and fish in lotus fields.

MATERIALS AND METHODS

Lotus field facilities and experimental design

This study was conducted using a randomized complete block design in 15 fields with surface areas ranging from 0.100 to 0.187 ha in Wugang City, Hunan Province, southern China, from March to November 2014 (Fig. 1). Five fish stocking density treatments—1 control (no fish) and 4 treatments (1500, 3000, 4500 and 6000 ind. ha^{-1}), each in triplicate—were used in the study (Table 1). Prior to the experiment, the ridge of each field was heightened and consolidated. For those fields stocked with fish, peripheral trenches (1 × 0.6 m, width × depth), linked with a cross-shaped trench in the field (0.8 × 0.5 m, width × depth) with an area of 10% of the field area, were dug to serve as fish refuges. In the field, screened inlets and outlets (2 mm mesh size) were installed to prevent indigenous fish and aquatic predators from entering the fields and to prevent stocked fish from escaping.

At the beginning of the experiment, the fields were harrowed, puddled and then leveled. Seedlings of the 'Elian 7' lotus variety, purchased from the Wuhan Vegetable Research Institute, were transplanted to fields of the 5 treatments at a constant density of 4500 kg ha^{-1} on 4 April 2014. The average length and weight of the transplanted lotus seedlings were 0.95 m and 2.50 kg, respectively. After the lotus seedlings were transplanted, water was added weekly to all fields and the water depth was increased as the height of lotus increased. Identical amounts of fertilizer were used as a basal dressing 1 d prior to transplanting and as top-dressing 50 d after transplanting. No herbicides or pesticides were applied to any of the treatments during the experiments.

Healthy fingerlings of triploid Xiangyun crucian carp *Carassius auratus* (L.) (initial mean weight: 93.5 ± 8.8 g, mean ± SD), triploid Xiangyun common carp *Cyprinus carpio* L. (117.2 ± 11.7 g) and grass carp *Ctenopharyngodon idellus* (Valenciennes) (413.7 ± 16.9 g), obtained from a nearby carp hatchery, were stocked at densities for the 3 species combined of 0, 1500, 3000, 4500 and 6000 ind. ha^{-1} on 22 May 2014 (Table 1). No supplemental food was provided to the

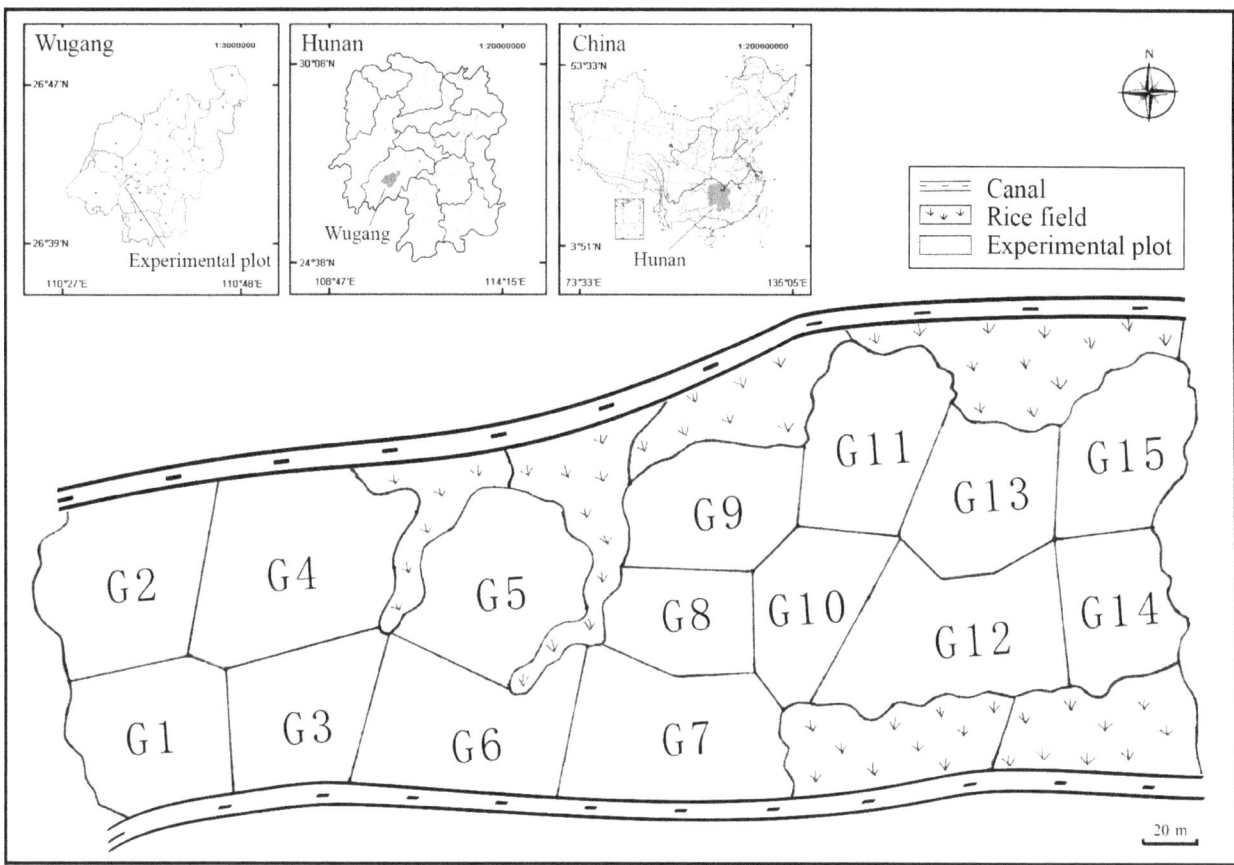

Fig. 1. Map of the experimental site in Wugang City, Hunan Province, southern China (insets), showing the distribution and area size of each lotus field (G1–G15)

fish during the experiment. The fish were reared for 210 d and then harvested after draining the fields.

The investigation of pests was conducted using the 5-point sampling method on each field once per week

(Ma et al. 2016). The fields were sampled regularly from 23 May (when fish were stocked) to 19 September 2014 (when lotus plants withered). The diameter of the sampling point in the field was 1 m, where all

Table 1. Fish stocking scheme in lotus fields

Treatment	Stocking density (ind. ha⁻¹)			Field number	Field size (ha)	Stocking amount (individuals)		
	Crucian carp	Common carp	Grass carp			Crucian carp	Common carp	Grass carp
I	0	0	0	G2	0.100	0	0	0
				G5	0.080	0	0	0
				G8	0.073	0	0	0
II	600	600	300	G7	0.173	104	104	52
				G9	0.120	72	72	36
				G11	0.107	64	64	32
III	1200	1200	600	G6	0.187	224	224	112
				G10	0.073	88	88	44
				G13	0.120	144	144	72
IV	1800	1800	900	G12	0.167	300	300	150
				G14	0.107	192	192	96
				G15	0.107	192	192	96
V	2400	2400	1200	G1	0.080	192	192	96
				G3	0.100	240	240	120
				G4	0.187	448	448	224

lotus plants within the 1-m circle were thoroughly surveyed for pest individuals. All pest individuals in the sampling sites were recorded using the following investigation method. The plants were checked from left to right, from top to bottom and from the upper surface of the blades to the lower surface of the blades. When a species was difficult to identify in the field, a photo was taken and the specimens were collected. Every photo and/or specimen was given a unique record number and identified in the laboratory. Weather conditions were recorded during the entire sampling period. The collected pest specimens were identified using the Huang & Li (2013) identification guide.

At the end of the experimental period, the number and individual weight of fish was recorded. Growth performance of the fish was evaluated using survival, relative weight gain (WG), specific growth rate (SGR), absolute growth rate (AGR) and total yield (Hossain et al. 2013). Before determining the total lotus harvest yield of each field, the lotus roots were placed in ditches and then the adhering mud was removed from each lotus plant. Temperature, dissolved oxygen and pH were recorded daily in each field between 07:00 and 08:00 h using portable electronic probes. All measurements were found to be in the acceptable range for carp culture. The water levels were measured daily using a graduated stick placed in each lotus field.

Economic benefit analysis

The economic analysis followed that of Gomes et al. (2006). Total net income equaled total revenue minus total cost. The total revenue included that of harvested lotus and fish, sold at 6 yuan kg^{-1} for lotus, 12 yuan kg^{-1} for crucian carp and common carp, and 14 yuan kg^{-1} for grass carp on the local market. The total costs consisted of the cost of the lotus seedlings (8 yuan kg^{-1}), fish fingerlings (1.2, 1.3 and 8 yuan $ind.^{-1}$ for crucian carp, common carp and grass carp, respectively), pond rent (7500 yuan ha^{-1} yr^{-1}), labor for daily management (6 mo, including fish stocking and harvesting, lotus planting, fertilizing, routine work, etc.) and labor for harvesting lotus (1.5 yuan kg^{-1} harvested lotus).

Statistical analysis

All data were expressed as mean ± SD. One-way ANOVA was used to determine the differences among treatments followed by a Tukey test to evalu-ate differences among treatment means for post hoc comparisons. Differences were considered significant at $p < 0.05$. All analyses were performed using the SPSS 16.0 statistical package.

RESULTS

Lotus harvesting

The per unit yield of the lotus, ranging from 20 905 kg ha^{-1} in treatment I to 24 581 kg ha^{-1} in treatment IV (with 20 905 ± 5338, 21 110 ± 3527, 22 950 ± 3520, 24 581 ± 4957, 22 662 ± 3549 kg ha^{-1} in treatments I, II, III, IV and V, respectively; see Table 1 for treatment specifics), were not significantly different among the 5 stocking density treatments ($F_4 = 0.375$, p = 0.822), although a slight increase trend in yield was observed with increasing stocking density except in the highest density treatment.

Survival and growth of the fish

Performance parameters (mean ± SD) of the 3 carp species reared for a 6 mo period in lotus fields at 5 stocking densities are shown in Table 2. The survival rates of the crucian carp, common carp and grass carp in the different treatments ranged from 46.8–64.2%, 48.4–58.2% and 45.3–60.4%, respectively. No significant differences were found in the 4 fish stocking densities for the 3 species (crucian carp: $F_3 = 0.641$, p = 0.610; common carp: $F_3 = 0.851$, p = 0.504; grass carp: $F_3 = 1.483$, p = 0.291). At the end of the study, however, there were significant differences among the stocking density treatments in the WG for each fish species (crucian carp: $F_3 = 6.340$, p = 0.017; common carp: $F_3 = 10.857$, p = 0.003; grass carp: $F_3 = 13.915$, p = 0.001), with the lowest values observed in the highest stocking density treatment. The AGR and SGR values showed a similar trend, with each of their values significantly decreasing with increasing stocking density in each of the 3 species (AGR: crucian carp: $F_3 = 5.349$, p = 0.036; common carp: $F_3 = 12.566$, p = 0.002; grass carp: $F_3 = 4.911$, p = 0.041; SGR: crucian carp: $F_3 = 5.704$, p = 0.032; common carp: $F_3 = 10.396$, p = 0.004; grass carp: $F_3 = 4.890$, p = 0.042). The total yield for each species showed significant differences between treatments (crucian carp: $F_3 = 10.933$, p = 0.002; common carp: $F_3 = 12.596$, p = 0.002; grass carp: $F_3 = 6.785$, p = 0.014) with the highest and lowest values found in treatments V and II, respectively.

Table 2. Performance parameters of 3 carp species reared for 6 mo at 5 stocking density treatments (I to V: 0, 1500, 3000, 4500 and 6000 ind. ha^{-1}) in lotus fields. WG: relative weight gain; AGR: absolute growth rate; SGR: specific growth rate. Data are presented as means ± SD; different superscripts in each row indicate significant differences (p < 0.05) between stocking densities

Parameter	Fish species	Treatment I	Treatment II	Treatment III	Treatment IV	Treatment V
Survival (%)	Crucian carp	–	64.2±5.9[a]	60.2±13.5[a]	46.8±9.5[a]	57.3±8.3[a]
	Common carp	–	56.4±3.3[a]	58.2±5.9[a]	48.4±4.9[a]	54.6±6.8[a]
	Grass carp	–	46.3±8.5[a]	60.4±7.8[a]	45.3±1.6[a]	45.3±7.7[a]
WG (g)	Crucian carp	–	413.2±29.6[a]	389.8±69.1[a]	306.5±28.9[a,b]	256.5±32.1[b]
	Common carp	–	489.5±6.7[b]	582.8±28.9[a]	399.5±33.3[c]	364.5±32.4[c]
	Grass carp	–	1053.0±240.4[a]	603.0±60.1[a,b]	619.6±185.6[a,b]	425.0±43.5[b]
AGR (g d^{-1})	Crucian carp	–	1.97±0.24[a]	1.86±0.49[a]	1.46±0.24[ab]	1.22±0.27[b]
	Common carp	–	2.33±0.05[b]	2.78±0.24[a]	1.90±0.28[c]	1.74±0.27[c]
	Grass carp	–	5.01±1.98[a]	2.87±0.49[a,b]	2.95±1.53[a,b]	2.02±0.36[b]
SGR (% d^{-1})	Crucian carp	–	0.80±0.05[a]	0.76±0.10[a]	0.69±0.06[a,b]	0.62±0.07[b]
	Common carp	–	0.78±0.01[a,b]	0.85±0.03[a]	0.70±0.05[b,c]	0.67±0.06[c]
	Grass carp	–	0.59±0.15[a]	0.43±0.05[a,b]	0.42±0.14[a,b]	0.34±0.04[b]
Total yield (kg ha^{-1})	Crucian carp	–	195.5±29.2[a]	367.8±46.4[b]	327.6±67.3[b]	493.7±88.3[c]
	Common carp	–	205.3±13.0[a]	492.9±39.6[b]	446.2±42.1[b]	639.8±83.5[c]
	Grass carp	–	198.7±76.5[a]	374.1±80.9[b]	421.0±90.3[bc]	462.0±88.9[c]

Economic benefit analysis

The comparative costs and return analysis for the 3 carp species cultured in lotus fields for 6 mo for each of the 5 different treatments are shown in Table 3. There were significant differences in total revenue, total costs and net income among density treatments (total revenue: $F_4 = 4.131$, p = 0.032; total costs: $F_4 = 5.405$, p = 0.014; net income: $F_4 = 3.488$, p = 0.041).

The total revenue was highest in treatment IV and lowest in treatment I, with the total costs significantly increasing with increased density. Total net income was highest in treatment IV and lowest in treatment I, while no significant differences was observed between treatments III and IV (p = 0.872). The revenue from lotus was the major contributor to the total net revenue realized in the 5 treatments. Compared with the control treatment, the revenue generated

Table 3. Economic benefit analysis of 3 carp species reared for 6 mo at 5 stocking density treatments in lotus fields. ROI: ratio of output and input. Values are means ± SE of 3 replicates; different superscripts in the same row indicate significant differences (p < 0.05)

Item	Amount and rate	Treatment I	Treatment II	Treatment III	Treatment IV	Treatment V
Total revenue	**10^4 yuan ha^{-1}**	**12.54±1.85[a]**	**13.43±1.16[a]**	**15.33±0.92[b]**	**16.27±1.67[b]**	**15.60±1.61[b]**
Lotus	6 yuan kg^{-1}	12.54±1.85[a]	12.67±1.22[a]	13.77±1.22[a]	14.75±1.71[a]	13.60±1.23[a]
Crucian carp	12 yuan kg^{-1}	–	0.23±0.02 [a]	0.44±0.15[b]	0.39±0.16 [b]	0.59±0.13[c]
Common carp	12 yuan kg^{-1}	–	0.25±0.01[a]	0.59±0.08[b]	0.54±0.03[b]	0.77±0.13[c]
Grass carp	14 yuan kg^{-1}	–	0.28±0.06[a]	0.52±0.09[b]	0.59±0.11[b,c]	0.65±0.14[c]
Total cost	**10^4 yuan ha^{-1}**	**8.96±0.46[a]**	**9.56±0.31[a,b]**	**10.22±0.30[a,b]**	**10.86±0.43[b]**	**10.96±0.31[b]**
Lotus seed	8 yuan kg^{-1}	36 000	36 000	36 000	36 000	36 000
Crucian carp fingerling	1.2 yuan ind.$^{-1}$	–	720	1440	2160	2880
Common carp fingerling	1.3 yuan ind.$^{-1}$	–	780	1560	2340	3120
Grass carp fingerling	8 yuan ind.$^{-1}$	–	2400	4800	7200	9600
Field rental cost	9000 yuan ha^{-1}	7500	7500	7500	7500	7500
Fertilizer	9000 yuan ha^{-1}	7500	7500	7500	7500	7500
Labor for daily management (6 mo)	1500 yuan ha^{-1} mo^{-1}	7200	9000	9000	9000	9000
Labor for harvesting lotus	1.5 yuan kg^{-1}	3.14±0.46[a]	3.17±0.31[a]	3.44±0.30[a]	3.69±0.43[a]	3.40±0.31[a]
Fish net income	**10^4 yuan ha^{-1}**	–	**0.37±0.07[a]**	**0.78±0.15[b]**	**0.34±0.06[a]**	**0.45±0.11[a]**
Total net income	**10^4 yuan ha^{-1}**	**3.59±1.38[a]**	**3.87±01.85[a]**	**5.10±0.62[b,c]**	**5.41±1.24[c]**	**4.64±1.30[a,b]**
ROI	–	1.39±0.13[a]	1.40±0.07[a]	1.50±0.05[a]	1.49±0.10[a]	1.44±0.04[a]

from the 3 carp species in the fish stocking treatments accounted for 5.65–12.86% of the total revenue. The major expenses (of the total costs) were for the lotus seed and the labor for harvesting the lotus, representing >60% of the total costs for all treatments. The ratios of output and input (ROI) were not significantly different among treatments, but the ROI of the fish stocking treatments were higher compared with that of the control treatment.

Pest abundance

Spodoptera litura (Fabricius) larvae (Lepidoptera: Noctuidae), *Rhopalosiphum nymphaeae* (L.) (Hemiptera: Aphididae) and Chironomidae spp. (Diptera) were the 3 most predominant lotus pests in the fields during our investigation (Fig. 2). The abundance of *S. litura* larvae showed similar temporal trends in the 5 density treatments by initially increasing and then decreasing between June and September. No significant difference was observed among treatments in any of the months (all p > 0.05; Fig. 2, Table 4), but their abundance was lower in the fish stocking treatments than in the control treatment from July to September. The abundance of *R. nymphaeae* showed similar variation trends in the 5 density treatments by increasing initially, then decreasing, and increasing again between May and September. At the beginning, there was no significant difference in abundance among treatments (p > 0.05; Fig. 2, Table 4), but significantly higher numbers were observed in the control compared to the 4 fish stocking treatments from June to September (all p < 0.05; Fig. 2, Table 4). The abundance of chironomid pests was highest in June, lowest in July and then increased again in subsequent months. Their abundance in the 4 fish stocking treatments was significantly lower than in the control treatment from June to September (all p < 0.05; Fig. 2, Table 4).

DISCUSSION

In the present study, the per unit yields of lotus were not significantly different among the 5 stocking density treatments, although it tended to increase slightly with increasing stocking density, except in the highest density treatment. This indicates that moderate fish stocking in lotus fields not only has no negative effects on the production of lotus, but also actually benefits the growth of the crop. Similar results have been observed in rice–fish and rice–crab

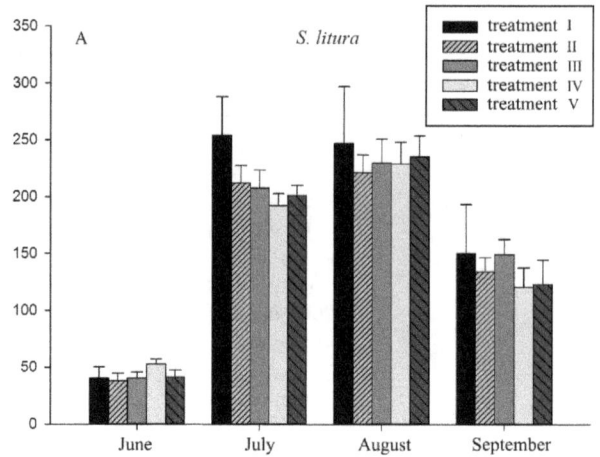

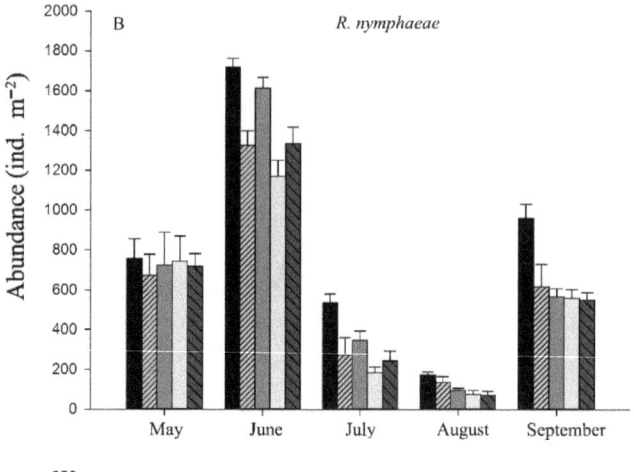

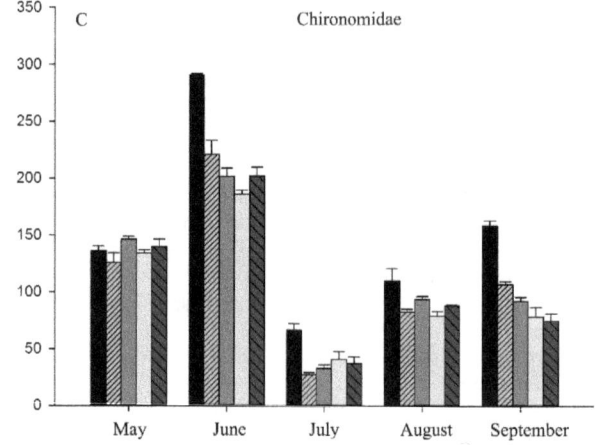

Fig. 2. Abundance of 3 main pests in 5 stocking density treatments in different months in the lotus fields: (A) *Spodoptera litura*, (B) *Rhopalosiphum nymphaeae*, (C) Chironomidae

culture systems, where fish or crabs reared in paddy fields were shown to have no negative effects on rice production (Cagauan et al. 2000, Li et al. 2007). In a review of 18 rice–fish studies, it was demonstrated that the tendency of integrating fish with rice was to

Table 4. Summary results comparing the abundance of the 3 most predominant lotus pests in different months in 5 stocking density treatments in lotus fields. Significant p-values ($p < 0.05$) are shown in **bold**

Source of variance	df	F	p
Spodoptera litura larvae			
June	4	1.222	0.361
July	4	2.463	0.114
August	4	0.508	0.731
September	4	0.943	0.478
Rhopalosiphum nymphaeae			
May	4	0.234	0.913
June	4	30.856	**0.000**
July	4	18.100	**0.000**
August	4	13.344	**0.001**
September	4	20.344	**0.000**
Chironomid pests			
May	4	0.828	0.537
June	4	30.146	**0.000**
July	4	8.596	**0.003**
August	4	5.142	**0.016**
September	4	18.976	**0.000**

improve rice yields (4.6–28.6%) (Lightfoot et al. 1992). In the present study, lotus production in fish stocking treatments increased 1–17.6% over the control treatment. This beneficial effect of fish stocking on lotus production is possibly related to increased nutrients (primarily nitrogen) availability in the soil as a result of fish activity in the lotus fields, as pointed out in similar reports involving rice–fish ecosystems (Panda et al. 1987, Lightfoot et al. 1992). Additionally, the reduction in the pest population in the lotus–fish culture system may be another factor responsible for the improved growth and production of lotus. The interaction between lotus and fish in lotus–fish system is complex, involving many factors, and needs to be studied further.

Stocking density is one of the most important factors that directly influence the survival, growth, behavior, feeding and production of many species. Thus, the determination of the most advantageous stocking density for cultured animals is essential in order to optimize production, profitability and sustainability (Imani et al. 2014). Although a positive effect of stocking density on growth is found in some species, an inverse relationship between these factors is more frequently reported (Weatherley 1976, Namukose et al. 1996, Hwang et al. 2014). Increasing stocking density results in stress (Leatherland & Cho 1985), which, in turn, leads to increased energy requirements and reduction in growth and food utilization (Hengsawat et al. 1997). In this study, the

WG, AGR and SGR showed similar trends in that their values significantly decreased with increasing stocking density in each of the 3 carp species. This indicates that fish growth is density-dependent during the grow-out phase of production in lotus fields. Density-dependent growth has been reported for fish (Sahoo et al. 2004, Schram et al. 2006), for penaeid shrimp species in tanks (Coman et al. 2004, Esparza-Leal et al. 2015), and for crabs in rice fields (Li et al. 2007). The main factor in the decrease of growth was probably food resource competition (Li et al. 2007, Esparza-Leal et al. 2015). The limited food resources available in the lotus fields in the present study, where supplemental food resources were not supplied, would likely impede fish growth at higher stocking densities.

Survival rates were not affected by stocking density in this study. Similar results have been reported for many fish species in cages or ponds (Hwang et al. 2014, Imani et al. 2014). However, many studies have found significantly negative effects of density on survival (Chakraborty & Mirza 2007, Li et al. 2007, Neal et al. 2010). The inconsistency of effects of stocking density on survival is probably attributable to experimental space, living habits and the capacity to adapt to the environment in different cultured species. Compared to tanks or cages, a lotus field is capable of providing an assortment of different food resources, such as zoobenthos, lotus pests, weeds and plankton, although their biomass was limited in this study. Crucian carp, common carp and grass carp, with different spatial and nutritive niches, high adaptability to varying environments and a lack of aggressive behavior toward each other, could, therefore, alternatively use limited food resources to maintain survival rates similar to the densities found in our study (1500–6000 ind. ha^{-1}).

In our study, the higher net income in 4 of the 5 lotus–fish culture treatments indicates that fish stocking is effective in producing beneficial results on production performance in lotus fields. Net income was higher in the 3000 and 4500 ind. ha^{-1} treatments, mainly due to the higher lotus yield and lower cost of fish fingerlings compared to other treatments. It can be concluded that reasonable stocking density of fish was rather important, not only in terms of the improvement of growth and economic points of view, but also considering the ecological benefits resulting from fewer lotus pests. The fish species stocked is also an important consideration in the total net income from lotus fields. In this study, the net income of 3 different carp species accounted for 6.43–15.21% of the total net income, with the apparent difference due to the fish species used. The net income from using grass carp was lower in each fish stocking treatment compared with that from using crucian and common carp, with the cost even surpassing the revenue for grass carp in the 4500 and 6000 ind. ha^{-1} treatments. This does suggest that crucian and common carp are preferable to grass carp for stocking in lotus fields, and that the stocking density of grass carp needs to be <600 ind. ha^{-1}.

In the present study, we found that the abundance of the 3 major pests (*Spodoptera litura* larvae, *Rhopalosiphum nymphaeae* and Chironomidae spp.) in lotus–fish culture systems were lower than those in lotus monoculture systems. This is the first validation study of pest control in lotus fields as a result of fish stocking. The decreased abundance of pests on the lotus plants is likely attributable mostly to fish feeding. It is also quite possible that pests on the stems and leaves of the lotus are physically knocked into the water and drowned as a result of increased agitation of the lotus plants due to the activity of the fish. The abundance of *R. nymphaeae* and chironomid individuals was significantly lower in the fish stocking treatments compared to the control treatment. There was no significant difference in the abundance of *S. litura* larvae between the non-fish and fish stocking treatments, although fewer of the pests were observed in lotus–fish culture systems from July to September. The difference was likely related to their microhabitat. *R. nymphaeae* and chironomid species feed mainly on the floating leaves, or the tender leaves and shoots of lotus plants close to the surface of the water (Wang & Ye 1986, Chen et al. 2013). In contrast, *S. litura* larvae are mainly found on emergent lotus leaves, or lotus flowers and ovaries above the waterline (Zhu et al. 2013).

CONCLUSIONS

Fish stocking is shown to have beneficial effects on pest control and plant performance in lotus fields, allowing for optimal utilization of the available space in lotus fields, and contributing substantially to improving the economic output and pest control in such an integrated system. Stocking densities of the 3 carp species tested had a significant effect on the WG, AGR and SGR values, but did not negatively affect fish survival rates or lotus production. The abundance of lotus pests declined with increasing stocking density. Considering the net profits as well as the growth performance of the 3 carp species, a stocking density of 3000–4200 total ind. ha^{-1} (crucian

carp: 1200–1800 ind. ha^{-1}; common carp: 1200–1800 ind. ha^{-1}; grass carp: <600 ind. ha^{-1}) is considered optimal in lotus–fish culture systems.

Acknowledgements. The authors thank Prof. Brendan Hicks and Dr. Casey Clark for their constructive comments and critically reading of the manuscript, and Dr. Cecil L. Smith (University of Georgia, USA) for editing and revising the English language. The present research was supported by a grant from the National Scientific and Technological Supporting Program of China (no. 2012BAD27B02).

LITERATURE CITED

Cagauan AG, Branckaert RD, van Hove C (2000) Integrating fish and azolla into rice-duck farming in Asia. Naga 23: 4–10

Chakraborty BK, Mirza MJA (2007) Effect of stocking density on survival and growth of endangered bata, *Labeo bata* (Hamilton–Buchanan) in nursery ponds. Aquaculture 265:156–162

Chen YS, Li ZJ (2007) Analysis of environmental and economical effectiveness in planting lotus in typical fish ponds in areas along the middle reaches of the Yangtze River. Resour Environ Yangtze Basin 16:609–614

Chen Q, Ma L, Zhu J, Wei Y, Huang GH (2013) Biological characteristics and control of *Rhopalosiphum nymphaeae* (Linnaeus, 1761). J Changjiang Veg 18:116–118

Coman GJ, Crocos PJ, Presto NP, Fielder D (2004) The effects of density on the growth and survival of different families of juvenile *Penaeus japonicus* Bate. Aquaculture 229:215–223

Edwards P (1987) Use of terrestrial vegetation and aquatic macrophytes in aquaculture. In: Moriarty DJW, Pullin RSV (eds) Detritus and microbial ecology in aquaculture. ICLARM Conference Proceedings, Vol. 14. International Center for Living Aquatic Resources Management, Manila, Philippines, p 311–335

Esparza-Leal HM, Cardozo AP, Wasielesky W (2015) Performance of *Litopenaeus vannamei* postlarvae reared in indoor nursery tanks at high stocking density in clearwater versus biofloc system. Aquacult Eng 68:28–34

Follett JM, Douglas JA (2003) Lotus root: production in Asia and potential for New Zealand. Combined Proc Int Plant Prop Soc 53:79–83

Gomes LC, Chagas EC, Martins-Junior H, Roubach R, Ono EA, Lourenço JNP (2006) Cage culture of tambaqui (*Colossoma macropomum*) in a central Amazon floodplain lake. Aquaculture 253:374–384

Hengsawat K, Ward FJ, Jaruratjamorn P (1997) The effect of stocking density on yield, growth and mortality of African catfish (*Clarias gariepinus* Burchell 1822) cultured in cages. Aquaculture 152:67–76

Hossain A, Sarker MAZ, Saifuzzaman M, Silva JATD, Lozovskaya MV, Akhter MM (2013) Evaluation of growth, yield, relative performance and heat susceptibility of eight wheat (*Triticum aestivum* L.) genotypes grown under heat stress. Int J Plant Prod 7:615–636

Huang GH, Li JH (2013) Color handbook of insect pests of aquatic vegetables in China. Changjiang Publishers Group, Wuhan

Hwang HK, Son MH, Myeong JI, Kim CW, Min BH (2014)

Effects of stocking density on the cage culture of Korean rockfish (*Sebastes schlegeli*). Aquaculture 434:303–306

Imani K, Enock M, Nasser K (2014) Effect of stocking density on the growth performance of sex reversed male Nile tilapia (*Oreochromis niloticus*) under pond conditions in Tanzania. World J Fish Mar Sci 6:156–161

Ke WD, Huang XF, Li JH, Yan SL, Liu YM, Li F (2015) Development and research of aquatic vegetables in China. J Changjiang Veg 14:33–37

Leatherland JF, Cho CY (1985) Effect of rearing density on thyroid and interrenal gland activity and plasma hepatic metabolite levels in rainbow trout, *Salmo gairdneri* Richardson. J Fish Biol 27:583–592

Li X, Dong S, Lei Y, Li Y (2007) The effect of stocking density of Chinese mitten crab *Eriocheir sinensis* on rice and crab seed yields in rice–crab culture systems. Aquaculture 273:487–493

Lightfoot CA, van Dam A, Costa-Pierce BA (1992) What's happening to the rice yields in rice–fish systems? In: Dela Cruz CR, Lightfoot C, Costa-Pierce BA, Carangal VR, Bimbao MP (eds) Rice–fish research and development in Asia. International Center for Living Aquatic Resource Management Conference Proceedings, Vol. 24. International Center for Living Aquatic Resource Management, Manila, p 177–183

Liu YM, Fu XF (2004) How to improve the effect for planting lotus. Shanghai Veg 4:13–14

Ma L, Zhu J, Chen Q, Li W, Huang GH (2016) Effect of fish stocking density on diversity of arthropod community in lotus field. J Hunan Agr Univ Nat Sci 42:64–69

Namukose M, Msuya FE, Ferse SCA, Slater MJ, Kunzmann A (2016) Growth performance of the sea cucumber *Holothuria scabra* and the seaweed *Eucheuma denticulatum*: integrated mariculture and effects on sediment organic characteristics. Aquacult Environ Interact 8:179–189

Neal RS, Coyle SD, Tidwell JH (2010) Evaluation of stocking density and light level on the growth and survival of the Pacific white shrimp, *Litopenaeus vannamei*, reared in zero-exchange systems. J World Aquacult Soc 41:533–544

Nguyen QV (2001) Lotus for export to Asia—an agronomic and physiological study. RIRDC Publication, Barton

Panda MM, Ghosh BC, Sinhababu DP (1987) Uptake of nutrients by rice under rice-cum-fish culture in intermediate deep water situation (up to 50-cm water depth). Plant Soil 102:131–132

Sahoo SK, Giri SS, Sahu AK (2004) Effect of stocking density on growth and survival of *Clarias batrachus* (Linn.) larvae and fry during hatchery rearing. J Appl Ichthyol 20:302–305

Schram E, Heul JWVD, Kamstra A, Verdegem MCJ (2006) Stocking density-dependent growth of Dover sole (*Solea solea*). Aquaculture 252:339–347

Tan Y, Li S, Sun L (2003) The pollution of pesticides to the water environment. Pesticides 42:12–14

Tian DK, Tilt KM, Woods FM, Sibley JL, Dane F (2006) Summary of development, introduction and marketing strategy to share lotus in the Southeast United States. Proc 13th Annu Conf Int Plant Prop Soc, Wakayama 56: 151–154

Tian DK, Tilt KM, Sibley JL, Woods FM, Dane F (2009) Response of lotus (*Nelumbo nucifera* Gaertn.) to planting time and disbudding. HortScience 44:656–659

Wang S, Ye Y (1986) Occurrence characteristics of *Stenichironomus nelumbus* (Tokunaga & Kuroda) on lotus and its related control technology. Entomol Knowledge 2:73–74

Weatherley AH (1976) Factors affecting maximization of fish growth. J Fish Res Board Can 33:1046–1048

Wu W (2005) Production in lotus-fish co-culture field. Fujian Agr 2:28–29

Xiong J, Wu LJ, Du J, Huang GH (2010) Control methods and damage characteristics of the main insect pest of lotus. J Changjiang Veg 14:94–98

Yi Y, Lin CK, Diana JS (2002) Recycling pond mud nutrients in integrated lotus-fish culture. Aquaculture 212:213–226

Zhu J, Niu C, Zhang Z, Zhao J, Li J, Gong Z (2013) Occurrence characteristics of *Spodoptera litura* (Fabricius) on lotus and its related green control technology. J Changjiang Veg 18:113–115

13

Impact of prawn farming effluent on coral reef water nutrients and microorganisms

Cynthia Becker[1,2], Konrad Hughen[1], Tracy J. Mincer[1], Justin Ossolinski[1],
Laura Weber[1], Amy Apprill[1,*]

[1]Department of Marine Chemistry and Geochemistry, Woods Hole Oceanographic Institution, Woods Hole, MA 02543, USA
[2]Department of Biology, Ithaca College, Ithaca, NY 14850, USA

ABSTRACT: Tropical coral reefs are characterized by low-nutrient waters that support oligotrophic picoplankton over a productive benthic ecosystem. Nutrient-rich effluent released from aquaculture facilities into coral reef environments may potentially upset the balance of these ecosystems by altering picoplankton dynamics. In this study, we examined how effluent from a prawn (*Litopenaeus vannamei*) farming facility in Al Lith, Saudi Arabia, impacted the inorganic nutrients and prokaryotic picoplankton community in the waters overlying coral reefs in the Red Sea. Across 24 sites, ranging 0–21 km from the effluent point source, we measured nutrient concentrations, quantified microbial cell abundances, and sequenced bacterial and archaeal small subunit ribosomal RNA (SSU rRNA) genes to examine picoplankton phylogenetic diversity and community composition. Our results demonstrated that sites nearest to the outfall had increased concentrations of phosphate and ammonium and elevated abundances of non-pigmented picoplankton (generally heterotrophic bacteria). Shifts in the composition of the picoplankton community were observed with increasing distance from the effluent canal outfall. Waters within 500 m of the outfall harbored the most distinct picoplanktonic community and contained putative pathogens within the genus *Francisella* and order *Rickettsiales*. While our study suggests that at the time of sampling, the Al Lith aquaculture facility exhibited relatively minor influences on inorganic nutrients and microbial communities, studying the longer-term impacts of the aquaculture effluent on the organisms within the reef will be necessary in order to understand the full extent of the facility's impact on the reef ecosystem.

KEY WORDS: Aquaculture · *Litopenaeus vannamei* · Oligotrophic · Microbial community · Coral reef · SSU rRNA gene · *Francisella* spp.

INTRODUCTION

Aquaculture facilities contribute substantially to the world fish supply and are increasing in abundance worldwide (Naylor et al. 2000). In 2014, over half of the fish that humans consumed were produced through aquaculture, highlighting the importance of aquaculture as a reliable food source for increasing populations (FAO 2016). Aquaculture is important, but the wastes and nutrients produced by these facilities can pose threats to coastal environments (Nogales et al. 2011, Jiang et al. 2013). As a result, monitoring efforts are routinely needed to understand potential impacts. Most aquaculture monitoring efforts are part of required environmental impact assessments that may include collecting measurements of water quality, nutrient concentrations, organic matter outputs, and residual chemicals and bacteria that affect the product quality (Hambrey 2009). In addition, indicator species assays are commonly used to detect a finite number of known pathogens (Hernández et al. 2009, Zhou et al. 2014).

*Corresponding author: aapprill@whoi.edu

There has been much less effort towards monitoring how entire bacterial and archaeal communities respond to the influx of aquaculture waste. A more inclusive method that uses comparative phylogenetic analyses to examine how picoplanktonic communities respond can help deepen this understanding.

Microorganisms, including *Bacteria*, *Archaea*, and phytoplankton, are critical to the coastal marine environment and actively participate in biogeochemical cycling of nutrients and organic matter (Kirchman et al. 2007, DeLong 2009). Microbes are sensitive to eutrophication and pollution-based changes in seawater, with some cells exhibiting enhanced or reduced growth under these conditions (dos Santos et al. 2011, Xiong et al. 2015). Studies have shown that in oligotrophic waters, continual enrichment of nitrate and phosphate alters the native bacterial community composition (Chen et al. 2016, Dong et al. 2017). As a result, microorganisms may be able to serve as sensitive indicators of change, especially in oligotrophic environments where the magnitude of nutrient and organic matter inputs from aquaculture could be much higher than natural inputs to the environment. For example, the naturally oligotrophic coral reef waters of the Red Sea feature very low nutrient concentrations of 0.05–0.1 µM phosphate, <0.1 µM ammonium, and <0.3 µM nitrate+nitrite (Furby et al. 2014). Therefore, water-column samples from coral reefs located within the Red Sea are ideal for identifying how anthropogenic inputs from aquaculture facilities perturb the low-nutrient system and impact the bacterial and archaeal picoplankton communities.

Nutrient pollution and eutrophication from aquaculture are important problems affecting tropical coral reef ecosystems that reside in oligotrophic waters. Specifically, aquaculture runoff introduces ammonium, phosphate, and organic solids into the surrounding water. Studies have shown that fish farm runoff can reduce the survivorship of juvenile corals (Villanueva et al. 2005) and impair coral reproduction (Loya et al. 2004), as well as impact coral-associated microbes (Garren et al. 2008, 2009, Campbell et al. 2015). Studies also demonstrate that the planktonic microbial community, which is easier to sample and analyze than coral-associated microbes (e.g. Apprill et al. 2016), shows similar community changes in response to environmental aquaculture pollution (Sousa et al. 2006, Garren et al. 2008, Fodelianakis et al. 2014, Xiong et al. 2015). For example, in aquaculture environments such as a coastal shrimp pond in southeast China (Wei et al. 2009) and a fish farm north of Crete, Greece (Fo-

delianakis et al. 2014), the seawater microbial communities had reduced phylogenetic diversity and experienced overall shifts in community composition. Increases in *Proteobacteria*, free-living bacteria, and virus-like particles were also noted in the oligotrophic reef waters of the Philippines, correlating with organic nutrient enrichment caused by a fish farm (Garren et al. 2008). These often site-specific alterations to the planktonic microbial community by aquaculture speak to the complexity of these oligotrophic ecosystems. It also exemplifies the need for continued assessments of these impacts, especially in understudied areas such as the eastern Red Sea.

The National Aquaculture Group, formerly known as the National Prawn Company, is a coastal aquaculture facility located next to coral reefs in the eastern Red Sea, near the city of Al Lith, Saudi Arabia. It is one of the largest desert aquaculture facilities in the world and covers approximately 250 km². From 2008 to 2010, our team examined the Red Sea marine environment to identify potential sources of putative pathogens that were associated with the presence of harmful coral lesions (Apprill et al. 2013, Furby et al. 2014). As part of this study, we examined whether the effluent from the National Prawn Company aquaculture facility could be a potential source of nutrients and pathogens to the surrounding environment. At the time of the sampling in 2009, the facility was hatching and rearing approximately 15 000 tons of white prawn *Litopenaeus vannamei* annually in ponds covering a total surface area of 2800 ha (https://web.archive.org/web/20091130220933/http: //www.robian.com.sa:80/home.html; see also www. naqua.com.sa). Since the time of sampling, the facility expanded to include multiple sea cages and many more shrimp farms and produces upwards of 100 000 tons of marine products annually, including shrimp, sea cucumber, and sea bass (www.naqua. com.sa). Furthermore, Saudi Arabia has plans to expand its aquaculture industry, with multiple investments in new facilities (Mon Chalil 2015). In the context of continuous expansions of the aquaculture industry, it is important to understand the extent to which this industry may threaten the adjacent delicate reef ecosystem. In this study, we aimed to assess the impact of the effluent from this aquaculture facility on the surrounding oligotrophic reef environment by examining the inorganic nutrient concentrations and microbial communities in an area extending nearly 22 km from the outfall. We hypothesized that concentrations of inorganic nitrogen and phosphorus would be elevated at the outfall, a distinct microbial community would reside there, including known

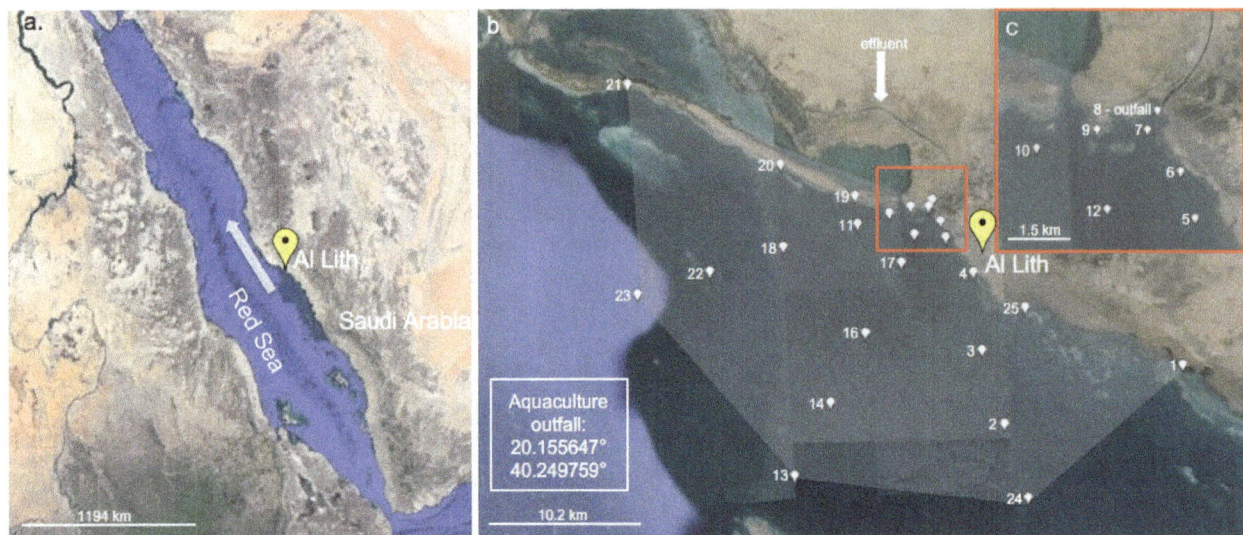

Fig. 1. (a) Location of the aquaculture facility in Al Lith, Saudi Arabia (yellow pin), in the eastern Red Sea. Gray arrow: the prevailing surface currents that move S to N, primarily along the eastern coast. (b) The 24 sites that were sampled along 5 transects extending from the aquaculture outfall up to 21.7 km from the outfall. (c) Inset: a more detailed map of the outfall (Site 8) and nearby sites. Images reproduced according to Google permissions and incorporate data from NOAA, US Navy, National Geospatial-Intelligence Agency, General Bathymetric Chart of the Oceans, DigitalGlobe, and US Geological Survey

pathogens, and that this impact would diminish with increasing distance from the outfall.

MATERIALS AND METHODS

Sampling

Sampling occurred in the eastern Red Sea, surrounding the runoff from the National Prawn Company aquaculture facility in Al Lith, Saudi Arabia (Fig. 1a). Seawater was sampled from 24 stations surrounding the aquaculture facility using a directional-transect grid pattern (Fig. 1b,c) from October 9 to 12, 2009, during 10:00–16:00 h. The sampling pattern extended in 5 transects away from the outfall point source: north, northwest, west, southwest, and south along the coastline (Table A1 in the Appendix). Sampling occurred first at the mouth of the outfall, and then was spaced 0.5–21.7 km along the 5 transects (Fig. 1). Seawater was sampled at a 10 m depth, which was just above the coral reef for the shallower sites, into 20 l acid-washed carboys. Inorganic nutrients and microbial abundance samples were collected and processed as previously outlined in Furby et al. (2014). Briefly, for inorganic nutrients, 150 ml polypropylene acid-washed bottles were filled with seawater and frozen at −20°C. Samples (1 ml) for cell counts were fixed in 1% (v:v) paraformaldehyde (final concentration) and placed in cryovials in liquid

nitrogen for 3 wk, then frozen at −80°C. Seawater for DNA analyses was stored on ice for no more than 4 h before it was filtered onto 0.22 µm Durapore membrane filters (142 mm) (Millipore) with a peristaltic pump. Filters were then frozen in liquid nitrogen.

Nutrient analysis

Nutrients were analyzed as outlined in Furby et al. (2014). In summary, dissolved concentrations of NH_4^+, $NO_3^- + NO_2^-$, NO_2^-, PO_4^{3-}, and silicate were determined with a continuous segmented flow system including a Technicon AutoAnalyzer II (SEAL Analytical) and an Alpkem RFA 300 Rapid Flow Analyzer. $NO_3^- + NO_2^-$ and NO_2^- were measured according to Armstrong et al. (1967). An adjusted molybdenum blue method was used to measure PO_4^{3-} (Bernhardt & Wilhelms 1967). The indophenol blue method was used to measure concentrations of NH_4^+ (USEPA 1983), and validated using a method developed by Holmes et al. (1999).

Direct cell counts

Microbial cell counts were enumerated using the flow cytometry methods described in Furby et al. (2014). To summarize, the preserved samples were analyzed using stained and unstained methods to determine the number of pigmented and non-pigmented cells within each sample. The aliquots of sample for the unstained method were run on an EPICS ALTRA flow cytometer (Beckman Coulter) to determine abundance of Cyanobacteria (*Prochlorococcus* and *Synechococcus*) and small eukaryotic phytoplankton (picoeukaryotes). The aliquot of sample for staining was prepared by diluting the sample 1:10 into a 30 mM (final concentration) potassium citrate buffer solution, and staining with Sybr Green I (1:5000 final dilution of initial stock) (Molecular Probes) for a duration of 2 h in the dark at 4°C. Excitation of the fluorescent stain was accomplished using a laser operating at a wavelength of 488 nm on the same EPICS ALTRA flow cytometer, and this allowed for enumerating picoplankton (*Bacteria* and *Archaea*) on the basis of DNA staining (Sybr Green I green fluorescence), chlorophyll, phycoerythrin, forward scatter, and 90° side scatter signatures. *Prochlorococcus* cell counts from the unstained method were subtracted from total prokaryotic cells to obtain counts of non-pigmented picoplankton. FlowJo software (v.6.3.3, Tree Star) was used for off-line data analysis.

DNA analysis

The filters were cut into quarters, and DNA was extracted from one of the quarters using a bead beating method combined with a sucrose-lysis extraction and spin-column separation. Briefly, 0.1 mm glass beads were used to homogenize cellular biomass collected on the filters for 10 min in a solution of 875 µl sucrose-EDTA lysis buffer (0.75 M sucrose, 20 mM EDTA, 400 mM NaCl, 50 mM Tris) and 100 ml of 10% sodium dodecyl sulfate. This was followed by a Proteinase K digestion at 55°C for 4 h, and separation of the DNA using the spin columns supplied by the DNeasy kit (Qiagen) (Santoro et al. 2010). The DNA samples were individually amplified with the 515F (5 -GTG CCA GCM GCC GCG GTA A-3) and 806RB primers (5 -GGA CTA CNV GGG TWT CTA AT-3), with the modified reverse primer designed to enhance the detection of different SAR11 clades (Apprill et al. 2015). The primers were designed similar to Kozich et al. (2013) and each included a unique 8 bp barcode, 10 bp pad, and 2 bp link in addition to the primers described above. PCR reactions were carried out in a Bio-Rad thermocycler using triplicate 25 µl volumes for each sample. Each reaction included 1.25 U GoTaq Flexi DNA Polymerase (Promega), 5× Colorless GoTaq Flexi Buffer, 2.5 mM

$MgCl_2$, 200 µM dNTP mix (Promega), 200 nM of the corresponding barcoded primer, and 1–4 ng of genomic DNA template. The reaction conditions were as follows: initial denaturation for 2 min at 95°C, followed by an iteration of 20 s at 95°C, 15 s at 55°C and 5 min at 72°C for 25–28 cycles, and a final extension step for 10 min at 72°C. Five µl of the reaction products were run on a 1% agarose/TBE gel containing the HyperLadder 50 bp DNA ladder (generally 5 ng µl^{-1}) (Bioline). After pooling the triplicate reactions per sample, the PCR products were purified with the QIAquick Purification Kit (Qiagen). The samples were quantified using the Qubit 2.0 Fluorometer with the dsDNA High Sensitivity Assay (Life Technologies). The PCR products were then combined into equimolar ratios and processed for sequencing using 250 bp paired-end MiSeq (Illumina) at the W. M. Keck Center for Comparative and Functional Genomics at the University of Illinois. Raw sequence data were deposited into the National Center for Biotechnology Information (NCBI) Sequence Read Archive (SRA) under accession number PRJNA357506.

Sequence analysis

Sequence data were filtered and analyzed using mothur v.1.36.1 (Schloss et al. 2009). This included constructing contigs, filtering out long amplicons (over 255 bp), and removing chimeras detected using de novo UCHIME v.4.2.40 (Edgar et al. 2011). The taxonomic assignment for each sequence was performed in mothur using the SILVA SSU reference database (v.123) and the *k*-nearest neighbor algorithm with an 80% cutoff. The number of small subunit ribosomal RNA (SSU rRNA) sequences per sample varied between 266 and 55182 and the dataset was therefore subsampled to a depth of 10200 sequences per sample, resulting in the loss of only Sample 20 (Appendix 1). Operational taxonomic units (OTUs) for each sample were generated using the minimum entropy decomposition (MED) clustering algorithm (Eren et al. 2015). MED further reduced the sequence reads to between 8853 and 9538 sequences per sample (Appendix 1).

Graphical and statistical analyses

Ocean Data View software (v.4.7.8, 64 bit) (Schlitzer 2002) was used to create contour plots of nutrients and microbial abundances using data-interpolating variational analysis (DIVA) gridding (40 × 40 scale length). PRIMER-E 7 (v.7.0.11, Quest Research) was used to compare the composition of OTUs at each sampling site. The data were square-root transformed, compared using Bray-Curtis similarity, and plotted using non-metric multidimensional scaling (nMDS) to compare microbial community structure across samples. Information regarding proximity to the outfall was superimposed onto the symbols for each sample in the nMDS plot. In addition, bar plots were generated in Excel for Mac 2011 (v.14.6.9, Microsoft) to illustrate microbial community composition by plotting the relative percent abundance of taxonomic groups of *Bacteria* and total *Archaea*.

Alpha diversity metrics of microbial communities at each site were computed using the plot_richness function in the phyloseq (v.1.16.2) and ggplot2 (v.2.1.0) R packages in RStudio (v.0.99.902) (Wickham 2009, McMurdie & Holmes 2013). Specific metrics plotted include 'observed' (observed richness), 'Chao1' (estimator of richness), and 'invsimpson' (inverse Simpson index of diversity). Differentially abundant OTUs between the outfall and other sites were obtained using DESeq2 (v.1.12.4) in RStudio (Love et al. 2014). Specifically, OTU sequence count data was compared between the outfall site and 6 different distance groups away from the outfall that encompassed the other sites (0.5–1.5, 2.5–3, 4.5–5.3, 9–10.1, 14.5–15.2, and 20–21.7 km). Significant differential abundance was determined in DESeq2 using the adjusted p-value ($p < 0.05$). Box plots of relative abundance data for significantly enriched (at the outfall) OTUs were created in PRIMER-E 7 (v.7.0.11).

RESULTS

Nutrients

Nutrient concentrations were generally similar across most of the sites spanning 0–21.7 km from the outfall (Fig. 2), with the exception of ammonium and phosphate, which were elevated at the outfall by 5–75 times and 4–14 times, respectively, compared to the other sites (Fig. 2a,d). Ammonium was 0.60 µM at the outfall and below 0.12 µM (Site 19) at all other sites (Fig. 2a,j). The concentration of phosphate at the outfall was the highest measured in the study, at 0.67 µM, while all other sites were lower, below 0.17 µM (Site 7) (Fig. 2d,m). Concentrations of nitrite, nitrate+nitrite, and silicate were not elevated near the outfall, and were similar in concentration to the more distant sites (Fig. 2b,c,e,k,l,n). Over the study

area, the combined nitrate+nitrite concentrations ranged from undetectable to 0.87 µM, nitrite ranged from 0.02 to 0.11 µM, and silicate ranged from 0.08 to 1.64 µM (Fig. 2b,c,e).

Microbial abundances

Cell counts of microbial groups were measured at all 24 sites (Fig. 2f–i), and there were no clear patterns in cell abundances, either elevated or depleted, directly at the outfall. Concentrations of non-pigmented picoplankton, a proxy for heterotrophic *Bacteria*, *Archaea*, and non-pigmented photoheterotrophic *Bacteria*, were high at the outfall (1.5×10^6 cells ml^{-1}) as well as at Sites 19, 17, 10, and 11 (1.3 to 1.5×10^6 cells ml^{-1}), which are all between 3 and 5 km away from the outfall (Fig. 2f,o). Concentrations were up to 9.3×10^5 cells ml^{-1} lower at the remaining sites, which ranged from 5.8×10^5 to 1.2×10^6 cells ml^{-1}. *Synechococcus* spp. were depleted directly at the outfall (1.5×10^5 cells ml^{-1}) compared to nearby Sites 7 and 9 that are ≤1.5 km away (2.0–2.1 × 10^5 cells ml^{-1}) (Fig. 2q). *Synechococcus* spp. were most abundant at sites 2.5 km from the outfall (Sites 5 and 12; 3.0–3.2 × 10^5 cells ml^{-1}; Fig. 2q). Abundances of *Prochlorococcus* spp. varied from 2.4×10^4 to 6.4×10^4 cells ml^{-1}, with no major changes observed over the sampling area (Fig. 2g). The abundance of picoeukaryotes ranged from 2.1×10^3 to 1.2×10^4 cells ml^{-1} (mean: 4.6×10^3 cells ml^{-1}), and were also consistent across the sampling area (Fig. 2i).

Microbial community phylogenetic diversity

Alpha diversity indices of bacterial and archaeal OTUs based on SSU rRNA gene sequences showed some variability between the sites. The observed community richness, or number of observed OTUs, ranged between 225 and 357 OTUs, and richness at the outfall site fell near the middle of this range, at 291 (Fig. 3a). Two sites, Site 25 (9 km from outfall) and Site 6 (1.5 km from outfall), contained microbial communities of lower richness, at 225 and 240 OTUs, respectively. The Chao1 estimator of richness produced patterns similar to the observed richness (Fig. 3a,b), indicating that the depth of sequencing utilized was appropriate for the study. When the samples were organized by distance from the outfall, there appeared to be a potential increase in estimated richness with increasing distance to ~5 km. Again, Sites 6 and 25 had lower predicted richness

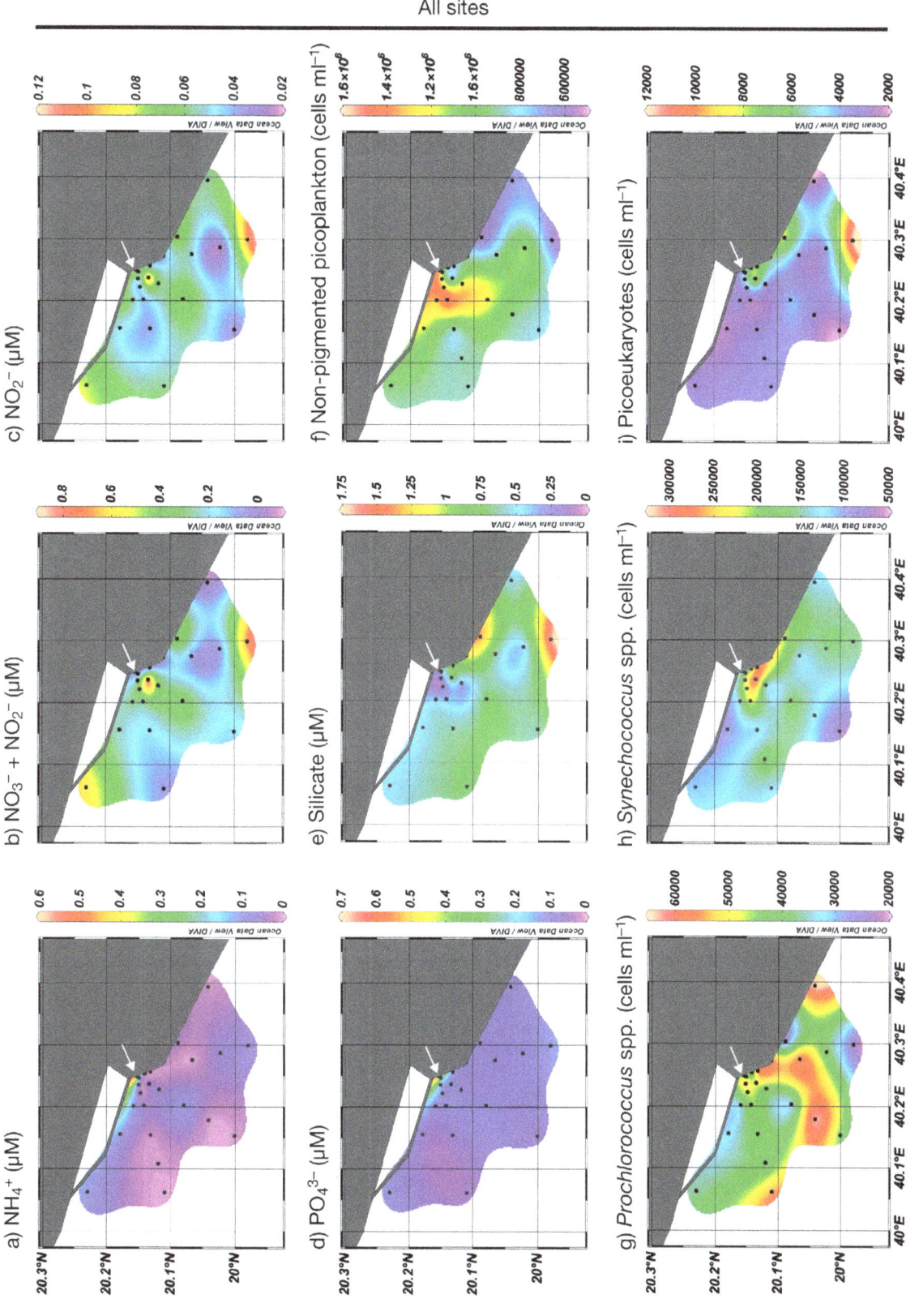

Fig. 2. Maps of surface concentrations of dissolved inorganic nutrients and microbial abundances at (a–i) all sample sites, and (j–r) sites <5.3 km from the aquaculture outfall. Inorganic nutrients: (a,j) NH$_4^+$, (b,k) NO$_3^-$ + NO$_2^-$, (c,l) NO$_2^-$, (d,m) PO$_4^{3-}$, and (e,n) silicate. Microbial abundances: (f,o) non-pigmented picoplankton, (g,p) *Prochlorococcus* spp., (h,q) *Synechococcus* spp., and (i,r) picoeukaryotes. White arrows: the outfall site. Northern coastline has been drawn in, to more accurately depict the geography

Sites nearest outfall

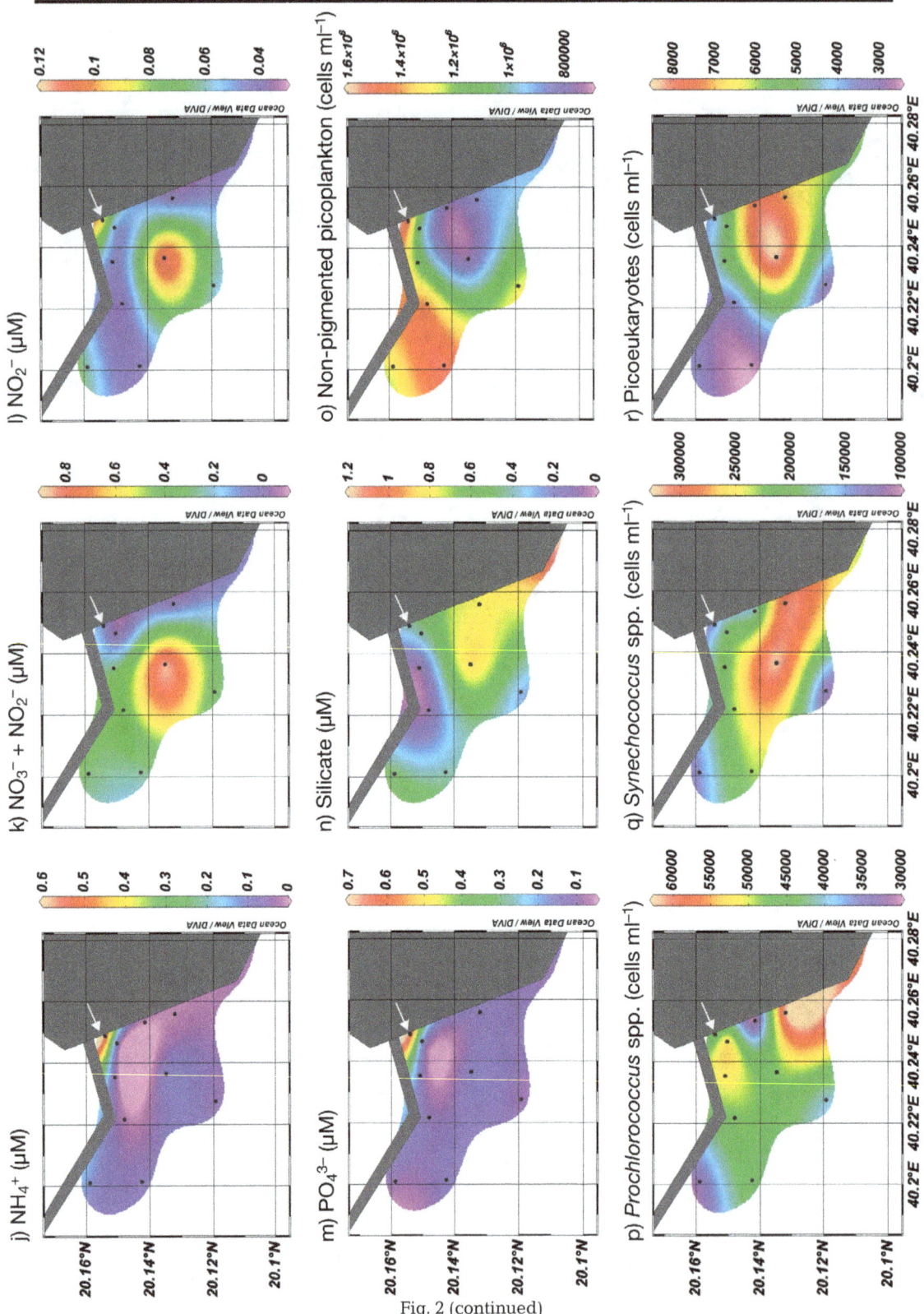

Fig. 2 (continued)

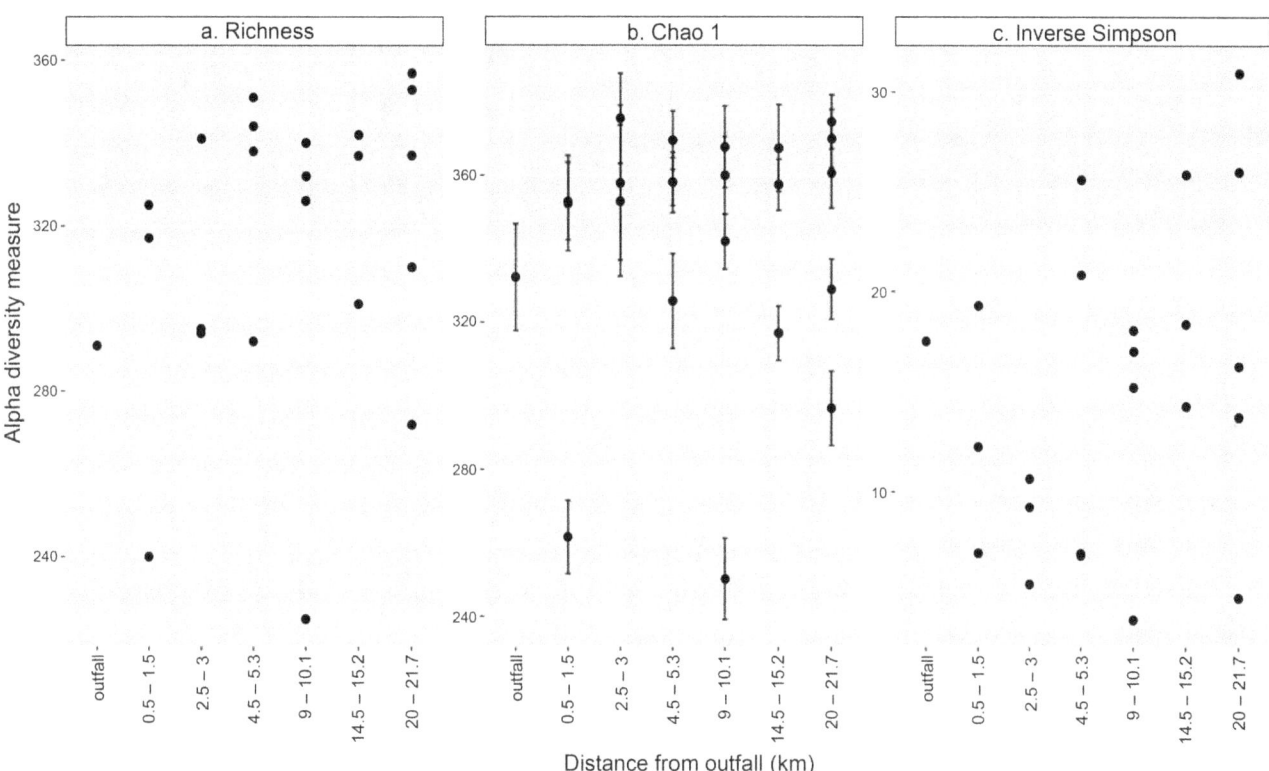

Fig. 3. Measures of alpha diversity based on bacterial and archaeal small subunit ribosomal RNA (SSU rRNA) genes at each sampling site: (a) community richness (e.g. number of operational taxonomic units), (b) Chao1 estimator (bars: 95% confidence intervals), and (c) inverse Simpson index. Data are organized into groups based on distance from the outfall

(Fig. 3b). The inverse Simpson index of diversity also indicates community evenness (Fig. 3c), and it ranged from low diversity and evenness, at 3.58 (Site 25, 9 km from the outfall), to higher diversity and evenness, at 31 (Site 13, 20 km from the outfall), with a mean of 15. Even though Site 6 was low in both observed and estimated richness (Fig. 3a,b), it was not low in diversity relative to the other sites, indicating that Site 6 had a more even microbial community. In terms of diversity, the outfall, with an inverse Simpson index of 17.5, was similar to the other sites.

Microbial community composition

The SSU rRNA gene sequences were dominated by *Cyanobacteria* and *Alphaproteobacteria* and showed similar trends in composition across the sites (Fig. 4). Two sites, 24 and 25, were made up of mostly *Cyanobacteria*, comprising 60 and 75% of total sequences, respectively. The outfall (Site 8) was similar to most sites, except that it had 3 times as much *Flavobacteria* as all other sites (15.6% compared to a mean of 4.7% for the other sites) (Fig. 4). An nMDS analysis provided a spatial representation of commu-

nity composition resemblance between sites (Fig. 5). Again, in this analysis, the southern Sites 24 and 25 were separated from the group, supporting how microbial communities at these sites were distinct from most of the other sites. Coastal Site 6 was also separated, both from the entire group of sites, and also from the other sites 0.5–1.5 km (Fig. 5) from the outfall. In general, the arrangement of sites in the nMDS plot corresponds to actual distance from outfall, demonstrating that the bacterial and archaeal communities shifted in composition with increasing distance away from the effluent outfall. The 0.5–3 km group plotted closer to the outfall than the 4.5–10.1 km distance group. In addition, the 14.5–21.7 km group was even farther from the outfall point in the nMDS plot (Fig. 5). The outfall site is distinctly separate in the plot, indicating that there are taxa within the community that are unique to the outfall.

Differentially abundant outfall bacteria

Microbes indicative of the outfall effluent were examined using the SSU rRNA gene sequence count data, and 22 OTUs were significantly elevated or de-

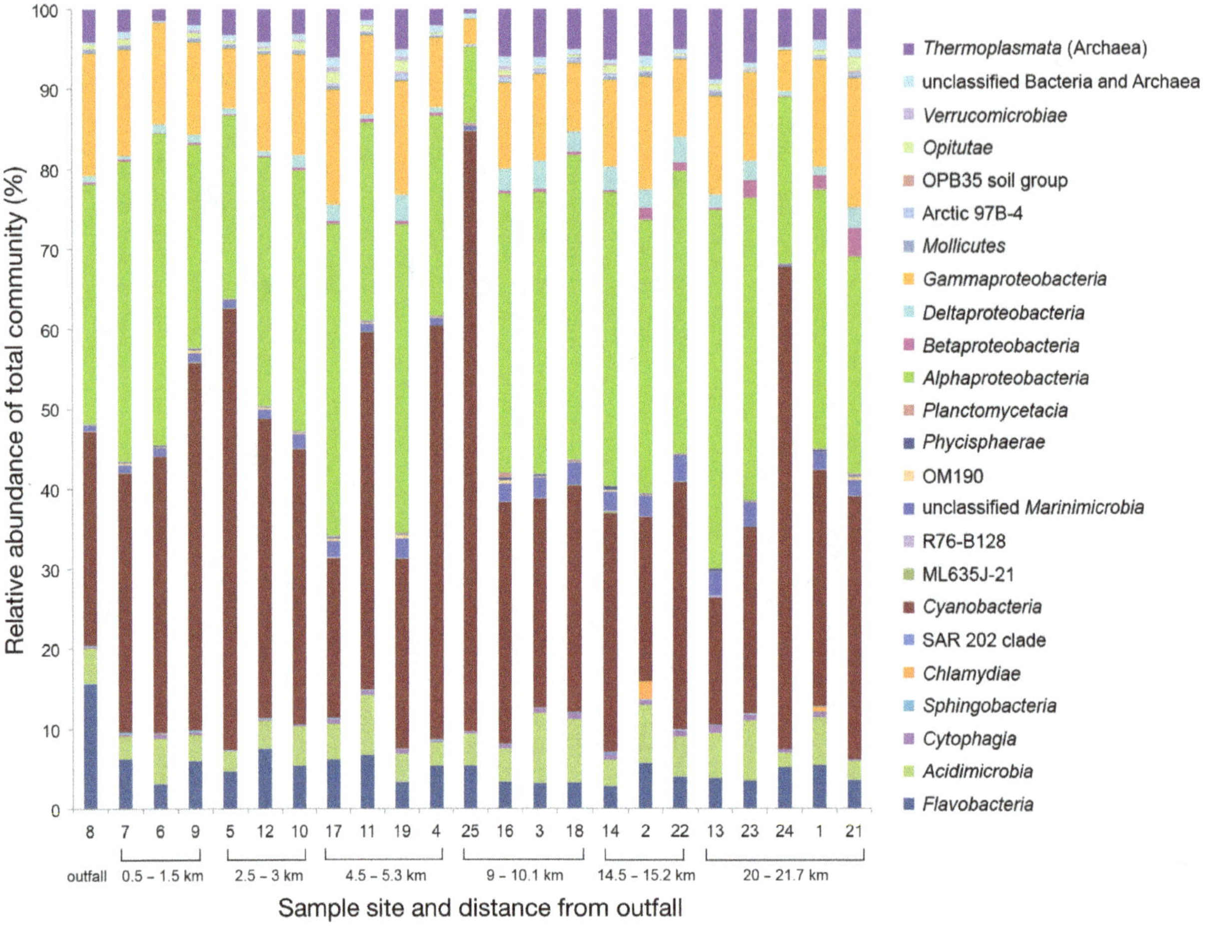

Fig. 4. Relative abundances of small subunit ribosomal RNA (SSU rRNA) gene sequences organized by taxonomic class for each site. Sites are organized by distance from outfall, with Sample 8 (outfall) on the far left and Sample 21 (21.7 km from Site 8) on the far right

pleted at the outfall when compared to the 6 distance groupings (0.5–1.5, 2.5–3, 4.5–5.3, 9–10.1, 14.5–15.2, and 20–21.7 km) (Fig. 6, Table 1). Of these, 7 OTUs were significantly elevated at the outfall (Fig. 6). These included sequences classifying as *Francisella* spp., *Caedibacter* spp., uncultured *Gammaproteobacteria*, members of the NS4 Marine Group, uncultured *Rickettsiales*, uncultured *Flavobacteriaceae*, and uncultured *Cryomorphaceae* (Fig. 6a–g). Sequences of those OTUs enriched at the outfall were found in lower abundances at the more distant site groupings, and were significantly different at many of these groups (p < 0.05; Fig. 6). For OTUs 622 (*Francisella* spp.), 705 (*Caedibacter* spp.), and 1653 (uncultured *Gammaproteobacteria*), sequence counts were significantly lower at all distances compared to the outfall (Fig. 6a–c). Sequence counts for OTUs 1743 (NS4 Marine Group) and 1342 (uncultured *Rickettsiales*) were significantly lower at all distance groups except for the closest one, 0.5–1 km from the outfall

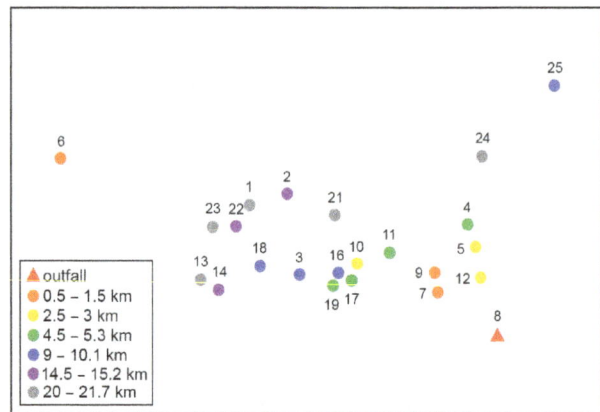

Fig. 5. Non-metric multidimensional scaling analysis of bacterial and archaeal community composition at each study site based on small subunit ribosomal RNA (SSU rRNA) gene sequences. Relative abundance data were square-root transformed and the resemblance of community composition was determined using Bray-Curtis similarity. Sites are grouped by distance from the outfall and the numbers indicate site. The 2-D stress of the solution is 0.06

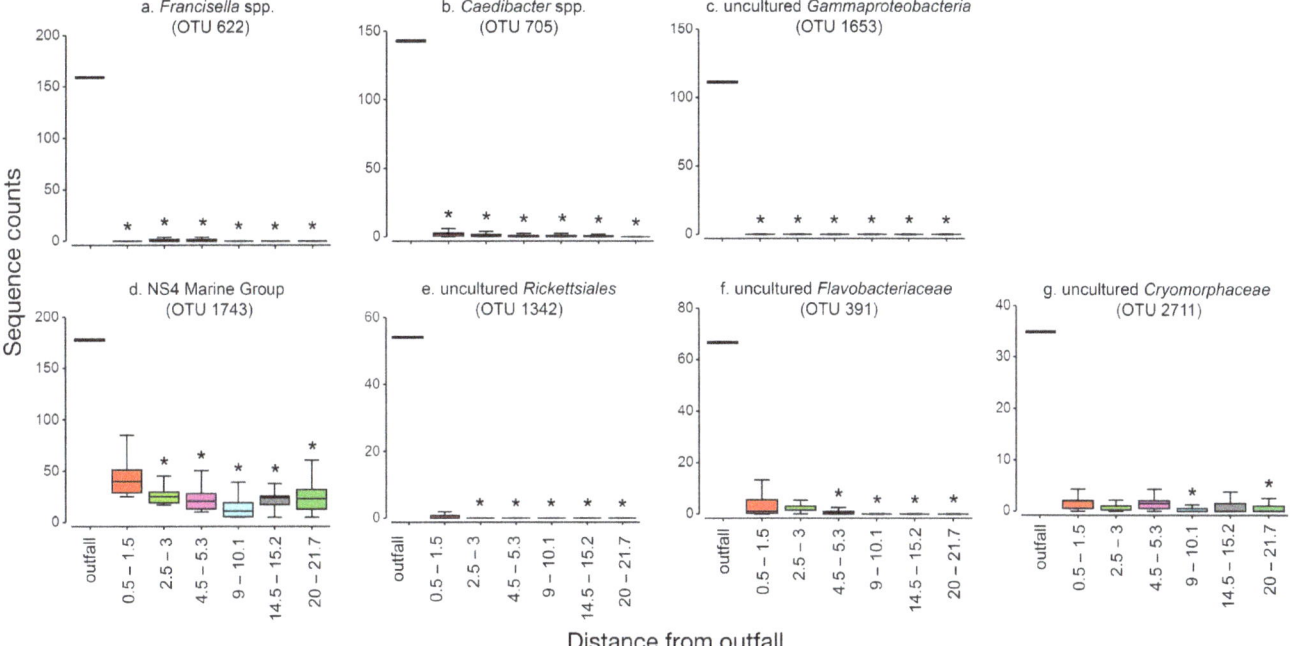

Fig. 6. Box-and-whisker plots of sequence counts for the 7 operational taxonomic units (OTUs) that were significantly enriched at the outfall site: (a) *Francisella* spp., (b) *Caedibacter* spp., (c) uncultured *Gammaproteobacteria*, (d) NS4 Marine Group, (e) uncultured *Rickettsiales*, (f) uncultured *Flavobacteriaceae*, and (g) uncultured *Cryomorphaceae*. *Distance groups with significantly depleted sequence counts compared to sequence counts at the outfall (p < 0.05). The outfall includes 1 sample (indicated by single line) and the other distance groups include 3–5 samples. In each plot, middle line denotes median, box outlines upper and lower 25 % quartiles, and whiskers denote maximum and minimum values

(Fig. 6d,e). Sequence counts for OTU 391 (uncultured *Flavobacteriaceae*) were significantly different from the outfall for the 4 farthest distance groups (4.5–5.3, 9–10.1, 14.5–15.2, and 20–21.7 km), but not the 2 closest distance groups (Fig. 6f). Sequence counts for OTU 2711 (uncultured *Cryomorphaceae*) were significantly distinct from the outfall at 2 of the farther distance groups (9–10.1 and 20–21.7 km) (Fig. 6g). The other 15 significantly differentially abundant OTUs were significantly enriched or depleted in 3 or fewer distant groups, when compared to the outfall, as indicated by the positive and negative log2 fold change (Table 1). Out of the 15 significantly enriched OTUs, 11 were from the taxonomic class *Flavobacteria* (Table 1), the class of *Bacteria* found at higher relative abundance at the outfall compared to the other sites (Fig. 4).

DISCUSSION

This study analyzes multiple factors, including nutrient concentrations, microbial cell abundances, and the fine-scale taxonomy and composition of the bacterial and archaeal seawater communities, to examine the impact of an effluent point source from a large aquaculture facility on oligotrophic coral reefs within the Red Sea. Although our results represent data from 2009 and were collected prior to recent expansions of the facility, they show a distinct elevation of ammonium and phosphate and an altered microbial community structure at the outfall, with subtler differences within microbial community composition as distance from the outfall increases.

The elevated levels of ammonium and phosphate at the outfall site were consistent with findings of aquaculture impacts in other studies. Biao et al. (2004) found elevated inorganic nutrient concentrations, including ammonium and phosphorus, in a shrimp farm outlet creek that flowed from shrimp ponds to the Yellow Sea. In that study, changes in ammonium reached 4.7 µM, much greater than in our study, whereas the magnitude of change in inorganic phosphorus concentrations was 0.26 µM, lower than that measured in our study (0.5 µM). In a study of fish cages in the eutrophic environment of the Xiangshan Bay, China, Xiong et al. (2015) observed overall increases in the concentrations of dissolved

Table 1. Operational taxonomic units (OTUs) that were significantly enriched or depleted at the outfall compared to the other distance groups, based on DESeq2 results

OTU	Classification	log2 fold change[a]	Enriched or depleted at outfall	Significant at distance groups (km) (p < 0.05)
622	Gammaproteobacteria, Francisella spp.	8.8; 6.6; 6.6; 9.2; 9.0; 9.5	Enriched	0.5–1.5; 2.5–3; 4.5–5.3; 9–10.1; 14.5–15.2; 20–21.7
705	Gammaproteobacteria, uncultured Caedibacter spp.	5.5; 6.2; 7.1; 7.2; 7.5; 9.4	Enriched	0.5–1.5; 2.5–3; 4.5–5.3; 9–10.1; 14.5–15.2; 20–21.7
1653	Uncultured Gammaproteobacteria	9.1; 9.2; 9.4; 9.3; 9.6	Enriched	0.5–1.5; 2.5–3; 4.5–5.3; 9–10.1; 14.5–15.2; 20–21.7
1743	Flavobacteria, NS4 Marine Group	2.9; 3.3; 4.0; 3.3; 3.2	Enriched	2.5–3; 4.5–5.3; 9–10.1; 14.5–15.2; 20–21.7
1342	Alphaproteobacteria, uncultured Rickettsiales	7.7; 8.0; 8.2; 8.2; 8.2	Enriched	2.5–3; 4.5–5.3; 9–10.1; 14.5–15.2; 20–21.7
391	Flavobacteria, uncultured Flavobacteria, Flavobacteriaceae	6.2; 8.2; 8.0; 8.5	Enriched	4.5–5.3; 9–10.1; 14.5–15.2; 20–21.7
2711	Flavobacteria, uncultured Cryomorphaceae	6.1; 5.7	Enriched	9–10.1; 20–21.7
2731	Flavobacteria, uncultured Cryomorphaceae	7.5	Enriched	14.5–15.2
1477	Flavobacteria, uncultured Flavobacteriaceae	6.7	Enriched	14.5–15.2
491	Flavobacteria, NS2b Marine Group	2.2	Enriched	14.5–15.2
2649	Flavobacteria, NS4 Marine Group	3.3	Enriched	14.5–15.2
1744	Flavobacteria, NS4 Marine Group	2.9	Enriched	14.5–15.2
1807	Flavobacteria, NS5 Marine Group	5.6	Enriched	14.5–15.2
2322	Flavobacteria, NS9 Marine Group	6.2	Enriched	14.5–15.2
2329	Flavobacteria, NS9 Marine Group	5.5	Enriched	14.5–15.2
1618	Gammaproteobacteria, Alteromonas spp.	−4.9; −3.7; −4.7	Depleted	0.5–1.5; 14.5–15.2; 20–21.7
2157	Cyanobacteria, uncultured Cyanobacteria	−6.5; −4.8	Depleted	0.5–1; 14.5–15.2
2055	Alphaproteobacteria, SAR116 clade	−3.2	Depleted	14.5–15.2
1018	Alphaproteobacteria, SAR116 clade	−4.7	Depleted	14.5–15.2
789	Alphaproteobacteria, SAR116 clade	−3.8	Depleted	14.5–15.2
2384	Euryarchaeota, Marine Group II	−4.4	Depleted	14.5–15.2
2765	Euryarchaeota, Marine Group III	−4.7	Depleted	14.5–15.2

[a]Order of values corresponds with respective distance group in last column

inorganic nitrogen and phosphate at fish cages relative to a reference site that was 8 km away. The magnitude of increase of dissolved inorganic nitrogen was approximately 2.4 µM, much higher than any concentration measured in our study. In contrast, while overall phosphate concentrations were higher within the eutrophic bay, the increase in phosphate at the fish farm (0.1 µM) was lower than the increase measured in our study (0.5 µM). Although the magnitude of changes varied among studies, the ubiquitous increase in seawater nutrient concentrations in areas supporting aquaculture facilities suggests that they can be a source of nutrient enrichment to both eutrophic and oligotrophic environments. This comparison further suggests that the elevated nutrient concentrations may be attributed to animal waste products generated at the aquaculture facility. Since the time of sampling in 2009, the Al Lith aquaculture facility has expanded considerably to include more shrimp farms and sea cages. If these inorganic nutrient concentrations were among the highest recorded in the Red Sea in 2009, we would hypothesize that nutrient levels will have only increased as a result of the expansions, further threatening the delicate reef ecosystem.

In the study presented here, we found it surprising that the effluent nutrient signal diminished so rapidly just 500 m away from the outfall. This could be related to the oligotrophic nature of the Red Sea waters, in which generally low-concentration or limiting nutrients are quickly mixed and diluted into the surrounding waters, or are rapidly assimilated by bacteria and other planktonic organisms that recycle these nutrients and incorporate them into microbial biomass. Although detailed circulation data for the Red Sea is scant, the study area features a buoyancy-driven boundary current that moves generally from south to north, which may also be contributing to the incorporation of nutrients into the surrounding waters (Eshel & Naik 1997). Another factor may be the length of the effluent canal. Biao et al. (2004) found diminishing concentrations of all measured inorganic nutrients along an outlet canal that extended from 3 shrimp ponds to the Yellow Sea, over a distance of 11 km. The effluent canal of the National Prawn Company facility runs the length of the shrimp pond area, over 20 km long, and may be acting as a potential biological buffer where higher levels of nutrients are consumed before reaching the oligotrophic waters.

Of the microbial cells enumerated, non-pigmented picoplankton were the only cell groups found to be elevated at the outfall, as well as at nearby sites. Non-pigmented picoplankton are primarily heterotrophic *Bacteria* and *Archaea* as well as non-pigmented photoheterotrophic *Bacteria*, and their abundance is often related to the organic carbon availability (Jumars et al. 1989), which was likely elevated as a result of shrimp organic wastes in the effluent. Although organic carbon was not measured here, preventing a complete assessment, a study of bacterial communities below a fish farm cage found increased abundances of heterotrophic bacteria as a result of total organic matter enrichment (La Rosa et al. 2004). This suggests the elevated abundances of non-pigmented picoplankton at the outfall may be due to populations of these cells residing in the organic-rich effluent, likely confirming the presence of increased organic matter. Indeed, future studies of aquaculture effluent should include quantification of particulate and dissolved organic carbon.

One particular strength of this study was the examination of the microbial SSU rRNA sequences from general bacterial and archaeal primers, which allowed for a more holistic examination of the communities compared to cultivation-dependent or indicator microbial assays, such as loop-mediated isothermal amplification assays that detect known bacterial pathogens (Zhou et al. 2014). An example of one of the benefits of using cultivation-independent techniques to study microbial communities within the environment is that it allowed us to produce new knowledge of potential indicator bacterial taxa that may be originating from shrimp aquaculture. We found alterations in the entire microbial community composition with increasing distance away from the aquaculture effluent point source. This finding is consistent with studies that found changes in the microbial communities between treated wastewater runoff and the seawater surrounding coral reefs in

Florida (Campbell et al. 2015) and in the Red Sea (Ziegler et al. 2016). It also confirms other study results that found alterations in bacterial communities in oligotrophic areas subject to aquaculture (fish farm) effluent (Garren et al. 2008, Fodelianakis et al. 2014). In our study, 22 groups of bacteria were identified as being significantly elevated or depleted at the outfall site. Of these, 7 were elevated at the outfall of the effluent canal when compared to multiple distances, suggesting that they potentially originated at the aquaculture facility. Of these enriched groups, 2 OTUs, classifying as *Francisella* spp. and uncultured *Rickettsiales*, are related to known pathogens. *Francisella* spp. have been implicated as pathogens for a wide taxonomic diversity of mammals and fish, as well as free-living cells (Sjödin et al. 2012). Pathogenic *Francisella* generally target marine organisms, causing disease in cod (Olsen et al. 2006), tilapia (Mauel et al. 2007), and giant abalone (Kamaishi et al. 2010). In addition, *Francisella* spp. can also impact a wide diversity of land mammals, including humans, through arthropod vectors or infected water (Pérez et al. 2016). It is likely that the infection breadth of *Francisella* spp. is highly underestimated (Birkbeck et al. 2011), and the significant occurrence of *Francisella* spp. at the shrimp aquaculture outfall may suggest that *Francisella* spp. originated in the shrimp ponds. Some pathogenic species, including *F. noatunensis* (cod pathogen) and *F. tularensis* (human pathogen), likely persist poorly in the environment without a host (Duodu & Colquhoun 2010). Despite this, *Francisella* spp. have been isolated in seawater samples from around the world (Barns et al. 2005, Berrada & Telford 2010, Duodu et al. 2012). Some pathogenic *Francisella* strains may interact with single-celled eukaryotes, allowing them to persist successfully in the environment (Verhoeven et al. 2010). While presence and abundance of single-celled eukaryotes were not measured in this study, they generally feed on bacteria, and increased levels of heterotrophic bacteria at the effluent outfall may have provided a favorable environment for eukaryotic *Francisella* spp. hosts. The other putative pathogen detected in this study was uncultured *Rickettsiales*. While not every *Rickettsiales* bacterium is pathogenic, these bacteria belong to an order that is characterized by having both obligate intracellular and free-living lifestyles. *Rickettsiales* can invade arthropod hosts (Darby et al. 2007), and in fact, the first marine *Rickettsiales* pathogen to be characterized, 'Candidatus Hepatobacter penaei', was originally isolated from *Penaeus vannamei*, and subsequently found in both an arthropod and crustacean as well as a commonly cultured shrimp in the western hemisphere (Nunan et al. 2013). *Caedibacter* spp. encompass another interesting group of bacteria elevated at the outfall. While this bacterium is not pathogenic, this genus comprises species that are unique endosymbionts of paramecia (Beier et al. 2002). Paramecia have been isolated from fresh, brackish, and seawater environments where they can prey on other bacteria and algae (Fokin et al. 1999, Fokin 2010). While little information on specific associations of paramecia within shrimp aquaculture exist, the outfall of the effluent canal was characterized by increased counts of heterotrophic bacteria, and may have hosted an environment favorable to the algae- and bacteria-consuming paramecia, which could account for the increase in *Caedibacter* spp. at the outfall.

Other bacteria elevated at the outfall included common seawater bacteria, including uncultured *Gammaproteobacteria*, uncultured *Flavobacteriaceae*, NS4 Marine Group, and uncultured *Cryomorphaceae*. The *Flavobacteriaceae* are a large family of marine bacteria, and this family includes some species that are associated with phosphorus and nitrogen enrichment as well as shrimp aquaculture (Maeda et al. 2002, Abell & Bowman 2005). NS4 Marine Group and uncultured *Cryomorphaceae* are also common bacteria in seawater, but less is known about their physiology (Alonso et al. 2007, Bowman 2014). While some bacteria were found at higher abundances at the outfall, others were significantly depleted at the outfall. For example, 3 OTUs belonging to the SAR116 clade of *Alphaproteobacteria* were depleted at the outfall. The SAR116 clade has been isolated from oligotrophic waters globally, such as the Sargasso Sea and the North Pacific Subtropical Gyre (Mullins et al. 1995, DeLong et al. 2006). While the Red Sea is naturally oligotrophic, the outfall site was characterized by nutrient enrichment. It is likely that the more eutrophic environment of the outfall was poorly suited for bacteria that thrive in oligotrophic waters, such as the SAR116 clade, but provided an ideal environment for other bacteria such as those from the class *Flavobacteria* and *Francisella* spp., as evidenced by the significant changes in sequence counts for those groups. In this way, the nutrient enrichment from the aquaculture facility appears to exert a bottom-up control on the microbial diversity near the effluent outfall.

Two limitations of this study included a lack of temporal resolution as well as site replication. These limitations prevented us from utilizing statistical approaches to compare the outfall nutrients and microbial cell abundances to the other sites. We were also not able to understand the consistency of the impact to the microbial communities over time, because one seawater sample was collected directly at the outfall. In the future, we recommend sampling farther up the effluent canal, or obtaining samples from the aquaculture pools to enhance our understanding of the microbial processes occurring within the effluent canal before the effluent reaches the Red Sea.

CONCLUSIONS

In our 2009 assessment of the impact of the National Aquaculture Group facility near Al Lith, Saudi Arabia on the microbial and biogeochemical composition of the eastern Red Sea, we found that the outfall site showed elevated concentrations of ammonium and phosphate, along with changes in the microbial communities, including the presence of the potential pathogens *Francisella* spp. and *Rickettsiales*. The impacts of the effluent were more localized than hypothesized, with elevated nutrients and statistically significant microbial community alterations restricted to only the outfall site and diminishing within 500 m, likely related to the highly oligotrophic nature of these waters. Aside from this seemingly localized impact, we also noticed a shift in the microbial communities with increasing distance from the outfall across the entire study area. The impact of the aquaculture facility on the entire coral reef ecosystem was not addressed here, but our data for the waters overlying the reef near the effluent canal suggests that this ecosystem may indeed be impacted by the aquaculture effluent, and these reefs and their water quality should be monitored in the future. This study expands on the existing knowledge of the impacts aquaculture imposes on oligotrophic coral reef ecosystems, although since 2009, the increased aquaculture output in the eastern Red Sea may have altered the observed impacts. As the global aquaculture industry continues to expand to meet the demands of population growth, studies of this nature will be critical in order to assess the impacts that this industry imposes on the surrounding marine environment, especially in oligotrophic areas harboring sensitive coral reef ecosystems.

Acknowledgements. We thank Whitney Bernstein, Kathryn Furby, Jesse Kneeland and the crew of the M/V 'Dream Island' for sample collections and logistical support. We thank Karen Selph of the University of Hawai'i School of Ocean and Earth Science and Technology (SOEST) flow cytometry facility for cell counts, Joe Jennings of Oregon State University for inorganic nutrient analysis, and C. Wright and the University of Illinois W. M. Keck Center for Comparative and Functional Genomics for sequencing. This research was supported by a Woods Hole Oceanographic Institution (WHOI) Ocean Life Institute postdoctoral scholar fellowship to A.A., the Semester at WHOI Program supporting C.B., and Award No. USA 00002 to K.H. made by King Abdullah University of Science and Technology (KAUST).

LITERATURE CITED

Abell GCJ, Bowman JP (2005) Ecological and biogeographic relationships of class Flavobacteria in the Southern Ocean. FEMS Microbiol Ecol 51:265–277

Alonso C, Warnecke F, Amann R, Pernthaler J (2007) High local and global diversity of Flavobacteria in marine plankton. Environ Microbiol 9:1253–1266

Apprill A, Hughen K, Mincer TJ (2013) Major similarities in the bacterial communities associated with lesioned and healthy *Fungiidae* corals. Environ Microbiol 15:2063–2072

Apprill A, McNally S, Parsons R, Weber L (2015) Minor revision to V4 region SSU rRNA 806R gene primer greatly increases detection of SAR11 bacterioplankton. Aquat Microb Ecol 75:129–137

Apprill A, Weber L, Santoro A (2016) Distinguishing between microbial habitats unravels ecological complexity in coral microbiomes. mSystems 1:e00143-16

Armstrong FAJ, Stearns CR, Strickland JDH (1967) The measurement of upwelling and subsequent biological process by means of the Technicon Autoanalyzer® and associated equipment. Deep-Sea Res Oceanogr Abstr 14:381–389

Barns SM, Grow CC, Okinaka RT, Keim P, Kuske CR (2005) Detection of diverse new *Francisella*-like bacteria in environmental samples. Appl Environ Microbiol 71:5494–5500

Beier CL, Horn M, Michel R, Schweikert M, Gortz HD, Wagner M (2002) The genus *Caedibacter* comprises endosymbionts of *Paramecium* spp. related to the *Rickettsiales* (Alphaproteobacteria) and to *Francisella tularensis* (Gammaproteobacteria). Appl Environ Microbiol 68:6043–6050

Bernhardt H, Wilhelms A (1967) The continuous determination of low-level iron, soluble phosphate and total phosphate with the Autoanalyzer. In: Scova NB (ed) Technicon symposia, Vol 1. Mediad, New York, NY, p 385–389

Berrada ZL, Telford SR (2010) Diversity of *Francisella* species in environmental samples from Martha's Vineyard, Massachusetts. Microb Ecol 59:277–283

Biao X, Zhuhong D, Xiaorong W (2004) Impact of the intensive shrimp farming on the water quality of the adjacent coastal creeks from Eastern China. Mar Pollut Bull 48:543–553

Birkbeck TH, Feist SW, Verner-Jeffreys DW (2011) *Francisella* infections in fish and shellfish. J Fish Dis 34:173–187

Bowman JP (2014) The family Cryomorphaceae. In: Rosenberg E, DeLong EF, Lory S, Stackebrandt E, Thompson F (eds) The prokaryotes. Springer, Berlin, p 539–550

Campbell AM, Fleisher J, Sinigalliano C, White JR, Lopez JV (2015) Dynamics of marine bacterial community diversity of the coastal waters of the reefs, inlets, and

wastewater outfalls of southeast Florida. MicrobiologyOpen 4:390–408

Chen X, Wang K, Guo A, Dong Z, Zhao Q, Qian J, Zhang D (2016) Excess phosphate loading shifts bacterioplankton community composition in oligotrophic coastal water microcosms over time. J Exp Mar Biol Ecol 483:139–146

Darby AC, Cho N, Fuxelius H, Westberg J, Andersson SGE (2007) Intracellular pathogens go extreme: genome evolution in the Rickettsiales. Trends Genet 23:511–520

DeLong EF (2009) The microbial ocean from genomes to biomes. Nature 459:200–206

DeLong EF, Preston CM, Mincer T, Rich V and others (2006) Community genomics among stratified microbial assemblages in the ocean's interior. Science 311:496–503

Dong Z, Wang K, Chen X, Zhu J, Hu C, Zhang D (2017) Temporal dynamics of bacterioplankton communities in response to excessive nitrate loading in oligotrophic coastal water. Mar Pollut Bull 114:656–663

dos Santos HF, Cury JC, do Carmo FL, dos Santos AL and others (2011) Mangrove bacterial diversity and the impact of oil contamination revealed by pyrosequencing: bacterial proxies for oil pollution. PLOS ONE 6:e16943

Duodu S, Colquhoun D (2010) Monitoring the survival of fish-pathogenic *Francisella* in water microcosms. FEMS Microbiol Ecol 74:534–541

Duodu S, Larsson P, Sjödin A, Forsman M, Colquhoun DJ (2012) The distribution of *Francisella*-like bacteria associated with coastal waters in Norway. Microb Ecol 64:370–377

Edgar RC, Haas BJ, Clemente JC, Quince C, Knight R (2011) UCHIME improves sensitivity and speed of chimera detection. Bioinformatics 27:2194–2200

Eren AM, Morrison HG, Lescault PJ, Reveillaud J, Vineis JH, Sogin ML (2015) Minimum entropy decomposition: unsupervised oligotyping for sensitive partitioning of high-throughput marker gene sequences. ISME J 9:968–979

Eshel G, Naik NH (1997) Climatological coastal jet collision, intermediate water formation, and the general circulation of the Red Sea. J Phys Oceanogr 27:1233–1257

FAO (Food and Agriculture Organization of the United Nations) (2016) The state of the world fisheries and aquaculture 2016. FAO, Rome

Fodelianakis S, Papageorgiou N, Pitta P, Kasapidis P, Karakassis I, Ladoukakis ED (2014) The pattern of change in the abundances of specific bacterioplankton groups is consistent across different nutrient-enriched habitats in Crete. Appl Environ Microbiol 80:3784–3792

Fokin SI (2010) *Paramecium* genus: biodiversity, some morphological features and the key to the main morphospecies discrimination. Protistology 6:227–235

Fokin S, Stoeck T, Schmidt H (1999) *Paramecium duboscqui* Chatton, Brachon, 1933. Distribution, ecology and taxonomy. Eur J Protistol 35:161–167

Furby KA, Apprill A, Cervino JM, Ossolinski JE, Hughen KA (2014) Incidence of lesions on *Fungiidae* corals in the eastern Red Sea is related to water temperature and coastal pollution. Mar Environ Res 98:29–38

Garren M, Smriga S, Azam F (2008) Gradients of coastal fish farm effluents and their effect on coral reef microbes. Environ Microbiol 10:2299–2312

Garren M, Raymundo L, Guest J, Harvell CD, Azam F (2009) Resilience of coral-associated bacterial communities exposed to fish farm effluent. PLOS ONE 4:e7319

Hambrey JB (2009) Global review and synthesis of reviews

of EIA and monitoring in aquaculture in four regions and for salmon aquaculture. In: Barg U, Soto D, Aguilar-Manjarrez J, Hambrey J, Castilla JL (eds). Environmental impact assessment and monitoring in aquaculture. FAO Fish Aquacult Tech Pap No. 527. FAO, Rome, p 3–57

Hernández E, Figueroa J, Iregui C (2009) Streptococcosis on a red tilapia, *Orechromis* sp., farm: a case study. J Fish Dis 32:247–252

Holmes RM, Aminot A, Kérouel R, Hooker BA, Peterson BJ (1999) A simple and precise method for measuring ammonium in marine and freshwater ecosystems. Can J Fish Aquat Sci 56:1801–1808

Jiang Z, Liao Y, Liu J, Shou L and others (2013) Effects of fish farming on phytoplankton community under the thermal stress caused by a power plant in a eutrophic, semi-enclosed bay: induce toxic dinoflagellate (*Prorocentrum minimum*) blooms in cold seasons. Mar Pollut Bull 76: 315–324

Jumars PA, Penry DL, Baross JA, Perry MJ, Frost BW (1989) Closing the microbial loop: dissolved carbon pathway to heterotrophic bacteria from incomplete ingestion, digestion and absorption in animals. Deep-Sea Res A 36: 483–495

Kamaishi T, Miwa S, Goto E, Matsuyama T, Oseko N (2010) Mass mortality of giant abalone *Haliotis gigantea* caused by a *Francisella* sp. bacterium. Dis Aquat Org 89:145–154

Kirchman DL, Elifantz H, Dittel AI, Malmstrom RR, Cottrell MT (2007) Standing stocks and activity of Archaea and Bacteria in the western Arctic Ocean. Limnol Oceanogr 52:495–507

Kozich JJ, Westcott SL, Baxter NT, Highlander SK, Schloss PD (2013) Development of a dual-index sequencing strategy and curation pipeline for analyzing amplicon sequence data on the MiSeq Illumina sequencing platform. Appl Environ Microbiol 79:5112–5120

La Rosa T, Mirto S, Mazzola A, Maugeri TL (2004) Benthic microbial indicators of fish farm impact in a coastal area of the Tyrrhenian Sea. Aquaculture 230:153–167

Love MI, Huber W, Anders S (2014) Moderated estimation of fold change and dispersion for RNA-seq data with DESeq2. Genome Biol 15:550

Loya Y, Lubinevsky H, Rosenfeld M, Kramarsky-Winter E (2004) Nutrient enrichment caused by *in situ* fish farms at Eilat, Red Sea is detrimental to coral reproduction. Mar Pollut Bull 49:344–353

Maeda M, Nogami K, Kanematsu S, Kotani Y (2002) Manipulation of microbial communities for improving the aquaculture environment. US-Japan Cooperative Program in Natural Resources (UJNR) Tech Rep 24:125–130

Mauel MJ, Soto E, Moralis JA, Hawke J (2007) A Piscirickettsiosis-like syndrome in cultured Nile tilapia in Latin America with *Francisella* spp. as the pathogenic agent. J Aquat Anim Health 19:27–34

McMurdie PJ, Holmes S (2013) phyloseq: an R package for reproducible interactive analysis and graphics of microbiome census data. PLOS ONE 8:e61217

Mon Chalil G (2015) The new investment wave into aquaculture in Middle East countries: opportunities and challenges. FAO, Rome. www.fao.org/in-action/globefish/fishery-information/resource-detail/en/c/338614/

Mullins TD, Britschgi TB, Krest RL, Giovannoni SJ (1995) Genetic comparisons reveal the same unknown bacterial lineages in Atlantic and Pacific bacterioplankton communities. Limnol Oceanogr 40:148–158

Naylor RL, Goldburg RJ, Primavera JH, Kautsky N and others (2000) Effect of aquaculture on world fish supplies. Nature 405:1017–1024

Nogales B, Lanfranconi MP, Piña-Villalonga JM, Bosch R (2011) Anthropogenic perturbations in marine microbial communities. FEMS Microbiol Rev 35:275–298

Nunan LM, Pantoja CR, Gomez-Jimenez S, Lightner DV (2013) '*Candidatus* Hepatobacter penaei,' an intracellular pathogenic enteric bacterium in the hepatopancreas of the marine shrimp *Penaeus vannamei* (Crustacea: Decapoda). Appl Environ Microbiol 79:1407–1409

Olsen AB, Mikalsen J, Rode M, Alfjorden A and others (2006) A novel systemic granulomatous inflammatory disease in farmed Atlantic cod, *Gadus morhua* L., associated with a bacterium belonging to the genus *Francisella*. J Fish Dis 29:307–311

Pérez N, Johnson R, Sen B, Ramakrishnan G (2016) Two parallel pathways for ferric and ferrous iron acquisition support growth and virulence of the intracellular pathogen *Francisella tularensis* Schu S4. MicrobiologyOpen 5: 453–468

Santoro AE, Casciotti KL, Francis CA (2010) Activity, abundance and diversity of nitrifying archaea and bacteria in the central California Current. Environ Microbiol 12: 1989–2006

Schlitzer R (2002) Interactive analysis and visualization of geoscience data with Ocean Data View. Comput Geosci 28:1211–1218

Schloss PD, Westcott SL, Ryabin T, Hall JR and others (2009) Introducing mothur: open-source, platform-independent, community-supported software for describing and comparing microbial communities. Appl Environ Microbiol 75:7537–7541

Sjödin A, Svensson K, Öhrman C, Ahlinder J and others (2012) Genome characterisation of the genus *Francisella* reveals insight into similar evolutionary paths in pathogens of mammals and fish. BMC Genomics 13:268

Sousa OV, Macrae A, Menezes FGR, Gomes NCM, Vieira RHSF, Mendonça-Hagler LCS (2006) The impact of shrimp farming effluent on bacterial communities in mangrove waters, Ceará, Brazil. Mar Pollut Bull 52: 1725–1734

USEPA (US Environmental Protection Agency) (1983) Nitrogen, ammonia. Method 350.1 (colorimetric, automated, phenate). In: Methods for chemical analysis of water and wastes. USEPA, Cincinnati, OH, p 350.1-1 – 350.1-4

Verhoeven AB, Durham-colleran MW, Pierson T, Boswell WT, Van Hoek ML (2010) *Francisella philomiragia* biofilm formation and interaction with the aquatic protist *Acanthamoeba castellanii*. Biol Bull (Woods Hole) 219: 178–188

Villanueva RD, Yap HT, Montaño MNE (2005) Survivorship of coral juveniles in a fish farm environment. Mar Pollut Bull 51:580–589

Wei C, Zeng Y, Tang K, Jiao N (2009) Comparison of bacterioplankton communities in three mariculture ponds farming different commercial animals in subtropical Chinese coast. Hydrobiologia 632:107–126

Wickham H (2009) Ggplot2: elegant graphics for data analysis. Springer, New York, NY

Xiong J, Chen H, Hu C, Ye X, Kong D, Zhang D (2015) Evidence of bacterioplankton community adaptation in response to long-term mariculture disturbance. Sci Rep 5:15274

Zhou Q, Wang L, Chen J, Wang R and others (2014) Development and evaluation of a real-time fluorogenic loop-mediated isothermal amplification assay integrated on a microfluidic disc chip (on-chip LAMP) for rapid and simultaneous detection of ten pathogenic bacteria in aquatic animals. J Microbiol Methods 104:26–35

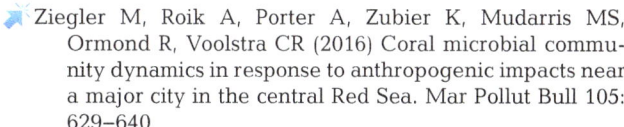 Ziegler M, Roik A, Porter A, Zubier K, Mudarris MS, Ormond R, Voolstra CR (2016) Coral microbial community dynamics in response to anthropogenic impacts near a major city in the central Red Sea. Mar Pollut Bull 105: 629–640

APPENDIX

Table A1. Summary of sampling sites, sequence reads, and alpha diversity metrics. MED: minimum entropy decomposition, na: not applicable, OTU: operational taxonomic unit

Sample site	Distance to outfall (km)	No. of reads	Sub-sampled reads	No. of reads after MED	No. of OTUs	Chao1	Inverse Simpson
8	0	23149	10200	9059	291	332	17.5
7	0.5 W	27653	10200	9132	317	352	12.3
6	1.5 S	15503	10200	8853	240	262	19.3
9	1.5 N	34380	10200	9473	325	353	6.92
5	2.5 S	30509	10200	9538	295	358	5.35
12	2.5 SW	23761	10200	9161	294	353	9.23
10	3 NW	18933	10200	9369	341	375	10.7
17	4.5 SW	19276	10200	9025	351	370	25.7
11	5 NW	27524	10200	9406	338	358	6.91
19	5 N	16656	10200	9075	344	358	20.9
4	5.3 S	15755	10200	9327	292	326	6.80
25	9 S	18132	10200	9205	225	250	3.58
16	9.5 SW	23822	10200	9115	326	342	15.2
3	10 S	27600	10200	9336	332	360	18.1
18	10.1 NW	28537	10200	9035	340	369	17.1
20[a]	10.7 N	266	na	na	na	na	na
14	14.5 SW	24107	10200	9258	301	317	18.4
2	15 S	33282	10200	9000	342	357	25.9
22	15.2 NW	34679	10200	8908	337	367	14.3
13	20 SW	22209	10200	9157	310	329	31.0
23	20.3 NW	55182	10200	9032	337	361	26.1
24	20.5 S	10200	10200	9365	272	297	4.68
1	21.5 S	24170	10200	8894	353	370	16.3
21	21.7 N	45407	10200	8881	357	375	13.8

[a]If the number of sequence reads was <10200, the site was not included in microbial community analyses

14

Effects of atmospheric cold fronts on stratification and water quality of a tropical reservoir: implications for aquaculture

Carlos A. S. Araújo[1,6,*], Fernanda G. Sampaio[2], Enner Alcântara[3], Marcelo P. Curtarelli[4], Igor Ogashawara[5], José L. Stech[1]

[1]Remote Sensing Division, National Institute for Space Research, 12227-010 São José dos Campos, SP, Brazil

[2]Embrapa Environment, Brazilian Agricultural Research Corporation, 13820-000 Jaguariúna, SP, Brazil

[3]Department of Environmental Engineering, São Paulo State University, 12247-004 São José dos Campos, SP, Brazil

[4]Green Economy Center, Reference Center Foundation for Innovative Technologies, 88040-970 Florianópolis, SC, Brazil

[5]Department of Earth Sciences, Indiana University–Purdue University Indianapolis, Indianapolis, IN 46202, USA

[6]*Present address:* Département de Biologie, Chimie et Géographie, Université du Québec à Rimouski, Rimouski, QC G5L 3A1, Canada

ABSTRACT: Stratification and mixing patterns of a water body are influenced by the variability of atmospheric systems, which can also modify their biogeochemical properties. The primary goal of this study was to analyze the effect of atmospheric cold fronts (CFs) on thermal stratification and water quality parameters in 4 embayments of the Furnas Hydroelectric Reservoir (FHR) (southeastern Brazil), a warm monomictic water body. A secondary goal was to evaluate the implications of this effect on the aquaculture of Nile tilapia *Oreochromis niloticus* (L.) in net cages. A 2 yr dataset of meteorological and water quality parameters was used to compute heat flux balance components and buoyancy frequency. These parameters were used to evaluate the influence of CFs on FHR water column stability and water quality. It was observed that the passing of CFs increased net heat loss and wind velocity, resulting in a partial mixture of surface waters with deeper layers. These changes in the physical structure of the water column altered the diel cycle of water temperature, led to a slight decrease in dissolved oxygen concentrations and pH values during the stratification period, and contributed to the increase in dilution power at the fish cultivation sites. However, following CF passages, no significant changes were observed in water quality parameters that influenced Nile tilapia cultivation in the FHR. Nevertheless, the understanding of meteorological systems and their influence on the physical and biogeochemical properties of an aquatic system is important for optimal management of aquaculture activities.

KEY WORDS: Tropical reservoir · Cold fronts · Water quality parameters · Aquaculture · Nile tilapia

INTRODUCTION

Worldwide consumption of fish is increasing, and aquaculture has been identified as the main supplier for this demand, considering that wild fisheries captures have not increased in the past 10 yr (FAO 2016).

In Brazil, the increase in aquaculture production follows the worldwide trend. The production of freshwater aquaculture grew at a mean rate of approximately 27 % yr^{-1} between 2009 and 2011 (MPA 2013). Brazilian aquaculture activities produced around 483 241 tons of fish in 2015 (IBGE 2016). One of the

*Corresponding author:
carlosalberto.sampaiodearaujo@uqar.ca

major products is the exotic Nile tilapia *Oreochromis niloticus* (L.), making up 45.4 % of total Brazilian finfish aquaculture production in 2015 (IBGE 2016).

To increase the freshwater aquaculture production for food supply, aquatic systems such as lakes and reservoirs have been hosting fish farming activities. Brazil has >250 reservoirs that could host such activities. However, these activities may lead to environmental problems such as eutrophication and changes in ecological succession. Although the increase in aquaculture production is important to assure food security, an environmental monitoring protocol is necessary to ensure environmental and human health as well as support policymakers' decisions. An analysis of the possible impacts of aquaculture on these reservoirs is required for the management of aquaculture activities.

Changes in the physical and biogeochemical properties of a water column, as well the associated implications to aquatic life caused by the succession of different meteorological systems over tropical reservoirs, have been studied in different regions of Brazil (Tundisi et al. 2004, 2010, Curtarelli et al. 2013, 2014a,b). These changes occur mainly because of variation in the heat budget of the aquatic system, which is primarily modulated by seasonal cycles of incoming shortwave radiation and other meteorological variables along with mesoscale atmospheric disturbances. Among the meteorological systems over South America, cold front (CF) incursions are one of the most recurrent atmospheric systems. CFs influence atmospheric circulation, precipitation, and temperature regime (Garreaud 2000). Moreover, they are also responsible for quick variations in the heat budget of aquatic systems (Liu et al. 2009, Alcântara et al. 2010, Curtarelli et al. 2013).

Several studies have shown a relationship between CF passages and changes in the physical properties of the water column at continental shelves (Stech & Lorenzzetti 1992) and reservoirs (Tundisi et al. 2004, Alcântara 2012, Curtarelli et al. 2013, 2014a). These changes are important for the vertical distribution of chemical constituents, which are crucial factors to estimate primary productivity in aquatic systems (Ganf 1974, Reynolds 1992, Serra et al. 2007, Vidal et al. 2010). Thus, processes that alter water column stability such as stratification and mixing are important for understanding phytoplankton population dynamics, diversity, and succession (Reynolds 1992, Calijuri et al. 2002, Becker et al. 2009a,b).

CF effects on water column stability and phytoplankton succession have been studied in a Brazilian tropical reservoir by Tundisi et al. (2004). They showed that alterations in the periods of vertical stratification and mixing were related to the passage and dissipation of CFs. Based on these observations, a framework was proposed for the management of lakes and reservoirs in southeastern Brazil. This framework relates the instability in the water column caused by the mixing process during a CF passage to the dominance of diatoms and green algae. Water column stability, which is caused by the stratification process in the absence of CFs, is related to cyanobacterial dominance periods (Tundisi et al. 2010).

The succession of mixing and stratification processes, often related to meteorological systems, is an important driver for environmental parameters such as phytoplankton abundance, water transparency, and dissolved oxygen (DO) levels (van Rijssel et al. 2016). Even still, the use of these stability processes in understanding their relation to higher-ranking levels of the trophic chain has not yet been totally explored. In addition, little or no information is available regarding CFs' effects on water quality parameters (WQPs) in aquaculture fields.

Since atmospheric radiation and air temperature are controlling factors for water column temperature, it is important to understand the role of CF passages over aquaculture fields. Such water column changes can affect the productivity and growth of fisheries (Kapetsky 2000). Therefore, the identification of water column changes can equip aquaculture management toward improving production and preventing environmental degradation.

From an aquaculture perspective, the maintenance of an ideal range for parameters such as temperature, DO, and pH are fundamental. Variations in WQPs affect fish by triggering their stress disruption, which can increase susceptibility to disease and cause a decrease in production. One example is the relationship between water temperature and DO consumption, which shows that, in warmer waters, DO consumption is higher (Boyd & Pillai 1985). It has also been found that low DO levels cause changes in the behavior, physiology, and morphology of fish (Pollock et al. 2007), and temperature also affects the chemical and biological processes of aquatic organisms' metabolism (Boyd & Pillai 1985). Changes in water temperature modify fish metabolism, resulting in a high-energy demand for maintenance and less energy available for growth (Jobling 1994).

We hypothesized that changes in wind pattern, heat flux balance, and the stratification-to-mixing conditions caused by a CF passage will affect WQPs to be outside the optimal range of environmental conditions for aquaculture. The aim of this work was to understand the role of CF passages over inland water fish farms located in a region of the Furnas Hydroelectric Reservoir (FHR; see Fig. 1), Guapé, Brazil. The specific objectives were to (1) identify the seasonal patterns of stratification and mixing processes in the FHR, (2) identify the frequency of the passage of CFs over the FHR, (3) quantify the changes in water temperature and in heat flux balance based on normal meteorological conditions, (4) understand how the water column stability changes with CF passages, and (5) compare changes in critical WQPs to the optimal ranges for Nile tilapia farming.

MATERIALS AND METHODS

Study site

FHR is in the middle course of the Grande River, Minas Gerais State, Brazil (Fig. 1), and it is formed mainly by 2 rivers: the Grande River, flowing in an east–west direction, and the Sapucaí River, flowing in a southeast–northwest direction. The flooded area and the storage capacity at its maximum level at 768 m above mean sea level (a.m.s.l.) are 1440 km^2 and 22.95 billion m^3 of water, respectively. FHR's useful volume is 17.22 billion m^3 of water and its minimum operational level is at 750 m a.m.s.l. (Furnas 2016). Based on its bathymetry, FHR reaches a maximum depth of 90 m and has a mean depth of 16 m.

FHR is surrounded by 23 municipalities, 19 of which produced 4683–5130 tons yr^{-1} of Nile tilapia

from 2013 to 2015. In Guapé municipality, Nile tilapia production was 2430, 1500, and 790 tons yr^{-1} for the years 2013, 2014, and 2015, respectively (IBGE 2016). This intense decrease in fish production was related to a constant and pronounced reservoir level fluctuation and the difficulty of regularizing the aquaculture activity in the area. Fish diseases were not a significant cause for this decrease, since reported cases were usually caused by the transport of fingerlings. However, occasional occurrences of *Streptococcus* and columnaris were observed in fish under production (Cardoso et al. 2013).

Selected areas for this study were 4 small embayments (Fig. 1C), located near the municipality of Guapé where there was an intensive cage cultivation of Nile tilapia. At the time of this study, there were >200 Nile tilapia producers in the FHR, though we focused on 11 fish farms (with a total cage volume of about 817–818 m^3), which have been active for the last 7 yr. They produce around 1000 tons yr^{-1} in 1.4 cycles yr^{-1}, with an individual commercial size between 0.85 and 1.00 kg. Fish farms have newly hatched, young, and adult Nile tilapia, even though most of the producers (52.4%) do not adopt any management practices, such as measuring fish weight and length (Cardoso et al. 2013).

Moored platforms

Six platforms were anchored within the study area (Fig. 1C). One was equipped with meteorological

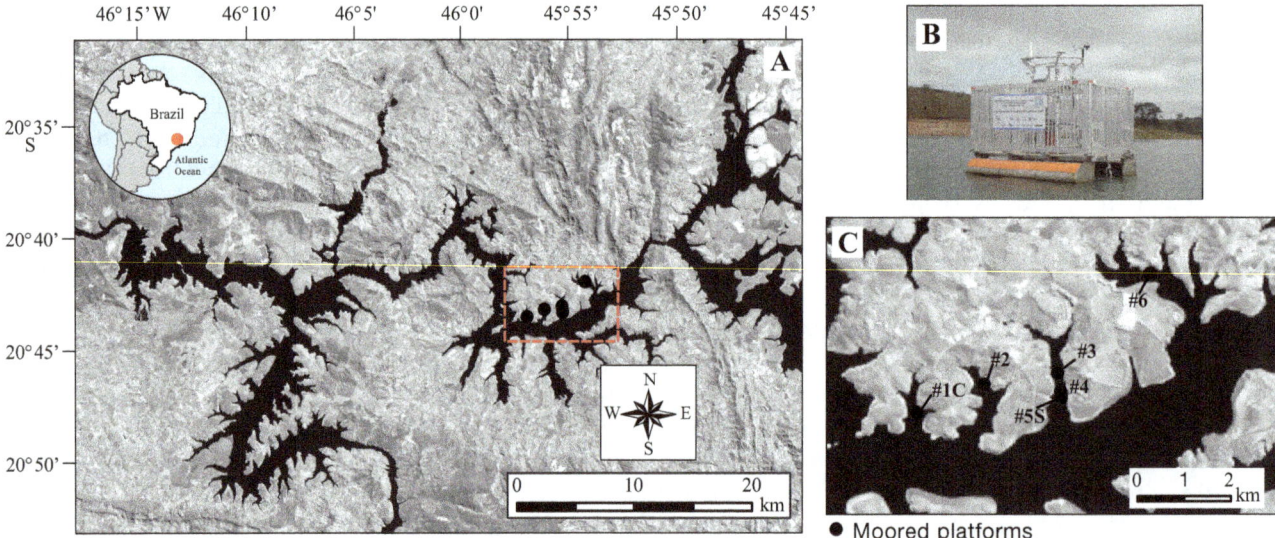

Fig. 1. (A) Furnas Hydroelectric Reservoir (red dot in inset) in Brazil and location of monitoring platforms (yellow dots). (B) Integrated System for Environmental Monitoring (SIMA) at platform #5S. (C) Zoom-in of red dashed box in (A) with the 6 moored platforms shown. Images in (A,C) from Landsat 8 OLI, near-infrared band, August 3, 2014

and water quality sensors (#5S; Fig. 1B), while the other 5 have only water quality sensors (#1C, #2, #3, #4, and #6). Platform #1C was placed in an area free of Nile tilapia cages to serve as a control site, while #2 was in an embayment in which a fish farm trader association produces around 150 tons yr^{-1}. Platforms #3, #4, and #5S were placed in an embayment where there were 9 producers totaling a production of around 500 tons yr^{-1}. Finally, platform #6 was in an embayment with 1 fish farm which has a yearly production of 550 tons.

At #5S, using a system called SIMA (an acronym, in Portuguese, for Integrated System for Environmental Monitoring; see Stech et al. 2006), the following meteorological parameters were measured at a height of 3 m: air temperature (T_a), relative humidity (R_h), atmospheric pressure (P_{atm}), wind speed (U_z) and direction, and shortwave radiation (S_w). From the SIMA (#5S) and SIMA-Aquaculture (#1C, #2, #3, #4, and #6), multiparameter sondes were installed, allowing measurement of water temperature (T_w), conductivity, pH, turbidity, and DO concentration at a depth of 1.5 m. Thermistor chains, which provided water temperature at different depths throughout the water column, were also installed in all systems. Each chain had 13 thermistors, distributed 0.25 m apart in the first meter, then 1 m apart in the next 5 m, and finally every 2 m until 1 m above the bottom of the water column. A distinct difference between SIMA and SIMA-Aquaculture is that the first can transmit hourly-acquired data telemetrically in near real-time (excluding temperature from the thermistor chain), while the second can only store the data until a backup is conducted. Telemetric data was only used when downloaded data was not available. Table A1 in the Appendix describes the instruments used.

All instruments were configured to store data at regular intervals of 10 min. Additionally, all wind direction data were corrected for magnetic declination for the FHR location using NOAA's magnetic field calculator (NOAA 2016).

SIMA and SIMA-Aquaculture operated from August 2013 to September 2015. However, due to battery recharging failure, from March 16 to May 10, 2014, no meteorological and WQP data were collected for #5S. Furthermore, from December 10, 2014 until the end of the entire period, data from SIMA was not backed up (on a 10 min basis), and therefore only the telemetrically transmitted data (hourly based) were used.

Maintenance on the deployed equipment was conducted approximately every 2 mo. All sensors from the multiparameter sondes were submitted to a thorough calibration procedure according to the manufacturer's instructions. In case of unsuccessful calibration, data were considered unreliable, and we also disqualified any data collected when the measurement was taken outside of the calibration's lifetime, which is sensor-dependent. For the thermistor chains, data were discarded in 2 cases: (1) if the end of the chain's rope reached the bottom, which mainly happened during the drought of 2014, and (2) if the chain's rope was wrapped into the anchor's cable. Although these imposed restrictions in the usage of the water column temperature data, this was not critical for our analysis since temperature profiles among platforms were highly correlated. For example, the relationship between the temperature time series for a depth of 0.5 m collected by platforms #1C and #6 in February 2014 showed a determination coefficient (R^2) of 0.85 and a root mean square error (RMSE) of 0.27°C. Thus, for an overview of the temporal variation in the thermal structure of the water column, we used data from platforms #1C, #5S, and #6, to improve the temporal coverage.

Rainfall and water level data

Rainfall and reservoir water level data were computed for 2 periods: historical (or climatological) and the studied period. Monthly-accumulated precipitation data were acquired from a station located near the reservoir (municipality of Machado; 21.68° S, 45.94° W), downloaded from the National Institute of Meteorology database (INMET 2016). Historical precipitation data from 1961 to 2015 were used to compute monthly means. Daily reservoir water level data, in meters (a.m.s.l.), were acquired for the studied period. The historical reservoir level was computed from 1963 to 2015, with the data from 1963 to 2001 acquired on a monthly basis (maximum and minimum levels), and data from 2002 to 2015 on a daily basis. To compare both periods, monthly means were linearly interpolated to daily means to make it comparable to the data for the analyzed period.

Surface heat flux balance and buoyancy frequency

If we exclude the inflow and outflow contributions to the hydrodynamics of a water column, water temperature in the mixed layer and vertical stratification of a reservoir are controlled essentially by a combination of heat and wind mixing (Imberger & Hamblin

1982). While the surface heat flux balance (S, W m^{-2}) acting over an aquatic system is one of the most important processes controlling the mixed layer dynamics (Henderson-Sellers 1986), the stability of the water column, or the buoyancy frequency (N, s^{-1}), has been widely used as a metric for quantifying the stratification force (Marcé et al. 2000, Calijuri et al. 2002, Rao et al. 2008, Liu et al. 2012).

In this study, the meteorological variables measured by the SIMA, T_a, T_w, R_h, P_{atm}, and U_z, were used as inputs for the calculation of the components of S, on a 10 min basis. S was computed following Eq. (1):

$$S = S_w (1 - \alpha) + L_w - (E + H) \qquad (1)$$

where S_w is shortwave radiation (directly measured by the SIMA), α is the albedo of the shortwave radiation (calculated as in Martin & McCutcheon 1999), L_w is the net longwave radiation, and E and H are the turbulent fluxes of latent and sensible heat fluxes, respectively. Positive values of S, or net heat gain, indicate warming of the surface layer, which leads to more stability, while negative values, or net heat loss, indicate cooling of the surface layer and can promote vertical mixing. A complete description of the procedures to calculate the heat budget is found in Lorenzzetti et al. (2015).

Both emitted ($L_{w(emi)}$) and incident ($L_{w(inc)}$) longwave radiation ($L_w = L_{w(inc)} - L_{w(emi)}$) were calculated following equations in Henderson-Sellers (1986). E and H were estimated using the bulk aerodynamic transfer method, considering the stability of the atmospheric boundary layer (Amorocho & DeVries 1980, Imberger & Patterson 1990, Verburg & Antenucci 2010).

Buoyancy frequency (N) was calculated based on data from the thermistors chains, interpolated for regular intervals of 1 m, as $N = [(-g/\bar{\rho})(\partial\rho/\partial z)]^{(1/2)}$, where g is gravitational acceleration (9.8 m s^{-2}), $\bar{\rho}$ (kg m^{-3}) is mean density of the water column, and $\partial\rho/\partial z$ (kg m^{-3} m^{-1}) is the density gradient. Water density was calculated as in Martin & McCutcheon (1999), and the effects of salinity on density were considered negligible, since conductivity values were <100 µS cm^{-1}. Positive values of N indicate stable conditions, while negative and near-zero values indicate unstable and neutral conditions, respectively.

CF identification

Meteorological systems were identified as atmospheric disturbances that change the normal pattern of meteorological variables near the surface, such as at- mospheric pressure, air temperature and relative humidity, wind speed and direction, and the radiative components of the heat balance. An example of a meteorological system identification based on meteorological variables and satellite images is shown in Fig. 2.

Wind direction and air temperature were 2 key parameters for the identification of a CF passage (Fig. 2A,B). It was observed that the normal pattern of wind direction, from northerly to easterly, changes to southerly-southwesterly during a CF passage. Likewise, air temperature with a normal diel variation of ~10°C was reduced to ~3°C.

Once changes in the normal pattern of the meteorological variables were identified, we used Geostationary Operational Environmental Satellite (GOES) images (Fig. 2C–E), previously processed by the Center for Weather Forecasting and Climate Research (CPTEC) of the National Institute of Space Research (CPTEC 2016), to confirm the CF passage, as shown in previous studies (Stech & Lorenzzetti 1992, Tundisi et al. 2004). To better evaluate this identification, we also used monthly bulletins of climatic monitoring and analysis, provided by CPTEC (Climanálise 2016), which contain detailed information on atmospheric disturbances over Brazil.

Critical conditions for Nile tilapia aquaculture

Fish under production should be maintained at ideal environmental conditions for optimal development and growth. Thus, the collected WQP values in FHR were compared with ranges considered ideal for Nile tilapia farming. However, optimal ranges of WQPs vary according to the developmental phase of the fish. Since in the FHR, there were different developmental phases of Nile tilapia, we only considered optimal ranges described for young Nile tilapia, until more information becomes available in the literature.

While the optimum water temperature for Nile tilapia growth has been reported to be ~30°C (Azaza et al. 2008), 27–30°C is also considered to be their thermal comfort zone (Boyd 1990, Ostrensky & Boeger 1998, El-Sayed 2006). Outside of this range, fish are more apt to suffer from thermal stress and to develop diseases (Marcusso et al. 2015).

The recommended ranges of pH vary from 6.0–9.0 (Popma & Masser 1999), 6.5–9.0 (Nandlal & Pickering 2004), and 6.5–8.0 (Özdemir et al. 2014). Although these are small variations for optimal pH, it was observed that the range 6.0–9.0 is important for proper fish development. Any pH value below or above this range will trigger biochemical and physi-

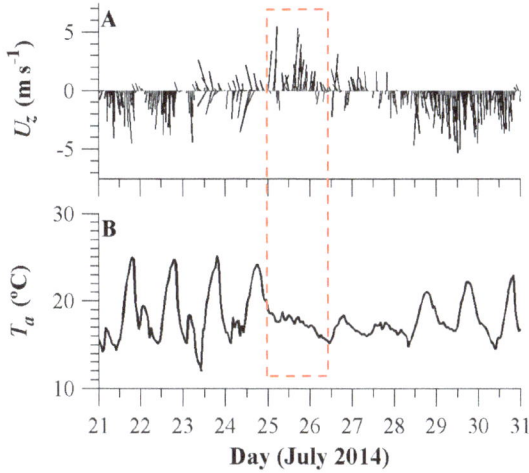

Fig. 2. Example of a cold front (CF) identified based on meteorological variables and satellite images. (A) The predominantly northerly wind direction changes to southerly (U_z: wind speed) and (B) the normal diel variability of air temperature (T_a) is altered on July 25, 2014 (red dashed box) during a CF passage. A sequence of GOES images, on July (C) 23, (D) 24, and (E) 25, 2014 at 00:00 h, showing the displacement of a CF over the Furnas Hydroelectric Reservoir (red square)

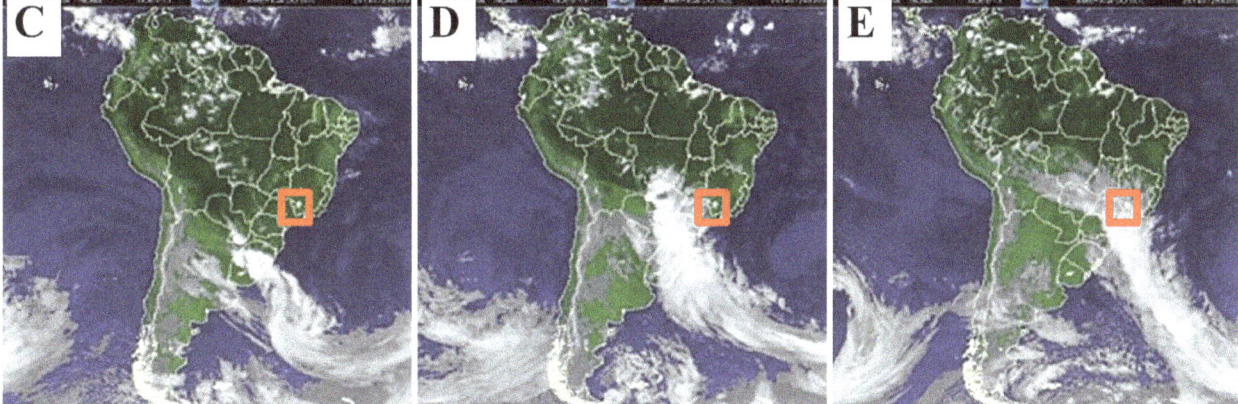

ological mechanisms to prevent death (Sampaio et al. 2010).

Nile tilapia can be considered a resistant fish in terms of adverse environmental conditions. They can survive at DO concentrations as low as 1.0 mg l^{-1} (Abdel-Tawwab et al. 2015), so DO is not a limiting factor for their development. However, fish growth and feed efficiency are affected by low DO concentrations (Bergheim et al. 2006, Duan et al. 2011, Abdel-Tawwab et al. 2015), and contaminant-induced toxicity could increase at lower DO levels (Sampaio et al. 2012). Therefore, even if low DO concentration is not a significant threat to Nile tilapia, adequate DO levels should be maintained to satisfy functions responsible for improving fish performance and health (Abdel-Tawwab et al. 2015). Experiments exposing Nile tilapia to different DO concentrations showed that performance is better when concentrations vary from 6.0 to 6.5 mg l^{-1} than at lower levels (Abdel-Tawwab et al. 2015).

It is important to highlight that producers in the monitored areas of the FHR do not use any fish management procedures during the production cycle. Because of the lack of management, there is no record of fish development. However, during the analyzed period, a survey of fish physiological parameters was conducted, and results showed that fish under production were healthy.

RESULTS

FHR water level, precipitation, and seasonal characterization of water column temperature and stability

Monthly-accumulated precipitation and the FHR's water level data are presented in Fig. 3A for both the historical and analyzed periods. Time series of water column temperature and buoyancy frequency are shown in Fig. 3B,C (depth isopleths are in parallel to FHR water level). The analyzed period started on August 9, 2013 and ended on September 24, 2015 (Fig. 3).

The water level decreased remarkably, comparing the historical and analyzed periods. While the seasonal historical fluctuation in the water level ranges to about 4.5 m, the fluctuation within the analyzed period ranged to about 11 m. The below-normal water

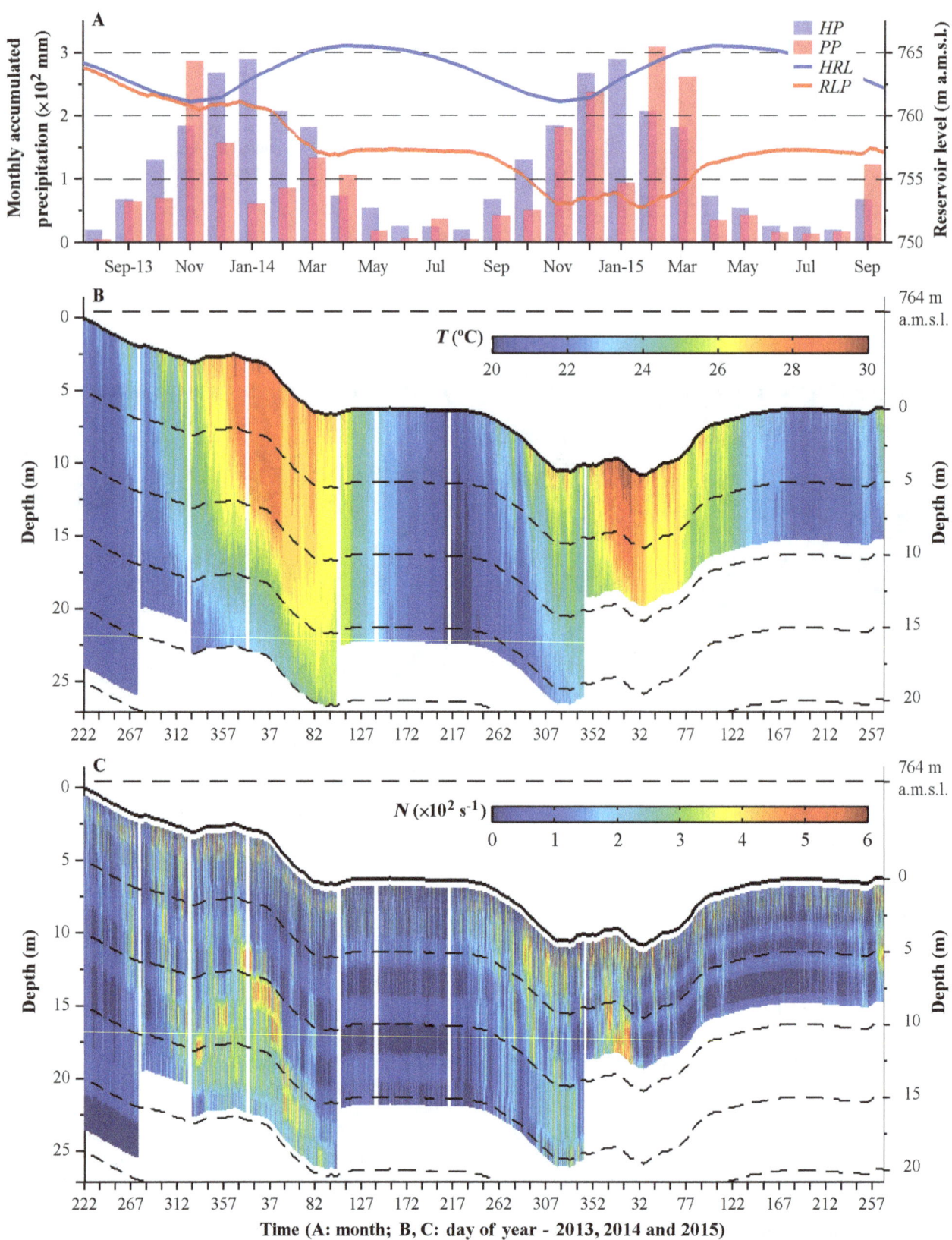

Fig. 3. Time series of (A) monthly mean accumulated precipitation (*HP*: averaged historical precipitation for 1961–2015, *PP*: precipitation in the analyzed period) and daily water level (*HRL*: averaged historical reservoir level for 1963–2015, *RLP*: reservoir level in the analyzed period); (B) water column temperature (*T*); and (C) buoyancy frequency (*N*). a.m.s.l.: above mean sea level

level is strongly related to a drought period, which can also be observed in Fig. 3A. This drought occurred in the austral summers of 2013–2014 and 2014–2015 and affected FHR's water level as well as many other reservoirs in the southeast region of Brazil. This drought caused a severe problem in water supply management, especially in São Paulo—the largest city in Brazil (Escobar 2015, Meganck et al. 2015, Nobre et al. 2016). The cause of this 2 yr drought has been related to large-scale and regional atmospheric circulation, as well as sea surface temperature anomalies from the South Atlantic and Pacific oceans (Seth et al. 2015, Coelho et al. 2016). This can be better illustrated by the precipitation data, which showed that the difference between monthly-accumulated precipitation, from August 2013 to January 2015, and the averaged historical precipitation was 850 mm.

Time series of the vertical thermal structure (Fig. 3B) revealed a strong seasonal pattern, with a well-marked period of stratification during the austral summer (approximately from day of the year [DOY] 300 to 100) and mixing during the austral winter (approximately from DOY 120 to 250). The buoyancy frequency data (Fig. 3C) confirmed these stratification-mixing patterns, with values reaching 0.07 s^{-1} in the stratification period. Based on this information and following Lewis' (1983) classification, the observed stratification-mixing pattern characterizes the FHR as a warm monomictic water body.

Water surface temperatures ranged from 31°C in the austral summer to 22°C in the austral winter, and at a depth of ~16 m, ranged from 26°C in summer to 20°C in winter. N values had a strong diel signal in the first 5 m, and they were observed in the entire time series. The seasonal behavior of N shows high values in austral summers of 2013–2014 and 2014–2015, varying from 15 to 10 m depth. Although N values >0.05 s^{-1} observed in the FHR are usually considered to be characteristic of a strong stratification, we observed neither a well-defined metalimnion nor a well-developed hypolimnion. Instead, a temperature gradient was present from the beginning of the metalimniom until the deepest parts of the water column.

WQPs in stratification and mixing periods

For practical purposes, an N value of 0.03 s^{-1} was established as a threshold for the identification of stratification and mixing periods. Thus, hereinafter, reference to stratification and mixing pertains to the periods from November to March (DOY 305 to 90) and May to September (DOY 121 to 273), respec-

tively. In these periods, water chemical property dynamics were driven partially by stratification and mixing regimes. Thus, time series of water pH, DO concentration, and turbidity were separated into these 2 periods (Fig. 4). Minimum and maximum values and mean ($\bar{\mu}$) and standard deviation (σ) for the same time series are shown in Table 1.

Mean water pH values were, in general, slightly higher when the water column was mixed rather than stratified. Comparing #1C with the other platforms, mean values of pH were lower in those where fish cultivation was present. Additionally, higher ranges (Min–Max) and higher values of σ were also observed at the platforms located in the aquaculture areas. These results indicate a distinction of processes acting naturally in the reservoir (#1C) in comparison with the areas where Nile tilapia cages were present. During the water column stratification period, pH >8 were observed for #1C and #3, while in the mixing period this was observed for #1C and #5S, with the mean pH also >8 on #1C.

DO concentrations during the stratification regime were lower than those of the mixing regime for all platforms. This is partially explained by higher DO saturation values found during the mixing regime, which had a mean temperature of ~22°C, while the mean temperature during the stratification regime was of ~28°C. Mean values of DO concentrations were lower at platforms #2, #3, #4, #5S, and #6 than at #1C. Following the pattern of the pH values, higher ranges and σ were observed in the cultivation areas (except σ for #2 in the mixing regime). Values <4 mg l^{-1} for DO concentrations were found only in #3 from both periods.

Specific conductance at 25°C values did not significantly vary between stratification-mixing periods or between platforms. The highest range of specific conductance was observed in #6, from 33 to 51 µS cm^{-1}, while for #1C it ranged from 35 to 45 µS cm^{-1}. Turbidity data presented several anomalous values, or spikes (e.g. Fig. 4K,L), which were associated with noise in the measurements. Therefore, the values obtained from all platforms were, in general, <10 NTU (excluding the spikes). The platforms installed in the cultivation area presented slightly higher values compared to the control site.

CF effects on mixing and stratification patterns and WQPs

Although several types of meteorological systems can affect water column stability, such as instability

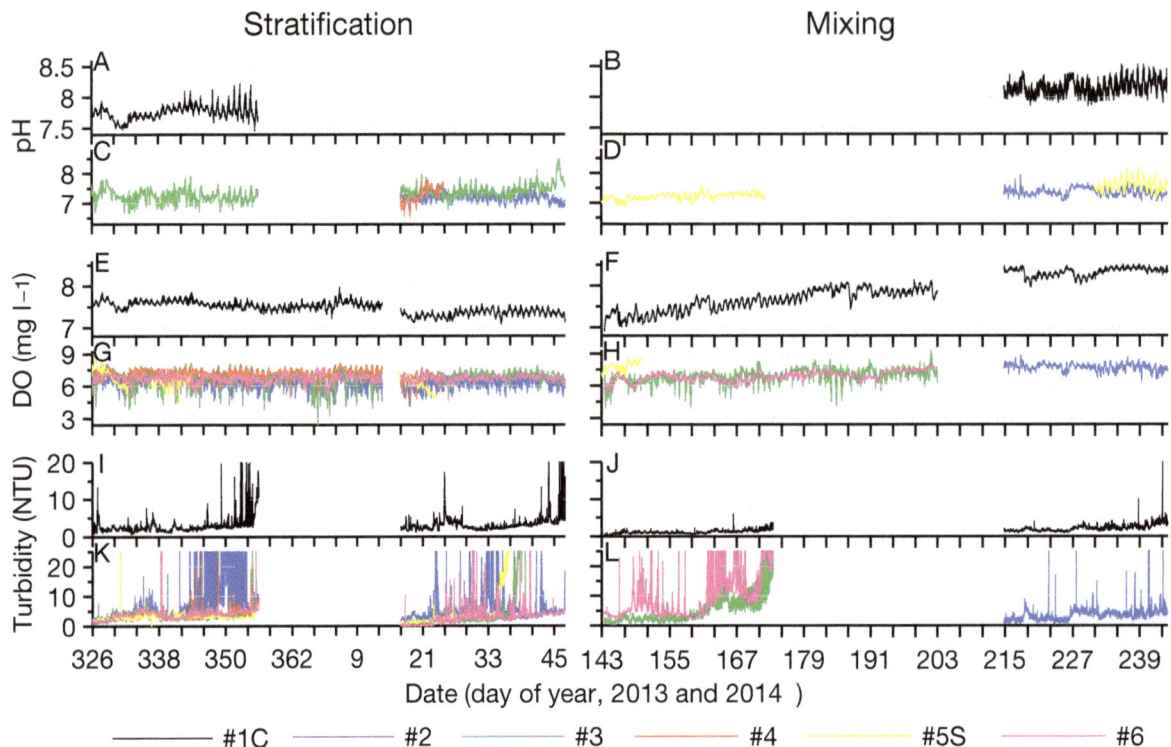

Fig. 4. Time series of (A–D) water pH, (E–H) dissolved oxygen (DO), and (I–L) turbidity for the periods of stratification (left) and mixing (right) regimes at Furnas Hydroelectric Reservoir, for (A,B,E,F,I,J) platform #1C and (C,D,G,H,K,L) platforms #2, #3, #4, #5S, and #6

lines and conflict between masses (as shown by Ogashawara et al. 2014, for Guarapiranga reservoir, also in southeast Brazil), CF passages were the main meteorological systems associated with disturbances in the water column of FHR.

Twenty-four CF passages were identified acting over FHR during the analyzed period. Fig. 5 shows the days when those events occurred, superimposed with a daily integration of S. The density of CF occurrences was higher during the mixing period, with 17 passages, while during the stratification period, only 4 passages were identified (3 CF passages occurred

in the transition between stratification and mixing periods). Daily integration values of S varied from −220 to 205 W m^{-2}, with positive values occurring with more frequency in the austral summer, and negative values occurring more frequently in the winter. The greatest daily net heat balance losses were associated with CF passages.

Wind direction and speed distribution, as well as the mean diel variability of S and its components, were analyzed considering days with and without CF passages (Fig. 6). In general, days without the passage of CF were characterized by easterly (and fewer

Table 1. Minimum, maximum, mean ($\bar{\mu}$), and standard deviation (σ) of pH and dissolved oxygen (DO) values of time series in Fig. 4. na: not available

Platform		pH						DO					
		#1C	#2	#3	#4	#5S	#6	#1C	#2	#3	#4	#5S	#6
Stratification	Min.	7.46	6.81	6.68	6.53	na	na	7.08	4.12	2.43	4.05	4.36	4.90
	Max.	8.22	7.69	8.48	7.77	na	na	7.97	7.67	8.16	7.99	8.32	7.67
	$\bar{\mu}$	7.75	7.17	7.33	7.26	na	na	7.47	6.30	6.89	7.07	6.44	6.68
	σ	0.11	0.14	0.23	0.25	na	na	0.14	0.52	0.62	0.43	0.73	0.32
Mixing	Min.	7.85	7.05	na	na	6.86	na	6.91	6.38	3.92	na	6.58	5.30
	Max.	8.52	7.93	na	na	8.19	na	8.51	8.8	9.26	na	8.52	7.97
	$\bar{\mu}$	8.13	7.35	na	na	7.32	na	7.86	7.72	6.89	na	7.78	6.84
	σ	0.11	0.13	na	na	0.23	na	0.39	0.33	0.61	na	0.42	0.43

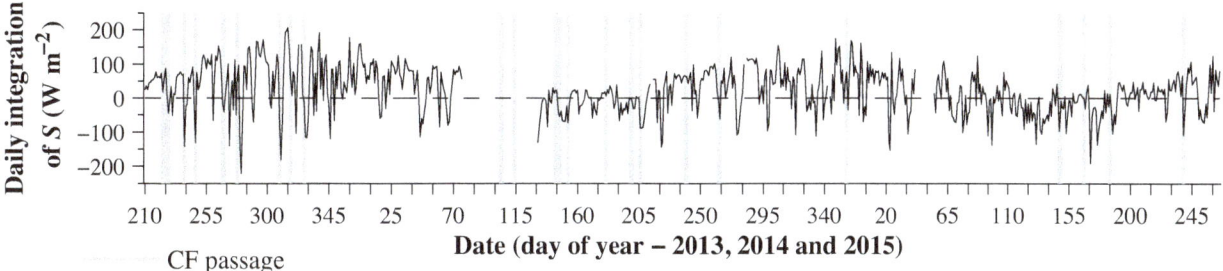

Fig. 5. Daily integration of surface heat flux balance (S) at Furnas Hydroelectric Reservoir from July 2013 to September 2015. CF: cold front

northerly) winds, with speeds <2 m s^{-1}. In contrast, days with CF passages were characterized by south-westerly and westerly winds, with speeds >3 m s^{-1}. The mean diel variability of S also showed different patterns for days with and without CF passages. On days without CF passages, the S peak of heat gain was 509 W m^{-2}, while on days with CF, this peak dropped to around 394 W m^{-2}. Integrating both diel

variability of S values, i.e. with and without CF passages (Fig. 6C,D), the obtained values were 25.22 and −21.36 W m^{-2} for days with and without CF passages, respectively.

Since strong wind speeds and heat balance losses favor vertical mixing of the water column, it was expected that CF would alter the mixed layer dynamics. Fig. 7 shows the effects of CF passages on the

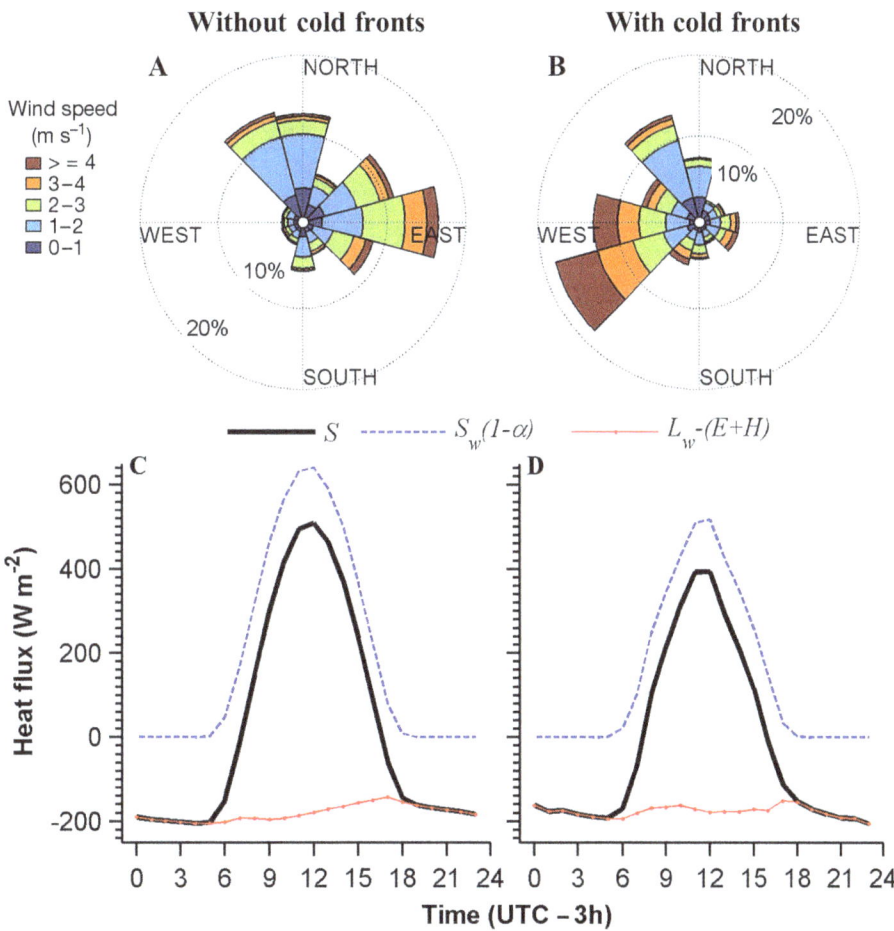

Fig. 6. (A,B) Wind direction and speed frequency distribution, and (C,D) diel variability of surface heat flux balance (S), shortwave radiation (S_w) multiplied by 1 minus the albedo (α), and net longwave radiation (L_w) minus latent (E) and sensible (H) heat fluxes, for days (A,C) without and (B,D) with cold front passages

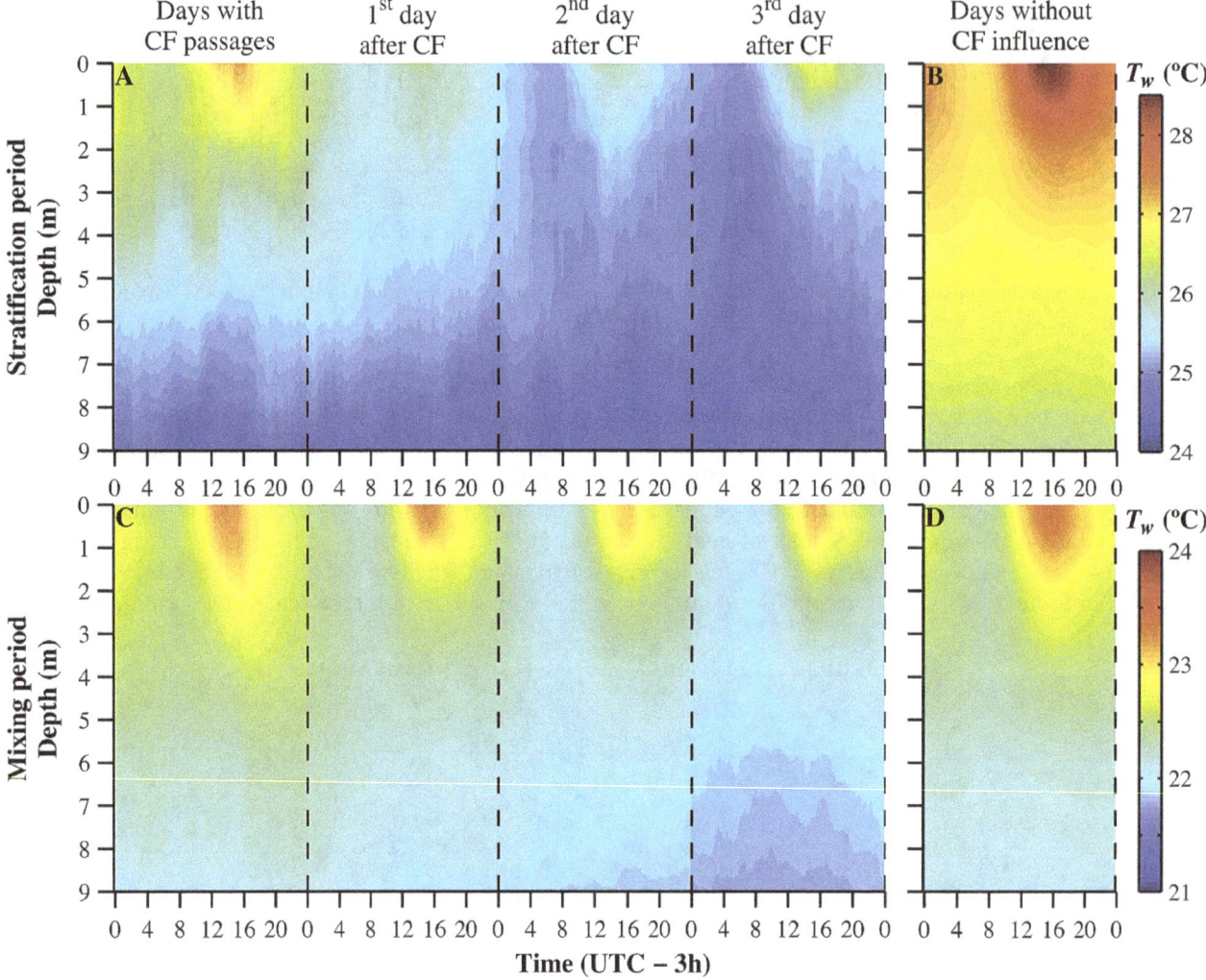

Fig. 7. Diel vertical distribution of water temperature (T_w) upon (A,C) occurrence of cold front (CF) passages and 3 d afterwards, and (B,D) without CF influence, in periods of (A,B) stratification and (C,D) mixing

vertical structure of temperature, up to 9 m depth. The data were analyzed for both stratification and mixing periods, also considering days with and without CF influences. Additionally, measurements up to 3 d after CF passages were also considered. Fig. 7 displays means of the diel variability of temperature of the water column.

The mean diel pattern of temperature changes in the mixed layer of FHR is strongly related to the mean diel pattern of the heat flux balance. Both patterns showed higher values at 16:00 h (UTC − 3 h) and minimum values at 07:00 h (UTC − 3 h), which correspond to the end times of net heat gain and losses, respectively. In the stratification period, during a day with a CF passage and continuing for 3 d after its occurrence, colder water rises from the deepest parts of the reservoir to the upper layers (Fig. 7A). For normal conditions (without CF influences) in the

stratification period, surface water temperature varied from 27 to 28.5°C, while after a CF passage, this variation was from 25°C (07:00 h on the 3rd day after the CF) to 26°C (16:00 h on the 2nd day after the CF).

Deep, colder waters reached surface layers after CF passages during the mixing period (Fig. 7C). For normal conditions, the temperature range was between 22.3 and 23.5°C, while after a CF passage, the variation ranged from 21.9°C (07:00 h on the 3rd day after the CF) to 23°C (16:00 h on the 2nd day after the CF). Differences in the mean values between days with and without the influence of CFs during the mixing period was around 0.5°C, in contrast to 2–2.5°C in the stratification period.

Vertical water temperature, buoyancy frequency, pH, and DO in the surface of the water column around a CF event, occurring on November 23, 2013 in the stratified period, are shown in detail in Fig. 8.

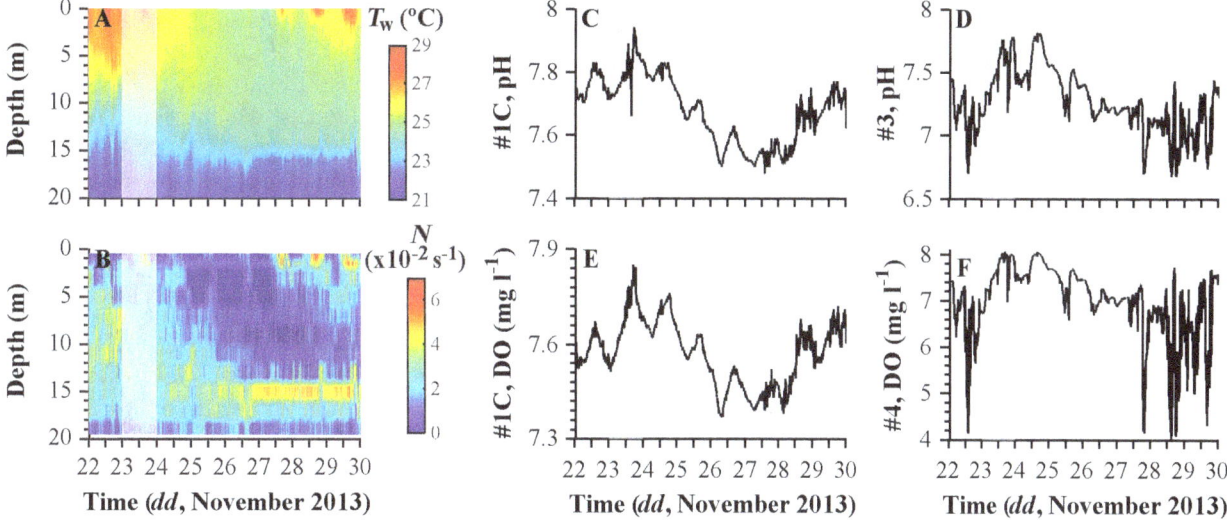

Fig. 8. Clipping of a short time-series of (A) water column temperature (T_w), (B) buoyancy frequency (N), (C,D) pH values at platforms #1C and #3, and (E,F) dissolved oxygen (DO) concentrations at platforms #1C and #4, around a cold front passage event on November 23, 2013 (highlighted in grey)

Before the CF passage, a strong gradient of tempera-ture was observed from the first few meters (~2–4 m) to 16–18 m. Buoyancy frequency in this depth range was almost homogeneous, with a value of 0.04 s^{-1}. During the passage of a CF, the stratification weak-ens and, after approximately 1 d, water temperature gradually becomes near-isothermal until approxi-mately 13 m in depth. About 3–4 d after the passage, a strong stratification develops at about 15 m (0.05 s^{-1}), while the diel variability of temperature in the upper layer develops concomitantly.

At #1C, the results of the mixing effect of the CF passage caused a slight decrease in pH values and DO concentrations (Fig. 8C,E). The same pattern was observed for the other platforms, as exemplified by #3 and #4 (Fig. 8D,F). However, at these platforms, peaks of lower values for both pH and DO concentra-tions were observed without the influence of the CF. Nevertheless, since no WQPs ranged beyond accept-able values for Nile tilapia, fish productivity was con-sidered normal, validated by the absence of mass mortality observations.

DISCUSSION

Water quality is a critical factor for any aquaculture activity, and the determination of the optimum range of WQPs for each species is important. Once these ranges are established, the monitoring of these WQPs is important to ensure growth and survival rates of the fishery. We discuss the relationships between changes in the monitored WQPs and CF

passages as well as the significance of these changes to Nile tilapia farming.

The analyzed period can be considered anomalous because of the low water levels caused by drought. Drastic reductions in water levels are considered a major problem for net cage fish farming in large reservoirs, requiring changes in location, reorganiza-tion of infrastructures, and sometimes changes in equipment, which all increase the cost of production. In FHR, these anomalous meteorological conditions led to changes in the operations of Nile tilapia farm-ing, such as changing of cage locations to promote the movement of water beneath them.

Although water level fluctuations may also lead to changes in nutrient concentration, plankton dynam-ics, and trophic state (Naselli-Flores & Barone 1994, Naselli-Flores 2000, Geraldes & Boavida 2005, da Costa et al. 2016), these were not observed to be a direct effect of the drought on the WQPs monitored in FHR.

A temperature gradient instead of a well-defined metalimnion in the stratification period, similar to the gradient that is characteristic for the FHR, was also observed in Foix Reservoir, Spain (Marcé et al. 2000), where a hypolimnion was not always present. In con-trast, in the Barra Bonita Reservoir (southeast Brazil), weaker N values (maximum of 0.027 s^{-1} in the sum-mer) were reported, indicating that this reservoir is polymictic (Calijuri et al. 2002). In Lake Palminhas (southeast Brazil), which is a mesotrophic warm mo-nomictic water body (Venturoti et al. 2015), during the mixing period (or dry season), the temperature and DO concentration differences between surface and deep layers (~20 m) are minimum, <1°C and 0.5 mg l^{-1} respectively. However, during the stratification period (or rainy season), these differences are about 4.5°C and 6.5 mg l^{-1} respectively, with the presence of an anoxic hypolimnion (Venturoti et al. 2014).

Water temperature is a crucial WQP for fish development, since it directly affects their growth (Martinez-Palacios et al. 1996, Azaza et al. 2008, 2010, El Sayed & Kawanna 2008). Based on our results, the ideal period for Nile tilapia cultivation in the FHR seems to be the austral summer (stratification period), although water surface temperatures >30°C may occur. On the other hand, during austral winter (mixing period), surface water temperatures <22°C are extremely stressing for Nile tilapia. Although it has been reported that Nile tilapia can maintain their normal activity and survive ~16°C (Sifa et al. 2002), at temperatures <27°C, fish will show reduced weight gain and increased disease susceptibility. Moreover, with the high frequency of CFs during this period, lower temperatures can be observed in the surface of the FHR. Therefore, it is recommended that workers avoid live fish handling and decrease fish density to prevent stress in this period of the year.

The increase of CF passages also increases FHR temperature fluctuation, which can also affect fish growth. Although the thermal fluctuation effect on Nile tilapia growth is not well understood, it has been demonstrated that it is size-dependent, allowing small juveniles to grow slightly faster but producing unsuitable conditions for fish of increasing size (Azaza et al. 2010). Therefore, the monitoring of meteorological systems can help to monitor aquaculture fields, since, with the higher density of CFs in the mixing period, thermal fluctuations will also increase.

Lower water pH levels and DO concentrations were found at all platforms located in the farming areas when compared to the control site. Oxygen depletion has a known effect on fish farming (Silvert 1992), caused by fish respiration as well as by the degradation process of organic matter, which can be higher within the cultivation areas due to uneaten food and excreta. Studies in Lake Palminhas (Venturoti et al. 2015) and FHR itself (Figueredo & Giani 2005) agree with our results, i.e. alterations in pH and DO concentrations were observed within cultivation areas of Nile tilapia when compared with areas without cultivation.

One of the major environmental problems of using net cages in a water body is eutrophication due to fish waste in the surrounding area (Edwards 2015). In 2 experiments conducted in the FHR (at the beginning of the rainy and dry season) with enclosures containing Nile tilapia fish, Figueredo & Giani (2005) verified that fish excretion increases the availability of nitrogen and phosphorus, which in turn promotes an increase in fast-growing algae. They also observed that selective feeding by Nile tilapia upon large algae (cyanobacteria and diatoms) controlled the relative abundance of phytoplankton species within the enclosures. In these 2 experiments, an initial decrease of both DO concentration and pH values was observed in relation to a control enclosure (no fish) and to another control point in the reservoir (no farming). Although the Nile tilapia contained in the enclosures in these experiments were not artificially fed (only naturally occurring plankton), the alterations in the water quality were noticeable. Therefore, the difference observed in water pH and DO concentration between #1C and the other platforms can be tied to the absence or presence of net cages. Despite the observed influence of net cages on pH fluctuation, the measured pH was still within the ideal range for Nile tilapia cultivation. However, DO concentrations <3.0 mg l^{-1} were observed in the FHR during the stratification period. Therefore, during this period, these low concentrations can be harmful for fish under production by decreasing their cost efficiency and increasing their stress susceptibility.

Low values of turbidity (excluding the noise) in this region of the FHR indicate that total suspended solids (TSS) concentrations are also low (since turbidity and TSS are highly correlated, e.g. Grayson et al. 1996, Bertrand-Krajewski 2004). TSS can be a limiting factor in light availability for primary production (e.g. Gikuma-Njuru & Hecky 2005); however, this is not the case for the studied embayments of the FHR. The observed noise in turbidity measurement in our study was also identified in a monitoring program of hard clam aquaculture using the same multiparameter sonde (Bergquist et al. 2009). Therefore, future studies and monitoring programs should focus on understanding the cause of this noise in the turbidity sensor.

The occurrences of CFs over the FHR were more frequent during the austral winter than during the summer, confirming the expected climatology shown by previous studies (Stech & Lorenzzetti 1992, Rodrigues et al. 2004). The combined effect of higher wind speeds and the increase in heat loss promoted by a CF acts in favor of mixing processes. Even though CFs occurred more intensively and more frequently during the austral winter, it is expected that changes in the environmental conditions of a water body due to a CF are more intense in the austral summer, when stratification is commonly established.

Mixing of chemically divergent layers during a stratification period, or atelomixis (Lewis 1973), may occur on diel scales in the epilimnion, reflecting the diel variability of heat flux balance. Stronger variations are expected during and after CF passages, since they will promote partial or entire mixing of the water column. Atelomixis processes act as a fundamental factor for the controlling of vertical phytoplankton distribution and succession (Barbosa & Padisák 2002). It is important to note that the physical processes associated with atelomixis may be related to basin-scale internal waves (Saggio & Imberger 1998, Hingsamer et al. 2014). Thus, the variation caused by the passage of a CF (Fig. 7) over the FHR can play a key role in primary production dynamics. Although CFs significantly altered the temperature distribution in the water column (Fig. 8), they did not completely erode the stratification of the water column.

Nevertheless, in some stratified lakes and reservoirs, CFs can drastically reduce oxygen concentration in the surface by the upwelling of anoxic waters from the hypolimnion (Tundisi et al. 2010). Faxinal Reservoir (south Brazil) (Becker et al. 2009b), Barra Bonita Reservoir (Calijuri et al. 2002), and Lake Palminhas (southeastern Brazil) (Venturoti et al. 2014) are examples of water bodies where a hypolimnetic hypoxia (or even anoxia) is present. Therefore, in such water bodies, CF passages have the potential to promote the upwelling of these deep waters and significantly affect the water quality in the cultivation areas.

In this study, we did not observe any critical changes in WQPs that would adversely affect Nile tilapia cultivation. This could be explained by the absence of a well-developed hypolimnion and because CFs did not completely erode the stratification (in the stratification period). Instead, higher variability of WQPs in the cultivation areas was observed in comparison to the control site (#1C) after a CF passage. Considering this effect, we observed that during the stratification period, 3 to 4 d after a CF passage, the water column had the ideal conditions for fish handling, since the effects of low pH and DO concentrations can be minimized by the dilution power promoted by the passage. The relatively low dilution power in the surface waters during the stratification period can lead to an increase in nutrient concentrations where the cages are present and, consequently, the phytoplankton ecological dynamics could be affected by these activities. Therefore, CF passages increase the embayments' dilution power by partially mixing the water column, benefiting the cultivation area in this region of the FHR.

The effects of CF passages over the FHR on water temperature (at 1.5 m depth), DO, and pH values are summarized in Fig. 9. To look at an area outside the influence of the farms, the data shown in Fig. 9 are from our control site (#1C). Although the mean temperature value in the stratification period decreases 2 or 3 d after the CF passages (~3°C), the mean of the other variables seems not to be highly affected, e.g. in a way that extrapolates the 1st and 3rd quartiles of days without CFs (exception of DO and pH values in the stratification periods).

The FHR, or at least the arm of the Grande River, can be considered suitable for Nile tilapia production. However, producers should consider that during the mixing periods, when temperatures are constantly <27°C, special management is needed. Since temperatures are outside the ideal range for the cultivation of Nile tilapia, producers should reduce fish density and fish manipulation to avoid fish stress. In this way, special management during mixing periods will allow them to maintain economic viability during the entire year.

We compare the WQPs in the FHR with the ideal range for 2 important native species in Brazilian aquaculture: tambaqui Colossoma macropomum and pacu Piaractus mesopotamicus. For tambaqui, the main challenge in the southeast and south regions of Brazil is its high sensitivity to low temperatures. In their natural habitat, tambaqui face temperatures varying from 25 to 40°C, and temperatures <20°C are adverse for their development. In terms of pH, tambaqui develop better in acidic water, with pH values varying from 4 to 6 (Aride et al. 2007). Based on these optimal ranges, the natural alkaline waters in the FHR and mean temperatures constantly >23°C during the mixing period may not be suitable for tambaqui cultivation.

On the other hand, pacu are more adapted to the southeast and south region of Brazil (Ferraz de Lima et al. 1988), since their ideal growing temperature is between 25 and 30°C. Although they can resist lower temperatures (15°C), the optimal range for their culture is 20–28°C (Saint-Paul 1989). An experimental study showed that a temperature of 7.5°C can be considered the lower limit for 1 yr old pacu survival (Milstein et al. 2000). The same study also stated that at temperatures under 16–18°C, fish feeding can be suspended to avoid wasting.

The main challenge in cultivating native species in the FHR is the relatively low temperature that occurs during winter, which will force a reduction in the production. Therefore, the farming of hybrid fish species in net cages in southeast Brazil is increasing. In

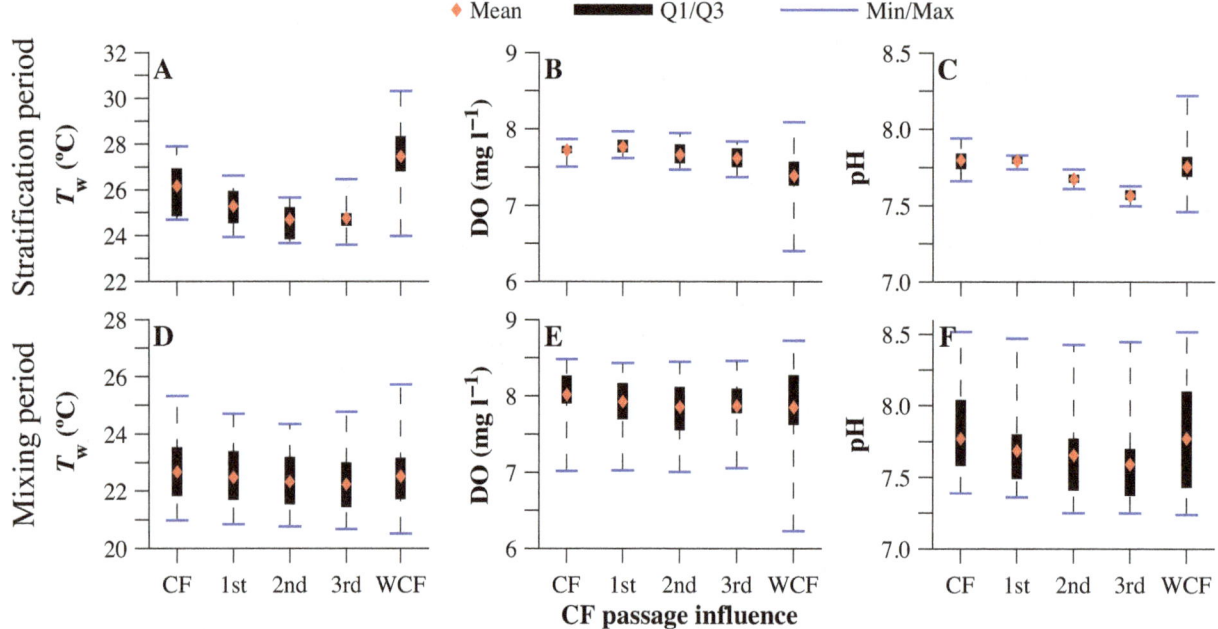

Fig. 9. Mean, 1st quartile, 3rd quartile, minimum, and maximum values of (A,D) water temperature (T_w), (B,E) dissolved oxygen (DO), and (C,F) pH, at platform #1C, for the (A–C) stratification and (D–F) mixing periods, on days without (WCF) and with (CF) the influence of cold fronts (including the 1st, 2nd, and 3rd days after its passage)

this region, an alternative could be the production of hybrids such as tambacu (female *C. macropomum* × male *P. mesopotamicus*), which are more resistant to lower temperatures, or the tambatinga (female *C. macropomum* × male *P. brachypomum*), which has been produced in central-west Brazil with reliable results. However, so far there is no data available about the optimal range of WQPs for tambatinga culture. It is important to highlight that the culture of hybrid species should comply with Brazilian legislation (IBAMA Order no. 145 from 1998) regulating the usage of species to be farmed in Brazilian hydrographic basins.

The understanding of stratification and mixing patterns of an aquatic system and their influences on WQPs are fundamental for the identification of the effects on aquaculture activities and vice versa. The economic and social benefit of fish farming is undisputed, but proper management is required for long-term sustainable activities (Beveridge et al. 1997, Abery et al. 2005). In this work, the combination of meteorology, limnology, and aquaculture were successfully applied to the development of a framework for the enhanced management of a fishery. The role of CFs in stratification and mixing processes and the implications for Nile tilapia-based aquaculture farming located in the FHR were analyzed. A dataset of meteorological data, WQPs, and water column temperature time-series was used on a high temporal

(continuous) acquisition basis, which was possible because of the use of moored platforms. Reliable data was able to be gathered due to the maintenance of the 6 moored platforms and their equipment. Despite the challenge to maintain moored systems for continuous acquisition of data (as previously noted in studies like Bergquist et al. 2009, Garel et al. 2009), the acquisition of high temporal resolution data for studies of lakes and reservoirs brings unprecedented knowledge of these environments in comparison to traditional fieldwork, where processes are addressed over short timeframes. For monitoring purposes, with an understanding of how meteorological systems affect heat flux balance and buoyancy frequency, the collection of meteorological variables could be reduced. Likewise, a thermistor chain could be deployed in only 1 representative location. Therefore, less effort would be needed to monitor WQPs in aquaculture fields.

The passage of CFs is a crucial factor to be considered in the monitoring of aquaculture activities in any water body. This may include, but is not limited to, south, southeast, and central-west regions of Brazil, as well as other countries in South America, like Argentina, Uruguay, and Paraguay. Although in this study, we highlighted the possible effects of CFs on aquaculture, it is important to note that other meteorological systems could also generate the same changes in the water column (Ogashawara et al.

2014). For example, Curtarelli et al. (2014b) showed that mesoscale convective systems are important meteorological forces that act over the Tucuruí Reservoir, northern Brazil. Thus, it is important to identify the main meteorological system, or systems, acting over a water body when considering the installation of an aquaculture enterprise.

The rapid increase in Brazilian aquaculture production in the last 2 decades indicates a growing market for exotic and native species (Pincinato & Asche 2016). Although the present study was restricted to Nile tilapia Oreochromis niloticus farming, which is considered a species that is highly tolerant to adverse environmental conditions, the proposed monitoring framework can be applicable to other fish cultures. The understanding and monitoring of the effects of CFs (and other meteorological systems) on different water bodies is highly important for the development of aquaculture and aquaculture policies worldwide.

Acknowledgements. This work is part of Furnas project, funded by the Ministry of Fisheries and Aquaculture of Brazil, and conducted in collaboration with the Brazilian Agricultural Research Corporation (Embrapa) and the National Institute for Space Research (INPE). C.A.S.A. thanks the National Council for Scientific and Technological Development (CNPq) for the PCI scholarship (Proc. no. 313071/2015-2). We thank Joaquim Leão and Geraldo Mendes for field assistance, and 3 anonymous reviewers for their suggestions to improve the manuscript.

LITERATURE CITED

Abdel-Tawwab M, Hagras AE, Elbaghdady HAM, Monier MN (2015) Effects of dissolved oxygen and fish size on Nile tilapia, Oreochromis niloticus (L.): growth performance, whole-body composition, and innate immunity. Aquacult Int 23:1261–1274

Abery NW, Sudaki F, Budhiman AA, Kartamihardja ES, Koeshendrajana S, Buddhiman, De Silva SS (2005) Fisheries and cage culture of three reservoirs in west Java, Indonesia; a case study of ambitious development and resulting interactions. Fish Manag Ecol 12:315–330

Alcântara EH (2012) Accessing the potential of satellite and telemetric data to evaluate the influence of the heat flux exchange in the water column mixing and stratification. Int J Geosci 3:899–907

Alcântara EH, Stech JL, Lorenzzetti JA, Bonnet MP, Casamitjana X, Assireu AT, Novo EMLM (2010) Remote sensing of water surface temperature and heat flux over a tropical hydroelectric reservoir. Remote Sens Environ 114:2651–2665

Amorocho J, DeVries JJ (1980) A new evaluation of the wind stress coefficient over water surfaces. J Geophys Res Oceans 85:433–442

Aride PHR, Roubach R, Val AL (2007) Tolerance response of tambaqui Colossoma macropomum (Cuvier) to water pH. Aquacult Res 38:588–594

Azaza MS, Dhraïef MN, Kraïem MM (2008) Effects of water temperature on growth and sex ratio of juvenile Nile tilapia Oreochromis niloticus (Linnaeus) reared in geothermal waters in southern Tunisia. J Therm Biol 33: 98–105

Azaza MS, Legendre M, Kraiem MM, Baras E (2010) Size-dependent effects of daily thermal fluctuations on the growth and size heterogeneity of Nile tilapia Oreochromis niloticus. J Fish Biol 76:669–683

Barbosa FAR, Padisák J (2002) The forgotten lake stratification pattern: atelomixis, and its ecological importance. Verh Int Ver Limnol 28:1385–1395

Becker V, Cardoso LS, Huszar VLM (2009a) Diel variation of phytoplankton functional groups in a subtropical reservoir in southern Brazil during an autumnal stratification period. Aquat Ecol 43:285–293

Becker V, Huszar VLM, Crossetti LO (2009b) Responses of phytoplankton functional groups to the mixing regime in a deep subtropical reservoir. Hydrobiologia 628:137–151

Bergheim A, Gausen M, Næss A, Hølland PM, Krogedal P, Crampton V (2006) A newly developed oxygen injection system for cage farms. Aquacult Eng 34:40–46

Bergquist DC, Heuberger D, Sturmer LN, Baker SM (2009) Continuous water quality monitoring for the hard clam industry in Florida, USA. Environ Monit Assess 148: 409–419

Bertrand-Krajewski JL (2004) TSS concentration in sewers estimated from turbidity measurements by means of linear regression accounting for uncertainties in both variables. Water Sci Technol 50:81–88

Beveridge MCM, Phillips MJ, Macintosh DJ (1997) Aquaculture and the environment: the supply of and demand for environmental goods and services by Asian aquaculture and the implications for sustainability. Aquacult Res 28:797–807

Boyd CE (1990) Water quality for pond aquaculture. Auburn University, Auburn, AL

Boyd CE, Pillai VK (1985) Water quality management in aquaculture. Spec Publ, Vol 22. CMFRI (ICAR), Cochin

Calijuri MC, Dos Santos ACA, Jati S (2002) Temporal changes in the phytoplankton community structure in a tropical and eutrophic reservoir (Barra Bonita, S.P.-Brazil). J Plankton Res 24:617–634

Cardoso EL, Gontijo VPM, Junior RMF, Morais ACR (2013) Ordenamento e monitoramento de áreas aquícolas do Reservatório de Furnas. Ser Doc 62. EPAMIG, Belo Horizonte

Climanálise (2016) Boletim de monitoramento e análise climática. www.climanalise.cptec.inpe.br/~rclimanl/boletim (accessed 1 July 2016)

Coelho CAS, Oliveira CP, Ambrizzi T, Reboita MS and others (2016) The 2014 southeast Brazil austral summer drought: regional scale mechanisms and teleconnections. Clim Dyn 46:3737–3752

CPTEC (Center for Weather Forecasting and Climate Research of the National Institute of Space Research) (2016) Collection BDI. www.cptec.inpe.br (accessed 1 Jul 2016)

Curtarelli MP, Alcântara EH, Rennó CD, Stech JL (2013) Effects of cold fronts on MODIS-derived sensible and latent heat fluxes in Itumbiara reservoir (Central Brazil). Adv Space Res 52:1668–1677

Curtarelli MP, Alcântara EH, Rennó CD, Stech JL (2014a) Physical changes within a large tropical hydroelectric reservoir induced by wintertime cold front activity. Hydrol Earth Syst Sci 18:3079–3093

Curtarelli MP, Ogashawara I, Araújo CAS, Alcântara EH, Lorenzzetti JA, Stech JL (2014b) Influence of summertime mesoscale convective systems on the heat balance and surface mixed layer dynamics of a large Amazonian hydroelectric reservoir. J Geophys Res Oceans 119: 8472–8494

da Costa MRA, Attayde JL, Becker V (2016) Effects of water level reduction on the dynamics of phytoplankton functional groups in tropical semi-arid shallow lakes. Hydrobiologia 778:75–89

Duan Y, Dong X, Zhang X, Miao Z (2011) Effects of dissolved oxygen concentration and stocking density on the growth, energy budget and body composition of juvenile Japanese flounder, Paralichthys olivaceus (Temminck et Schlegel). Aquacult Res 42:407–416

Edwards P (2015) Aquaculture environment interactions: past, present and likely future trends. Aquaculture 447: 2–14

El-Sayed AFM (2006) Tilapia culture. CABI Publishing, Wallingford

El-Sayed AFM, Kawanna M (2008) Optimum water temperature boosts the growth performance of Nile tilapia (Oreochromis niloticus) fry reared in a recycling system. Aquacult Res 39:670–672

Escobar H (2015) Drought triggers alarms in Brazil's biggest metropolis. Science 347:812

FAO (Food and Agriculture Organization ot the United Nations) (2016) The state of world fisheries and aquaculture 2016 (SOFIA). Contributing to food security and nutrition for all. FAO, Rome

Ferraz de Lima JA, Ferrari VA, Colares de Melo JS (1988) Comportamento do pacu em um cultivo experimental, no centro oeste do Brasil. Bol Tec CEPTA 1:15–28

Figueredo CC, Giani A (2005) Ecological interactions between Nile tilapia (Oreochromis niloticus, L.) and the phytoplanktonic community of the Furnas Reservoir (Brazil). Freshw Biol 50:1391–1403

Furnas (2016) Parque Gerador: usina hidrelétrica de Furnas. www.furnas.com.br (accessed 29 Sep 2016)

Ganf GG (1974) Diurnal mixing and the vertical distribution of phytoplankton in a shallow equatorial lake (Lake George, Uganda). J Ecol 62:611–629

Garel E, Nunes S, Neto JM, Fernandes R, Neves R, Marques JC, Ferreira Ó (2009) The autonomous Simpatico system for real-time continuous water-quality and current velocity monitoring: examples of application in three Portuguese estuaries. Geo-Mar Lett 29:331–341

Garreaud RD (2000) Cold air incursion over subtropical South America: mean structure and dynamics. Mon Weather Rev 128:2544–2559

Geraldes AM, Boavida MJ (2005) Seasonal water level fluctuations: implications for reservoir limnology and management. Lakes Reservoirs: Res Manage 10:59–69

Gikuma-Njuru P, Hecky RE (2005) Nutrient concentrations in Nyanza Gulf, Lake Victoria, Kenya: light limits algal demand and abundance. Hydrobiologia 534:131–140

Grayson RB, Finlayson BL, Gippel CJ, Hart BT (1996) The potential of field turbidity measurements for the computation of total phosphorus and suspended solids loads. J Environ Manage 47:257–267

Henderson-Sellers B (1986) Calculating the surface energy balance for lake and reservoir modelling: a review. Rev Geophys 24:625–649

Hingsamer P, Peeters F, Hofmann H (2014) The consequences of internal waves for phytoplankton focusing on the distribution and production of Planktothrix rubensces. PLOS ONE 9:e104359

IBGE (Brazilian Institute of Geography and Statistics) (2016) Produção da pecuária municipal 2015, Vol 43, IBGE, Produção da Pecuária Minicipal, Rio de Janeiro, p 1–49. https://biblioteca.ibge.gov.br/visualizacao/periodicos/84/ppm_2015_v43_br.pdf (accessed 3 Oct 2016)

Imberger J, Hamblin PF (1982) Dynamics of lakes, reservoirs, and cooling ponds. Annu Rev Fluid Mech 14:153–187

Imberger J, Patterson JC (1990) Physical limnology. Adv Appl Mech 27:303–475

INMET (National Institute of Meteorology) (2016) Banco de dados meteorológicos para ensino e pesquisa. www.inmet.gov.br (accessed 10 Jul 2016)

Jobling M (1994) Fish bioenergetics. Chapman & Hall, London

Kapetsky JM (2000) Present applications and future needs of meteorological and climatological data in inland fisheries and aquaculture. Agric For Meteorol 103:109–117

Lewis WM Jr (1973) The thermal regime of Lake Lanao (Philippines) and its theoretical implications for tropical lakes. Limnol Oceanogr 18:200–217

Lewis WM Jr (1983) A revised classification of lakes based on mixing. Can J Fish Aquat Sci 40:1779–1787

Liu H, Zhang Y, Liu S, Jiang H, Sheng L, Williams QL (2009) Eddy covariance measurements of surface energy budget and evaporation in a cool season over southern open water in Mississippi. J Geophys Res Atmos 114:D04110

Liu L, Liu D, Johnson DM, Yi Z, Huanh Y (2012) Effects of vertical mixing on phytoplankton blooms in Xiagxi Bay of Three Gorges Reservoir: implications for management. Water Res 46:2121–2130

Lorenzzetti JA, Araújo CAS, Curtarelli MP (2015) Mean diel variability of surface energy fluxes over Manso Reservoir. Inland Waters 5:155–172

Marcé R, Comerma M, Garcia JC, Gomà J, Armengol J (2000) Limnology of Foix Reservoir (Barcelona, Spain). Limnetica 19:175–191

Marcusso PF, Aguinaga JY, Claudiano GS, Eto SF and others (2015) Influence of temperature on Streptococcus agalactiae infection in Nile tilapia. Braz J Vet Res Anim Sci 52:57–62

Martin JL, McCutcheon SC (1999) Hydrodynamics and transport for water quality modeling. CRC Press, Boca Raton, FL

Martinez-Palacios CA, Chavez-Sanchez MC, Ross LG (1996) The effects of water temperature on food intake, growth and body composition of Cichlasoma urophthalmus (Günther) juveniles. Aquacult Res 27:455–461

Meganck R, Havens K, Pinto-Coelho RM (2015) Water: megacities running dry in Brazil. Nature 521:289

Milstein A, Zoran M, Peretz Y, Joseph D (2000) Low temperature tolerance of pacu, Piaractus mesopotamicus. Environ Biol Fishes 58:455–460

MPA (Ministry of Fisheries & Aquaculture of Brazil) (2013) Boletim estatístico da pesca e aquicultura 2011. MPA, Brasília. www.icmbio.gov.br/cepsul/images/stories/biblioteca/download/estatistica/est_2011_bol_bra.pdf (accessed 29 Sep 2016)

Nandlal S, Pickering T (2004) Tilapia hatchery Operation. In: Tilapia fish farming in Pacific Island countries, Vol 1. University of the South Pacific, Noumea

Naselli-Flores L (2000) Phytoplankton assemblages in twenty-one Sicilian reservoirs: relationships between species composition and environmental factors. Hydrobiologia

424:1–11

Naselli-Flores L, Barone R (1994) Relationship between trophic state and plankton community structure in 21 Sicilian dam reservoirs. Hydrobiologia 275/276:197–205

NOAA (2016) Magnetic field calculators. www.ngdc.noaa.gov/geomag-web (accessed 5 Jun 2016)

Nobre CA, Marengo JA, Seluchi ME, Cuartas AL, Alves LM (2016) Some characteristics and impacts of the drought and water crisis in southeastern Brazil during 2014 and 2015. J Water Resour Prot 8:252–262

Ogashawara I, Zavattini JA, Tundisi JG (2014) The climatic rhythm and blooms of cyanobacteria in a tropical reservoir in São Paulo, Brazil. Braz J Biol 74:72–78

Ostrensky A, Boeger W (1998) Piscicultura: fundamentos e técnicas de manejo. Agropecuária, Guaíba

Özdemir N, Demirak A, Keskin F (2014) Quality of water used during cage cultivation of rainbow trout (*Oncorhynchus mykiss*) in Bereket HES IV Dam Lake (Muğla, Turkey). Environ Monit Assess 186:8463–8472

Pincinato RBM, Asche F (2016) The development of Brazilian aquaculture: introduced and native species. Aquacult Econ Manage 20:312–323

Pollock MS, Clarke LMJ, Dubé MG (2007) The effects of hypoxia on fishes: from ecological relevance to physiological effects. Environ Rev 15:1–14

Popma T, Masser M (1999) Tilapia: life history and biology. SRAC Publ No. 283. Southern Regional Aquaculture Center, Stoneville, MS

Rao YR, Hawley N, Charlton MN, Schertzer WM (2008) Physical processes and hypoxia in the central basin of Lake Erie. Limnol Oceanogr 53:2007–2020

Reynolds CS (1992) Dynamics, selection and composition of phytoplankton in relation to vertical structure in lakes. Arch Hydrobiol Beih Ergeb Limnol 35:13–31

Rodrigues MLG, Franco D, Sugahara S (2004) Climatologia de frentes frias no litoral de Santa Catarina. Rev Bras Geofís 22:135–151

Saggio A, Imberger J (1998) Internal wave weather in a stratified lake. Limnol Oceanogr 43:1780–1795

Saint-Paul U (1989) Aquaculture in Latin America. Indigenous species promise increased yields. Naga IGLARM Q 12:3–5

Sampaio FG, Boijink CL, Santos LRB, Oba ET, Kalinin AL, Rantin FT (2010) The combined effect of copper and low pH on antioxidant defenses and biochemical parameters in neotropical fish pacu, *Piaractus mesopotamicus* (Holmberg, 1887). Ecotoxicology 19:963–976

Sampaio FG, Boijink CL, Santos LRB, Oba ET, Kalinin AL, Luiz AJB, Rantin FT (2012) Antioxidant defenses and biochemical changes in pacu, *Piaractus mesopotamicus*:

responses to single and combined copper and hypercarbia exposure. Comp Biochem Physiol C 156:178–186

Serra T, Vidal J, Casamitjana X, Soler M, Colomer J (2007) The role of surface vertical mixing in phytoplankton distribution in a stratified reservoir. Limnol Oceanogr 52:620–634

Seth A, Fernandes K, Camargo SJ (2015) Two summers of São Paulo drought: origins in the western tropical Pacific. Geophys Res Lett 42:10816–10823

Sifa L, Chenhong L, Dey M, Gagalac F, Dunham R (2002) Cold tolerance of three strains of Nile tilapia, *Oreochromis niloticus*, in China. Aquaculture 213:123–129

Silvert W (1992) Assessing environmental impacts of finfish aquaculture in marine waters. Aquaculture 107:67–79

Stech JL, Lorenzzetti JL (1992) The response of the South Brazil Bight to the passage of wintertime cold fronts. J Geophys Res Oceans 97:9507–9520

Stech JL, Lima IBT, Novo EMLM, Silva CM and others (2006) Telemetric monitoring system for meteorological and limnological data acquisition. Verh Int Ver Limnol 29:1747–1750

Tundisi JG, Matsumura-Tundisi T, Arantes JD Jr, Tundisi JEM, Manzini NF, Ducrot R (2004) The response of Carlos Botelho (Lobo, Broa) reservoir to the passage of cold fronts as reflected by physical, chemical and biological variables. Braz J Biol 64:177–186

Tundisi JG, Matsumura-Tundisi T, Pereira KC, Luzia AP and others (2010) Cold fronts and reservoir limnology: an integrated approach towards the ecological dynamics of freshwater ecosystems. Braz J Biol 70:815–824

van Rijssel JC, Hecky RE, Kishe-Machumu MA, Meijer SE and others (2016) Climatic variability in combination with eutrophication drives adaptive responses in the gills of Lake Victoria cichlids. Oecologia 182:1187–1201

Venturoti GP, Veronez AC, Salla RV, Gomes LC (2014) Phosphorous, total ammonia nitrogen and chlorophyll *a* from fish cages in a tropical lake (Lake Plaminhas, Espirito Santo, Brazil). Aquacult Res 2014:1–15

Venturoti GP, Veronez AC, Salla RV, Gomes LC (2015) Variation of limnological parameters in a tropical lake used for tilapia cage farming. Aquacult Rep 2:152–157

Verburg P, Antenucci JP (2010) Persistent unstable atmospheric boundary layer enhances sensible and latent heat loss in a tropical great lake: Lake Tanganyika. J Geophys Res Atmos 115:D11109

Vidal J, Moreno-Ostos E, Escot C, Quesada R, Rueda F (2010) The effects of diel changes in circulation and mixing on the longitudinal distribution of phytoplankton in a canyon-shaped Mediterranean reservoir. Freshw Biol 55:1945–1957

Appendix.

Table A1. Meteorological and water quality parameters collected by Integrated System for Environmental Monitoring (SIMA) and SIMA-Aquaculture at the 6 platforms (see Fig. 1C)

Data	Unit	Manufacturer and model	Range; accuracy	Platform
Air temperature (T_a)	°C	Rotronic MP 101A	−40 to 60; ±0.3	#5S
Relative humidity (R_h)	%		0 to 100; ±1	
Atmospheric pressure (P_{atm})	hPa	Vaisala PTB 110	800 to 1060; ±0.3 at 20°C	#5S
Wind speed (U_z)	m s^{-1}	RM Young 05106	0 to 100; ±0.3	#5S
Wind direction	degrees		0 to 360; ±3	
Shortwave radiation (S_w)	W m^{-2}	Novalynx 840-8102	0 to 1500; <1	#5S
Multiparameter sonde:	–	YSI 6600 V2-4 (or V2-2)[a]	–	
Water temperature (T_w)	°C	YSI 6560	−5 to 60; ±0.15	
Conductivity	mS cm^{-1}		0 to 100; ±0.5 % of reading + 0.001	#1C, #2, #3, #4, #5S, and #6
pH	–	YSI 6561	0 to 14; ±0.2	
Turbidity	NTU	YSI 6136	0 to 1000; ±2 % of reading or 0.3 (whichever is greater)	#1C, #2, #3, #4, #5S, and #6
Dissolved oxygen (DO)	mg l^{-1}	YSI 6150 (or YSI 6562)[a]	0 to 50; ±0.1 (0.2) or 1 % (2 %) of reading (whichever is greater)	#1C, #2, #3, #4, #5S, and #6
Thermistor chain:				
Water temperature (T_w)	°C	Onset HOBO U22-01 Water Temp Pro v2	−40 to 50; ±0.2	#1C, #2, #3, #4, #5S, and #6

[a]YSI 6600 V2-2 and YSI 6562 were only used in the SIMA

15

Vertical particle fluxes dominate integrated multi-trophic aquaculture (IMTA) sites: implications for shellfish–finfish synergy

R. Filgueira[1,*], T. Guyondet[2], G. K. Reid[3,4], J. Grant[5], P. J. Cranford[6]

[1]Marine Affairs Program, Dalhousie University, 1355 Oxford St., PO Box 15000, Halifax, NS B3H 1R2, Canada

[2]Department of Fisheries and Oceans, Gulf Fisheries Centre, Science Branch, PO Box 5030, Moncton, NB E1C 9B6, Canada

[3]Canadian Integrated Multi-Trophic Aquaculture Network (CIMTAN), University of New Brunswick, PO Box 5050, Saint John, NB E2L 4L5, Canada

[4]Department of Fisheries and Oceans, St. Andrews Biological Station, 531 Brandy Cove Rd., St. Andrews, NB E5B 2L9, Canada

[5]Department of Oceanography, Dalhousie University, Halifax, NS B3H 4R2, Canada

[6]Department of Fisheries and Oceans, Bedford Institute of Oceanography, 1 Challenger Dr., Dartmouth, NS B2Y 4A2, Canada

ABSTRACT: Maximizing the mitigation potential of open-water finfish–shellfish integrated multi-trophic aquaculture (IMTA) farms is complex in terms of co-locating the trophic components. Both the dispersal of finfish aquaculture wastes and biological processes are highly influenced by water circulation. Consequently, the evaluation of shellfish–finfish synergy requires a combined study of biological and physical processes, which can be achieved by the implementation and coupling of mathematical models. A highly configurable mathematical model was developed that can be applied at the apparent spatial scale of IMTA sites. The model tracks different components of the seston, including feed wastes, fish faeces, shellfish faeces, natural detritus and phytoplankton. Based on the characterization of these fluxes, a hypothetical IMTA site was used to explore different spatial arrangements for evaluating finfish–shellfish farm mitigation efficiency. The site was modelled following a factorial design, which tested 2 levels of background seston concentrations, 3 farm designs, 2 hydrodynamic conditions and 2 levels of aquaculture intensity. The model predicts that mitigation efficiency is highly dependent on the background environmental conditions, obtaining maximum mitigation under oligotrophic conditions that stimulate shellfish filtration activity. The dominance of vertical fluxes of particulate matter triggered by the high settling velocity of finfish aquaculture wastes suggests that suspended shellfish aquaculture cannot significantly reduce organic loading of the seabed. Consequently, this suggests that waste mitigation at IMTA sites should be best achieved by placing organic extractive species (e.g. deposit feeders) on the seabed directly beneath finfish cages rather than in suspension in the water column.

KEY WORDS: Settling velocity · Organic loading · Mitigation · Connectivity · Ecosystem model

INTRODUCTION

The projection of human population growth (United Nations, Department of Economic and Social Affairs, Population Division 2013) and the increasing demand for seafood per capita (FAO 2014) suggest an expansion in aquaculture activities in the coming decades. The rate of expansion will be driven by specific local pressures. For example, while social aspects such as aesthetics or conflict with other uses of the ocean will

*Corresponding author: ramon.filgueira@dal.ca

probably dominate European and North American expansion, physical limits (e.g. space) on total production will most likely determine the fate of this expansion in Asia, where maximizing the food supply is the paramount concern (Ferreira et al. 2013). In between social and production concerns, ecological drivers play a significant role in this expansion because adverse ecosystem interactions can cause a negative feedback on both society and on the production potential (Stigebrandt 2011). Therefore, developing ecologically sustainable aquaculture will reduce negative societal pressures on this industry with concomitant economic benefits.

In the case of fed finfish aquaculture, one of the major ecological concerns is related to benthic organic loading (e.g. Strain & Hargrave 2005). The vertical flux of uneaten feed and fish faeces can significantly alter sediment characteristics and communities, leading, in a worst-case scenario, to benthic oxygen depletion and local community impacts (Hargrave 2010). Dissolved nutrients are also released at fish farms and although they are quickly dispersed, they have been suggested to contribute to eutrophication (Sarà et al. 2011) and are the focus of regulations (Gillibrand et al. 2002). The ecologically engineered concept of integrated multi-trophic aquaculture (IMTA) uses a natural recycling approach across different trophic groups, where the by-products or wastes from one species become inputs for another within the same culture system (Chopin et al. 2012). Accordingly, the fed component (e.g. finfish) is combined with extractive species that recapture particulate organic matter (e.g. suspension- and deposit-feeders) and dissolved inorganic matter (e.g. seaweeds). In addition to potentially mitigating organic loading, the organic extractive species could benefit from the additional available food, potentially resulting in augmented growth and commercial benefits.

The capacity of an IMTA site to utilize wastes depends on: (1) the capacity of the extractive species to capture wastes, which is a function of its physiology, the amount of available wastes and the environmental conditions at the farm; and (2) the connectivity among the different components of the farm. Both factors are influenced by hydrodynamics, the spatial arrangement of extractive species and physicochemical properties of the wastes. Connectivity is crucial for increasing the probability of encounters between wastes and extractive species and is highly influenced by the settling velocity of the waste. For example, feed pellets sink directly beneath a fish cage but excreted ammonia remains in the water column. Regarding the capacity of the extractive species to

capture wastes, it is important to highlight the role of the individual physiological response to environmental conditions, including the diet to which the individual is exposed. The mitigation capacity of the extractive species is proportional to the contribution of wastes to the naturally available food. That is, if wastes are diluted within naturally abundant food, wastes will not be a significant part of the extractive species' diet, which results in low mitigation. For example, it has been stated that 15 to 35% of the organic diet of mussels must be from fish farm particulates (faeces or feed 'fines') to significantly reduce net benthic organic loading, otherwise the organic loading generated by mussel faeces would be higher than the mitigation by capturing fish faeces (Cranford et al. 2013).

One of the key components in the historical and present design of IMTA systems is the use of bivalves as the main organic extractive species (Soto 2009). Given that bivalves are widely distributed, can be cultured at high densities and can severely deplete suspended particulate matter at the ecosystem scale, they are ideal candidates to be used as waste biofilters (Cranford et al. 2013). Despite this theoretically good fit between finfish and mussels in the context of IMTA farming, literature data reveal contradictory results. For example, some studies reported higher growth of IMTA bivalves compared to controls (Wallace 1980, Jones & Iwama 1991, Stirling & Okumus 1995, Troell & Norberg 1998, Lander et al. 2004, 2012, Peharda et al. 2007, Sarà et al. 2009, 2012, Handå et al. 2012, Jiang et al. 2013) due to an increase in organic particles in the vicinity of finfish cages, but other studies showed little or no significant effects (Farias-Sánchez 1983, Taylor et al. 1992, Gryska et al. 1996, Parsons et al. 2002, Cheshuk et al. 2003, Navarrete-Mier et al. 2010, Irisarri et al. 2013). Rensel et al. (2011) highlighted interspecific differences under the same conditions in a case where oysters benefited from proximity to finfish cages whereas mussels did not. As Cranford et al. (2013) stated, it is difficult to compare the results of these studies to evaluate the ultimate cause of these discrepancies owing to the different experimental designs and environmental conditions. One of the key ways to evaluate whether mussels benefit from finfish wastes is to establish a reference site with identical environmental conditions so that IMTA mussels could be compared directly to reference mussels. However, natural environmental variability makes it difficult to define 2 identical sites that do not influence one another such that one of them could be used as a reference for comparison purposes (e.g. Brager et al. 2015).

Mathematical modelling provides an opportunity to theoretically explore finfish–shellfish IMTA performance by means of comparing different 'what if' scenarios. Models have been used to explore a range of sustainability and/or management alternatives at aquaculture sites, including polyculture and IMTA sites (e.g. Duarte et al. 2003, Nunes et al. 2003, Ferreira et al. 2012, Ren et al. 2012). Numerical models allow for a detailed description of organic deposition and associated biogeochemical fluxes in finfish aquaculture sites. A model commonly applied to cage aquaculture for this purpose is DEPOMOD (Cromey et al. 2002), a Lagrangian particle-tracking model that predicts the spatial distribution of carbon deposition in the vicinity of the farm based on biodeposit production rates, current velocities and bathymetry. Other Lagrangian models (e.g. Jusup et al. 2007) predict organic loading and have been successfully coupled to other models to predict biogeochemical fluxes around fish cages (Brigolin et al. 2014). One of the most powerful capabilities of numerical models is the possibility of exploring different scenarios, which constitutes a critical aspect for management purposes (Nobre et al. 2010). For example, Tsagaraki et al. (2011) explored the effects of fish farming under different production levels and hydrodynamic conditions. These scenarios can even be coupled to economic models to analyse the profitability of an IMTA operation under different hydrodynamic scenarios (Ferreira et al. 2012). While different methodological approaches can be used for this purpose (Grant & Filgueira 2011), a key aspect when evaluating the probability of encounter between waste and extractive species is spatial resolution. High spatial resolution is always desirable because it provides a more detailed description of the model domain. In addition, spatial resolution can also affect model predictions (Melbourne-Thomas et al. 2011), especially when the simulated processes are dependent on concentration (Fennel & Neumann 2004). Consequently, in this study we developed 2 high-resolution fully spatial modelling approaches with the specific objectives of: (1) evaluating the maximum mitigation efficiency under different finfish–shellfish IMTA configurations (ecosystem model); and (2) exploring the fate of particles (feed waste and finfish faeces) released in finfish cages (particle-tracking model).

MATERIALS AND METHODS

Model domain

A hypothetical aquaculture site was used as a testing ground for the numerical experiments. Although the models that have been developed in this study can be applied to real case locations, the use of a hypothetical and ideal site is aligned with the main goal regarding the exploration of the maximum mitigation efficiency. Accordingly, this hypothetical site was constructed to maximize mitigation. This hypothetical site was defined using a structured triangular mesh that includes 1360 triangles and 718 nodes (Fig. 1). The dimensions of the model domain are 510 m long × 300 m wide and 15 or 30 m deep for the ecosystem and particle-tracking models, respectively. Although the bathymetry is important for determining the footprint of organic loading in the benthos, it is not relevant for estimating the mitigation efficiency of organic particles from shellfish culture, which is typically suspended within the top 15 m of the water column. Accordingly, bathymetry was not included in the factorial design to test mitigation.

Ecosystem model

The following 5 sources of organic carbon were modelled: fish feed waste (FW), fish faeces (FF), shellfish faeces (ShF), background detritus (D) and background phytoplankton (P). Only the unsettleable fraction of FW (feed 'fines') was modelled; the remaining organic material in FW (uneaten feed pellets) was assumed to sink directly to the bottom due

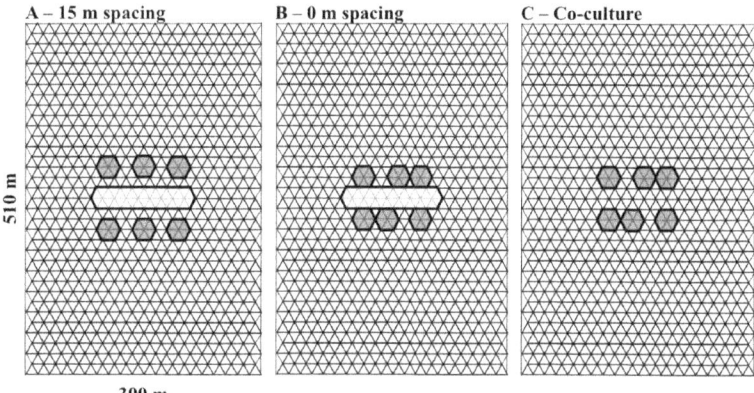

Fig. 1. Structured triangular mesh within the ecosystem model domain and the 3 farm designs studied: (A) 15 m separation, (B) 0 m separation and (C) co-culture. Dark and light grey hexagons represent finfish cages and shellfish longlines, respectively. In the co-culture design, finfish and shellfish are co-cultured together in the fish cages

to high settling velocities. It is assumed that all particles are within the filtering size range of the shellfish and consequently they can be effectively captured. The term 'background' makes reference to the concentration of detritus or phytoplankton that would exist in the model domain in the absence of any aquaculture activity. All of the stocks were characterized in mg C m^{-3}. The fluxes of the different particles were simulated according to the following equations coded in Matlab® (https://www.mathworks.com):

Table 1. Settling velocity (references indicated) and mussel absorption efficiency (AE, Reid et al. 2010) for each type of particle. na: not applicable

Particle type	Settling velocity (cm s^{-1})	Reference	Mussel AE (%)
Feed waste (FW)	0 (unsettleable)	na	85
Fish faeces (FF)	3	Robinson & Reid (2014)	85
Shellfish faeces (ShF)	1	Liutkus et al. (2012)	1
Phytoplankton (P)	5.8×10^{-5}	Herndl & Reinthaler (2013)	85
Detritus (D)	1.7×10^{-4}	McDonnell & Buesseler (2010)	45

$$\frac{dFW}{dt} = +FW_{production} - Sh_{ingestion} \pm FW_{mixing} \quad (1)$$

$$\frac{dFF}{dt} = +FF_{production} - FF_{sinking} - Sh_{ingestion} \pm FF_{mixing} \quad (2)$$

$$\frac{dShF}{dt} = +Sh_{egestion} - ShF_{sinking} - Sh_{ingestion} \pm ShF_{mixing} \quad (3)$$

$$\frac{dD}{dt} = +D_{production} - D_{sinking} - Sh_{ingestion} \pm D_{mixing}, \ D_{production} = D_{sinking} \quad (4)$$

$$\frac{dP}{dt} = +P_{production} - P_{sinking} - Sh_{ingestion} \pm P_{mixing}, \ P_{production} = P_{sinking} \quad (5)$$

where t is time, 'production' represents the production of particles by aquaculture activities (FW, FF and ShF) and natural processes (D and P); 'sinking' is the loss of particles due to sinking to the bottom following the average settling velocities depicted in Table 1; 'ingestion' is the ingestion of particles by shellfish; and 'mixing' is the exchange of particles among adjacent triangles. See the Supplement at www.int-res.com/articles/suppl/q009p127_supp.pdf for further details regarding the ecosystem model.

FW$_{production}$ and FF$_{production}$ rates were prescribed as constant forcing functions. This approach minimizes peaks in FW and FF, and better represents average conditions in the farm. This approach is less suitable for FW because feeding is not continuous through time. However, this simplification has a minimal impact on the available organic matter in the water column due to the small contribution of the unsettleable fraction of FW compared to other sources of organic matter such as FF (Table 2). Each finfish cage was characterized according to a typical mid-size farm in the Bay of Fundy (eastern Canada) area (see Reid et al. 2013a). The total feed was calcu-

lated by taking into account Atlantic salmon *Salmo salar* biomass, growth rate and a feed conversion ratio of 1.2 (Reid 2007, Reid et al. 2013b). Fifteen percent of consumed fish feed was assumed to be FF$_{production}$ and 3% FW (Reid et al. 2009), of which 1.5% was assumed to be unsettleable (FW$_{production}$). It was assumed that 77% of FF (93% of FW) is organic and 38% of that (57% for FW) is carbon (Reid et al. 2009, 2010). D$_{production}$ and P$_{production}$ were set up equal in magnitude to D$_{sinking}$ and P$_{sinking}$, respectively, in order to guarantee constant composition of the idealized ocean through time. Constant P concentration was established in terms of µg chl *a* l^{-1} (Table 2), and a C:chl *a* ratio of 50:1 was assumed to calculate organic carbon. Constant total seston concentration was established in terms of mg l^{-1} (Table 2), of which 65% was assumed to be organic and, in turn, 50% of the organic fraction was assumed to be carbon. D concentration was calculated by subtracting P concentration from total organic carbon. Constant D and P values are intended to mimic average conditions in farming areas in the Bay of Fundy area (e.g. Brager et al. 2015).

The terms Sh$_{ingestion}$ and Sh$_{egestion}$ represent the ingestion of particles by shellfish and the production of faeces, respectively. The shellfish used in this study is *Mytilus edulis*, whose ecophysiology has been simulated by coupling a previously published dynamic energy budget (DEB) model (Filgueira et al.

Table 2. Different scenarios (constant forcing conditions) of background seston in terms of phytoplankton (P, chl *a*) and detritus (D, total seston) concentration, as well as farm production rate of fish faeces (FF$_{production}$) and unsettleable feed waste (FW$_{production}$)

Background seston	Farm particles	chl-*a* (µg l^{-1})	Total seston (mg l^{-1})	FF$_{production}$ (kg d^{-1} farm^{-1})	FW$_{production}$ (g d^{-1} farm^{-1})
Low		1.0	1.0		
Average		2.5	3.0		
High		4.0	5.0		
	Low			15	35
	Average			25	70
	High			35	105

2011). See the Supplement for further details regarding the DEB model. DEB is coupled to the ecosystem model though $Sh_{ingestion}$ and $Sh_{egestion}$, that is, ingestion of all the potential food sources and faeces production, respectively. Pumping rate (PR, l h^{-1}) is derived from DEB and multiplied by the concentration of a given particle, which results in $Sh_{ingestion}$. For each particle type, the multiplication of $Sh_{ingestion}$ × (1 – absorption efficiency [Table 1]) provides $Sh_{egestion}$. The parameters of the DEB model follow Filgueira et al. (2011) with the exception of 'absorption efficiency' which varied depending on the food particle (our Table 1, following Reid et al. 2010) and the 'scaled functional response', which was prescribed here as 1. This parameterization of the scaled functional response ensures that the mussels will ingest at maximum capacity. Given that ingestion is maximum at all times, the results will represent the maximum mitigation capacity, which is aligned with the main goal of the study. No pre-ingestive selection is considered in this version of the DEB model, so the specific removal of each kind of particle is proportional to the relative abundance of the organic carbon of that particle in relation to the total organic carbon. PR was calculated by taking into account the mussel's organic carbon ingestion rate and the concentration of organic carbon in the surrounding water. Given that it was assumed that all of the particles are 100% retained in the gills, PR is equivalent to clearance rate. The average density of farmed mussels was prescribed as 30 kg m^{-2} in terms of wet weight. The total ingestion and egestion of the population was extrapolated from the individual rates estimated by the DEB model. Given the relevance of PR in mitigation efficiency, both were selected as response variables to evaluate the performance under each tested scenario.

Mixing among adjacent cells was calculated following the protocol described by Filgueira et al. (2012), in which the velocity field is used as an external forcing to calculate transport following a first-order upwind scheme. This results in an offline coupling scheme between the hydrodynamics and biogeochemical reactions. This protocol, also coded in Matlab®, was coupled to the biogeochemical model. In this case, the velocity field was theoretically predefined following a semidiurnal cycle (Fig. 2). The bi-directional current direction always followed the longitudinal axis of the model domain. This set up maximizes the potential mitigation efficiency of the IMTA site because it maximizes the connectivity among finfish–shellfish structures, which is coherent with the goal of estimating maximum mitigation efficiency under optimal conditions (see above). Two

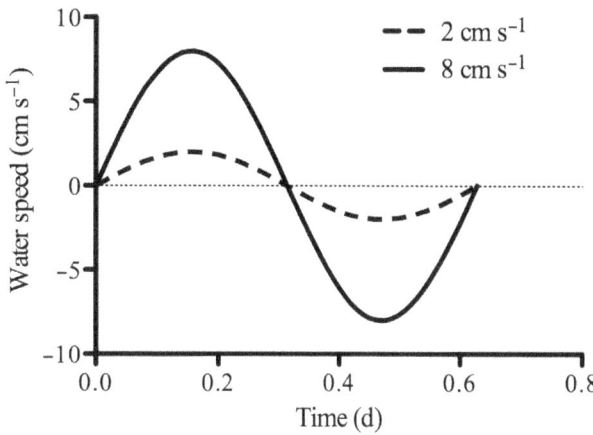

Fig. 2. Semidiurnal velocity patterns at the integrated multi-trophic aquaculture site tested in the factorial design experiment

potential scenarios with maximum horizontal water speed of 2 and 8 cm s^{-1} were defined, representing the minimum and maximum water speed at which a finfish aquaculture site can operate (values representative of typical Bay of Fundy facilities).

A factorial design was followed to explore different finfish–shellfish IMTA configurations. The following 3 farm designs were tested:

(1) 2 rows of 3 finfish cages separated by a central shellfish farm at 15 m distance from each row (15 m spacing; Fig. 1A);

(2) the same design but with no spacing between fish pens and mussel structures (0 m spacing, Fig. 1B); and

(3) co-culture of finfish and shellfish inside the fish pens (co-culture; Fig. 1C). Note that the total area available for mussel culture in this design is slightly larger than for the other 2 designs. Consequently, the density of mussels was reduced to have the same biomass of extractive species in all designs.

For each of the 3 farm designs, the 2 velocity fields (Fig. 2) were tested and 2 different levels of background P and D were used (Table 2, low and high), which represent oligo- and eutrophic conditions. In addition, 2 levels of particle production, $FF_{production}$ and $FW_{production}$, were considered (Table 2, high and low) that represent extreme husbandry practices. This factorial design gives a total of 24 scenarios. In addition, average conditions in terms of background and farm production levels were run for the different farm designs and hydrodynamic regimes (Table 2). Accordingly, the total number of tested scenarios reached 30.

The model was run for 2 tidal cycles (1.66 d) using a time step of 0.00001 d. Data were recorded every 0.01 d, but the first tidal cycle was discarded to minimize the impact of prescribing the homogeneous initial conditions across the model domain. The model was forced at the open boundaries with constant values of FW, FF, D and P, whose values change according to the different scenarios (Table 2). ShF was assumed to be negligible far from the mussel structures and consequently prescribed as 0 at the boundary. The velocity field was also prescribed as external forcing and assumed to be homogenous across the whole model domain. The velocity field followed Fig. 2 for the different hydrodynamic scenarios. The initial conditions of D and P were assumed to be homogeneous across the model domain (Table 2 for each scenario). Initial values of FF, FW and ShF across the domain were set up as 0 mg C m^{-3}.

A sensitivity test was carried out in the scenarios that reported maximum mitigation efficiency for each farm design to explore the impact of settling velocity on modelling outcomes. The sensitivity test was performed by increasing and decreasing the settling velocity of FF by 10% and recording the change in mitigation efficiency.

Particle tracking model

This model used the same grid as the ecosystem model and was run only for the 15 m spacing farm design scenario (Fig. 1A). The water depth was increased to 30 m to describe full particle trajectories up to and beyond the maximum depth of the mussel structures (15 m). The 3-dimensional finite-volume model FVCOM (Chen et al. 2003, 2007, Cowles et al. 2008) was applied to estimate water flows over the domain using a 1 m vertical resolution. The model was also modified according to Wu et al. (2014) to incorporate the effect of fish cage drag on water circulation. From the 2 water flow scenarios, only the 8 cm s^{-1} amplitude (Fig. 2) was tested here as it maximizes the connectivity between fish and shellfish farms for the 15 m spacing design. A 2 d spin-up period was used to ramp up the system from initial still conditions.

A Lagrangian particle tracking module was coupled to the hydrodynamic engine (Foreman et al. 2015) to reproduce fish waste trajectories in response to water flow and settling velocity. Thirty categories of settling velocity were considered, ranging from 0.1 to 3 cm s^{-1}. The different types of particles are identified by their average settling velocity (Table 1). The initial position of particles was randomly chosen inside each fish cage and both in the vertical and horizontal. Particles were released every 30 min over the first 12 h of an

experiment to cover the tidal cycle variability. Each particle was then followed for at least 24 h.

At each release time, 10 particles of each settling velocity category were introduced in each fish cage, amounting to a total of 43 200 particles per experiment or 1440 particles per settling velocity category. The experiment was repeated 3 times with different initial release locations such that probability analyses were performed on 4320 trajectories for each settling velocity category. Each trajectory was analysed to record whether the particle entered the mussel farm. In the affirmative, the depth at which the particle entered the farm was also recorded.

RESULTS AND DISCUSSION

Shellfish PR and mitigation efficiency

As noted in the introduction, the most relevant drivers of mitigation potential of suspended shellfish in an IMTA are (1) the connectivity between the supply of wastes from finfish cages and the location of the shellfish, and (2) shellfish physiology, namely PR, which ultimately will determine ingestion rate and consequently waste extraction capacity. The blue mussel *Mytilus edulis* was selected as the extractive species due to its wide distribution and filtration capacity. Mussels can be stocked at high densities, which, in combination with their PR, result in a high extraction capacity compared to other species, e.g. oysters (Filgueira et al. 2013). Therefore, mussels are ideal to test our main goal, that is, to evaluate the maximum mitigation efficiency under different finfish–shellfish IMTA configurations. The mussel DEB model determines the physiological response under different conditions in this numerical experiment. Mussel pumping activity was assumed to be physiologically controlled to maximize the ingestion of organic material (Winter 1976), with negligible food selection and rejection as pseudofaeces. Although food selection can take place at particle concentrations under the tested environmental conditions, more significant selectivity usually occurs at higher particle concentrations (>10 mg l^{-1}; Hawkins et al. 1998, 1999). These assumptions trigger maximum PRs under low background seston conditions (Fig. 3). Under these seston conditions, the contribution of finfish particles causes a minimal effect on pumping rate, although pumping rate is always slightly higher at low versus high production of finfish particles (Fig. 3). Therefore, low seston environments would promote the best conditions for mitigation potential through

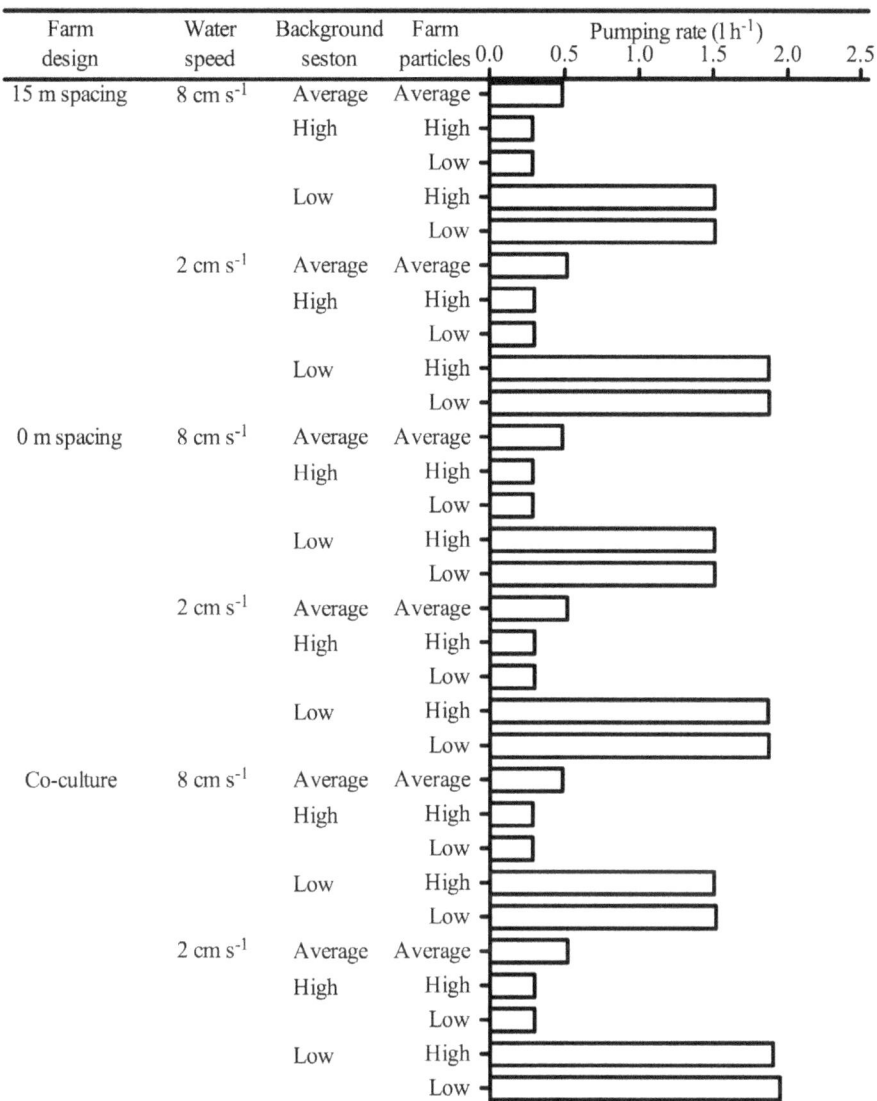

Fig. 3. Mussel pumping rate, standardized to a 60 mm shell length animal, in different scenarios defined in terms of farm design, water speed, background seston and production of particles in finfish cages

the stimulation of pumping activity and the relatively high contribution of wastes to the shellfish diet (Troell & Norberg 1998, Cranford et al. 2013, Irisarri et al. 2014).

For identical trophic conditions of background seston and farm particle load, and independently of farm design, the regime with low water speeds, 2 cm s^{-1}, promotes higher mussel PRs than high water speeds, 8 cm s^{-1} (Fig. 3). This is caused by the lower advection of background seston under the low water speed regime due to reduced horizontal flux and increased sinking, which decreases seston availability at the shellfish longline scale. This reduction in available organic material stimulates PRs in order to maximize organic ingestion. The design of the IMTA

farm is not a significant factor for mussel PR since similar pumping values are observed independently of the relative positioning of longlines to finfish cages (Fig. 3). This result suggests that environmental conditions, namely background seston levels, dominate the ecophysiological response of the extractive species rather than the design of the IMTA site. This is consistent with observations by Irisarri et al. (2014) in Ría de Ares-Betanzos (Galicia, Spain), which showed that *M. galloprovincialis* in 2 locations at different distances from finfish cages reported similar clearance rates, and similar scope for growth.

Waste mitigation efficiency was evaluated as the percentage of FF ingested by mussels, which is analogous to an exploitation efficiency used to describe resource acquisition in natural systems. This concept integrates the connectivity among IMTA structures and shellfish physiology, and it should not be understood as the net mitigation in terms of organic loading, which must include the contribution of ShF (Cranford et al. 2013). The combination of high PRs at low background seston, low production of particles by finfish cages (Fig. 3) and the strong connectivity that results from the co-culture of finfish and shellfish under a low current regime triggered a maximum mitigation efficiency of 6.6% (Fig. 4). The comparison of Figs. 3 & 4 highlights that high PRs are needed for greater mitigation efficiency, but the connectivity between finfish cages and shellfish longlines is also

critical. The use of a hypothetical bi-directional velocity field along the north-south axis maximizes the connectivity among IMTA structures and consequently mitigation efficiency. Accordingly, any deviation in the field from that velocity field would cause a decrease in mitigation efficiency. Nevertheless, the simulations suggest that the highest connectivity is obtained under the co-culture design at low water speed regimes, 2 cm s^{-1}. Under these conditions, FF are accessible for shellfish and consequently mitigation efficiency is increased. The increase of water speed to 8 cm s^{-1} in this farm configuration more rapidly flushes FF out of the cultivation area, reducing connectivity among IMTA species and consequently

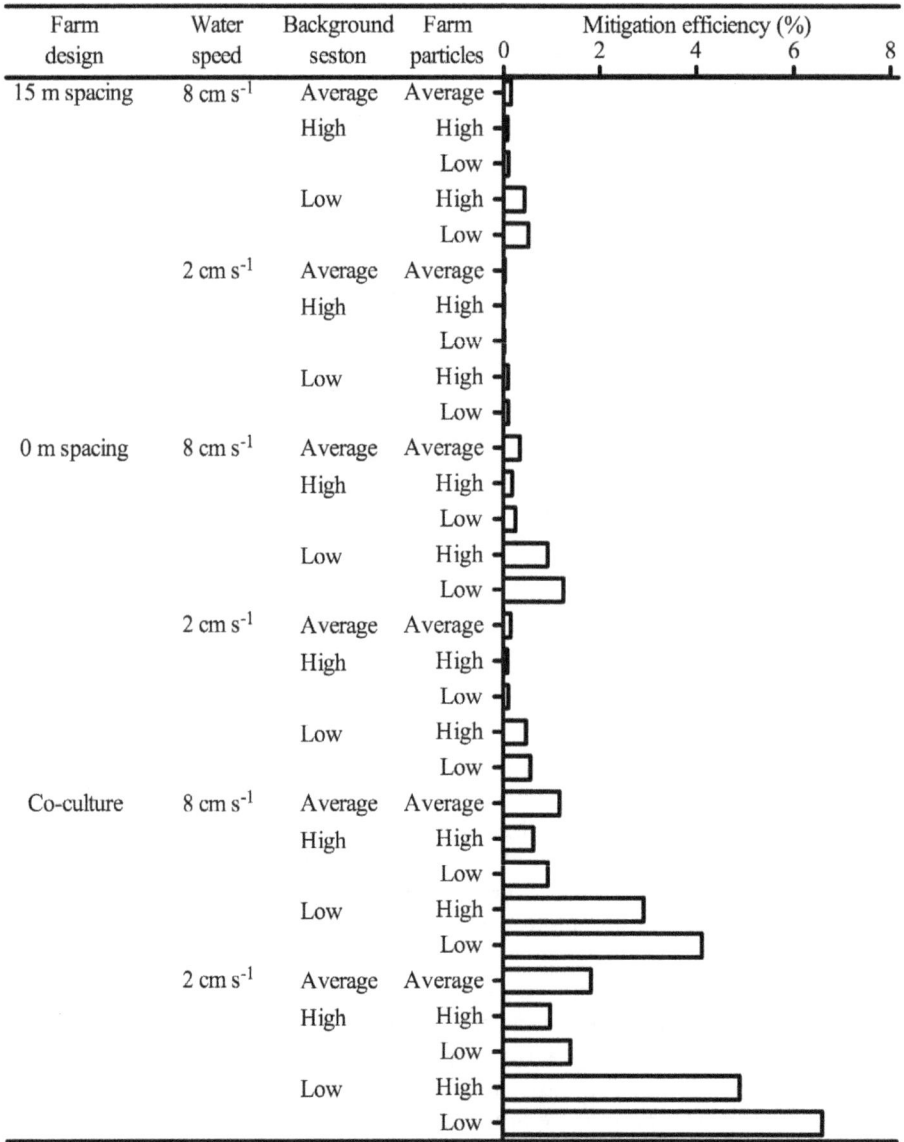

Fig. 4. Mitigation efficiency (percentage of produced fish faeces that have been ingested by mussels) in different scenarios defined in terms of farm design, water speed, background seston and production of particles in finfish cages

causing a reduction in mitigation efficiency (Fig. 4). The role of water currents in delivering particles is obvious when comparing the co-culture design with the other 2 configurations in which the mussels are farther from the source of faeces. In the 15 m and 0 m spacing designs, the increase in water currents enhances the delivery of wastes to the extractive species, consequently increasing mitigation efficiency, while the opposite occurs for co-culture (Fig. 4). These results are in good agreement with other observations and estimations that suggest that water currents have a direct impact on the concentration of wastes that exit finfish cages and consequently are available for integrated shellfish (Troell & Norberg 1998).

The effect of current speed on mitigation efficiency is strongly related to settling velocity of wastes. Due to the high settling velocity of FF, 3 cm s^{-1} (Table 1), the amount of time that the wastes are in suspension and consequently available as food source for mussels is limited. The relatively high current speed in the 15 m and 0 m spacing designs allow a greater proportion of FF to travel from cages to shellfish longlines before sinking and becoming unavailable. Despite the increase in mitigation efficiency at high current speeds, mitigation efficiency for 15 m and 0 m spacing designs is always below 0.55 and 1.25%, respectively, far from the 6.6% observed in co-culture, suggesting that settling velocity is key for determining the connectivity among IMTA structures and consequently mitigation potential. These low values match other predictions in the literature. Ferreira et al. (2012) suggested that the typical IMTA configurations in North America and Europe, equivalent to the one tested in this study, do not normally allow detection of concentration changes for particu-

late organic matter (POM) or chlorophyll in the field. Ferreira et al. (2012) studied these changes through modelling, and according to their Fig. 8, IMTA can reduce POM as much as 2.5% compared to a scenario with only fish (at the western boundary of the studied aquaculture park).

The sensitivity test demonstrated that the model is very sensitive to the settling velocity of FF. A change of ±10% in this parameter always produces changes greater than 10% in mitigation efficiency (Fig. 5). The response in mitigation efficiency follows the same pattern in the 3 farm designs (Fig. 5), that is, a reduction in settling velocity keeps the particles in suspension for longer, allowing the mussels more time to feed on FF. This effect is relatively more important when mussel farms are far away from the source of FF (Fig. 5), further demonstrating that settling velocity is key to understand the connectivity among IMTA structures. The reduction in settling velocity and the concomitant increase in mitigation efficiency matches the hypothesis of Cranford et al. (2013), who suggested that the available time to trap wastes is one of the critical limitations in finfish–shellfish IMTA mitigation potential.

Fluxes of matter in IMTA sites

The hydrodynamic–biogeochemical coupled model allows for the spatially explicit analysis of all of the IMTA particles in the water column at each time step. This facilitates the interpretation of fluxes of matter and understanding of spatial connectivity. For example, the normalized distribution of FF in the model domain at peak water speed in both hydrodynamic

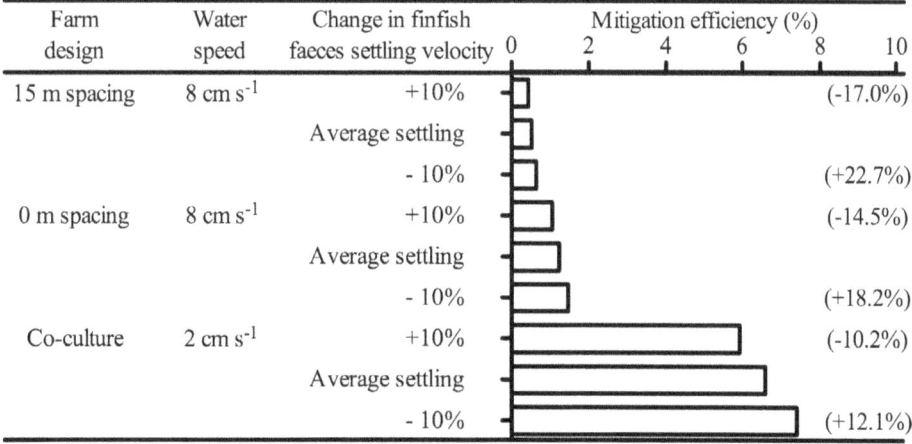

Fig. 5. Sensitivity test of mitigation efficiency to the settling velocity of finfish faeces evaluated in the scenarios that reported the highest mitigation efficiency for each farm design. The values in brackets represent the relative change in mitigation efficiency compared to the scenario with average settling velocity

regimes for the 15 m spacing scenario is presented in Fig. 6. Water speed in both tested hydrodynamic regimes ranges from −2 to +2 and −8 to +8 cm s⁻¹ (Fig. 2), but the time step presented in this figure represents the distribution at peak water speed. The distribution at peak velocity is the farthest distance that FF can travel from the cages before sinking to depths at which they are no longer accessible for mussel longlines (15 m). Given the effect of settling velocities on mitigation efficiency (Fig. 5), the distribution of FF with the settling velocities explored in the sensitivity test are also plotted in Fig. 6. The distribution of faeces explains the observations based on PRs regarding connectivity between finfish cages and mussel longlines: (1) high current regimes (Fig. 6A–C) favour the spreading of FF over a larger area than low current regimes (Fig. 6D–F), improving connec-

tivity; and (2) low settling rates maintain the highest FF concentrations in the water column, which is evident when comparing Fig. 6A vs. 6C and 6D vs. 6F, favouring connectivity.

Even when selecting the scenario with the highest degree of connectivity (high currents and low settling velocities) (Fig. 6A), the concentration of FF at the mussel longlines is negligible compared to the concentration at the center of the cages. However, the existence of a plume of material exiting the finfish cages is critical for the emplacement of extractive species and exploitation of IMTA synergies (Reid et al. 2010). The low concentration of particles at the shellfish farm predicted in this study (Fig. 6) suggests that this theoretical plume is weak, which would explain the low mitigation efficiency (Fig. 4). Several studies have focused on the identification of this

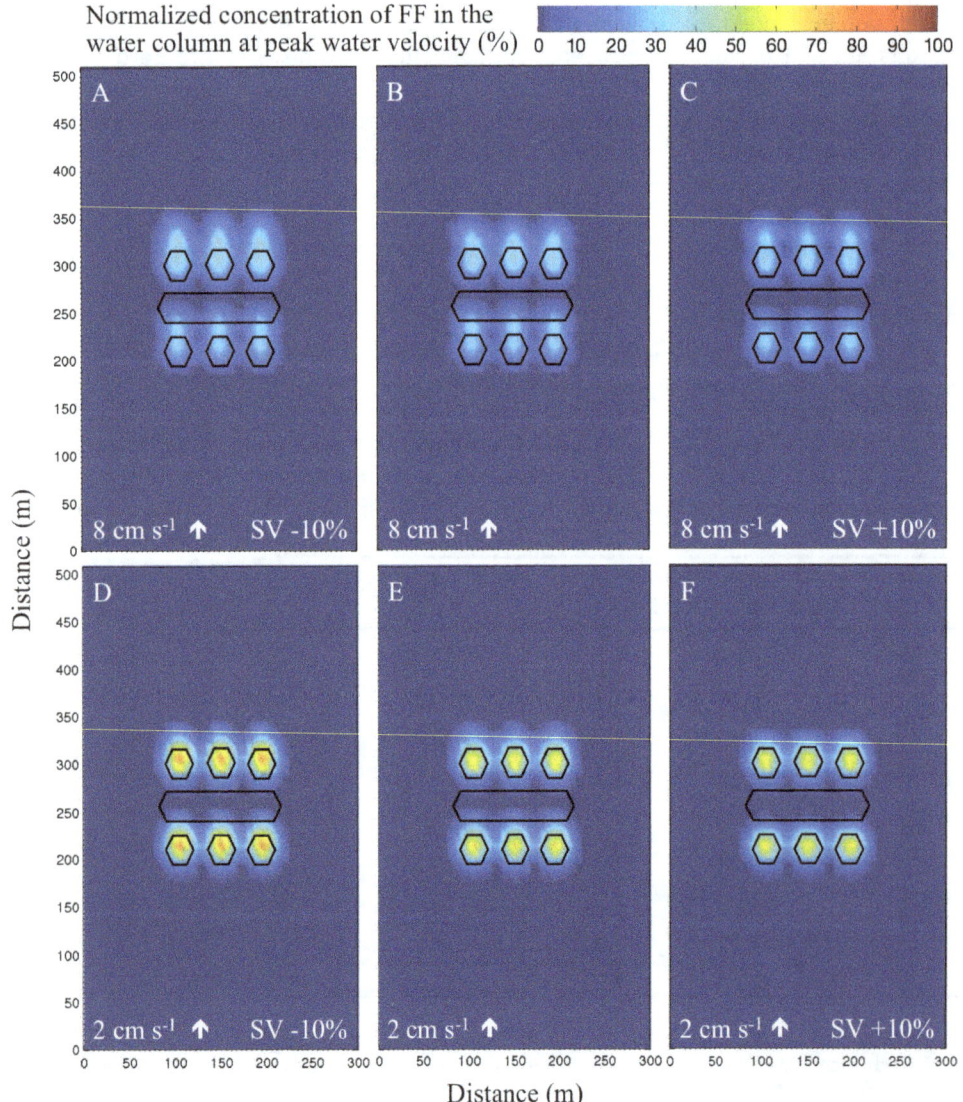

Fig. 6. Normalized concentration of finfish faeces (FF, %) in the water column at the time step in which velocity is at maximum in the farm design with 15 m spacing (see Fig. 1) in both hydrodynamic regimes (A–C: high current, D–F: low current) for the 3 tested FF settling velocities (SV). The concentration of FF has been normalized to the maximum concentration across the 6 simulations for comparative purposes. The white arrow represents the direction of water currents at peak velocity

waste plume, and the literature shows a degree of controversy. For example, Jones & Iwama (1991), Lefebvre et al. (2000) and MacDonald et al. (2011) observed elevated particle concentrations around finfish cages, but Pridmore & Rutherford (1992), Buschmann et al. (1996) and Cheshuk et al. (2003) did not observe significant enrichment. High-resolution suspended particle mapping around salmon pens in the Bay of Fundy occasionally detected low levels of particle enhancement (<1 mg l^{-1}), with any significant effect being highly localized and episodic (Brager et al. 2015). Despite these confounding observations, there is a general consensus that shellfish should be placed as close as possible to finfish cages (Brown et al. 1987, Gowen & Bradbury 1987, Gowen et al. 1988, Coyne et al. 1994, Findlay et al. 1995, Cheshuk et al. 2003, Lander et al. 2012, 2013, Brager et al. 2015) to avoid the strong dilution of the plume beyond the cage, which occurs beyond 10 m according to Lander et al. (2013). This placement requires a very specific design, such as the square cages used in British Columbia (Canada), where the bivalve extractive species can be located right beside the finfish cage, potentially benefiting from the additional organic matter (Weldrick & Jelinski 2016). Another alternative arrangement that could benefit IMTA is the spatial design developed in some Chinese bays, such as Sanggou Bay, in which most of the available area is occupied by finfish, shellfish and seaweed farms (Fang et al. 2016 and references therein). However, this placement or spatial arrangement is not usually practical for logistical reasons in most of the farms elsewhere.

The fact that FF are maintained in suspension for only a limited time due to high settling velocity suggests that vertical fluxes dominate IMTA sites. Brager et al. (2015) also highlighted that several processes at the cage level such as turbulence, reduced flow due to baffling by the structures, biofouling and the presence of extracellular polymers can enhance particle aggregation and the vertical flux of fine particles. The vertical fluxes of organic carbon have been computed at 15 m depth, which is assumed to be close to the average depth of a finfish cage (~13 m) and the maximum depth at which mussel culture is commonly carried out worldwide. Consequently, beyond that point, particles are assumed to be inaccessible to mussels. The normalized vertical flux of organic carbon over a tidal cycle is presented in Fig. 7 for the best scenarios in terms of mitigation efficiency for each farm design. The simulations strongly suggest that the highest vertical fluxes of organic matter in IMTA sites are located directly beneath finfish cages (Fig. 7). This is despite the fact that in these simulations the settleable fraction of feed wastes are not included because they cannot be mitigated by mussels given that (1) they sink directly to the bottom due to their even higher settling velocity, and (2) they are too big to be ingested by mussels. Although the final fate of the vertical flux of organic carbon at 15 m depth also depends on the total depth of the area (see below), these simulations strongly agree with other studies, which suggest that the impact of organic loading on finfish farms is restricted to the near-field (e.g. Brooks & Mahnken 2003, Wildish et al. 2004, Chang et al. 2014). These results high-

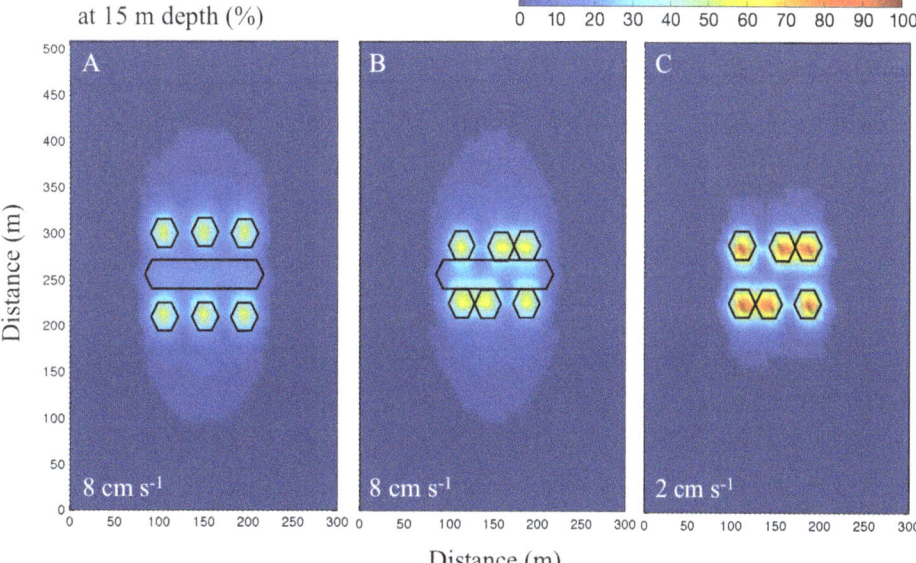

Fig. 7. Normalized vertical flux of carbon towards the bottom at 15 m depth averaged over a tidal cycle for the best scenarios in terms of mitigation efficiency in the 3 farm designs (see Fig. 1): (A) 15 m separation, (B) 0 m separation and (C) co-culture. The vertical flux of carbon was normalized to the maximum flux across the 3 simulations for comparative purposes

lighting the strong vertical fluxes are in good agreement with Cubillo et al. (2016) and suggest a limited usefulness of shellfish for mitigating solid wastes in IMTA sites using current technology and aquaculture practices.

The second significant conclusion that can be extracted from Fig. 7 is that despite the highest removal of FF in the co-culture scenario, the vertical flux of organic carbon is also highest in the co-culture scenario (Fig. 7C), suggesting that there is no net mitigation in terms of organic loading. The direct relationship between mitigation efficiency and the peak of vertical flux of organic carbon suggests that positive mitigation with FF close to mussels, is negative for organic loading due to faeces production by the extractive species. To overcome this limitation, Cranford et al. (2013) suggested that 15 to 35% of the organic matter that mussels consume should be from fish waste to significantly reduce net benthic organic loading and compensate for the new production of ShF. Following a similar approach, Reid et al. (2013a) suggested that depending on diet quality, 11 to 20% of total mussel diet must be comprised of fish farm solids, in order to reduce the site-wide organic load.

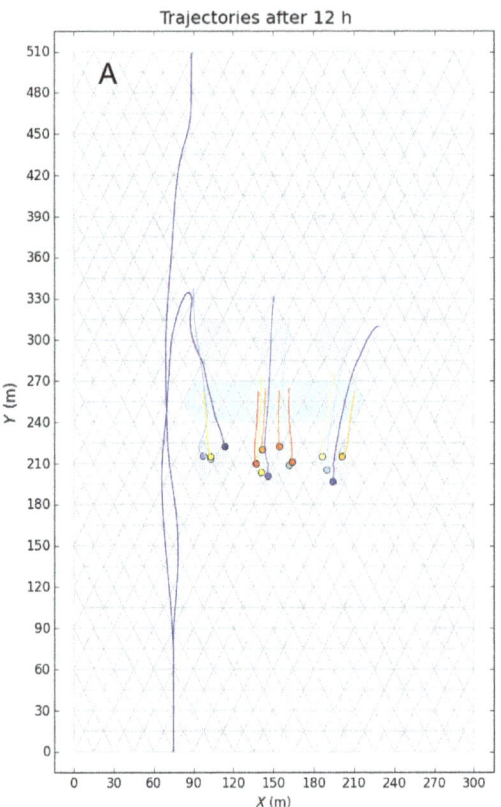

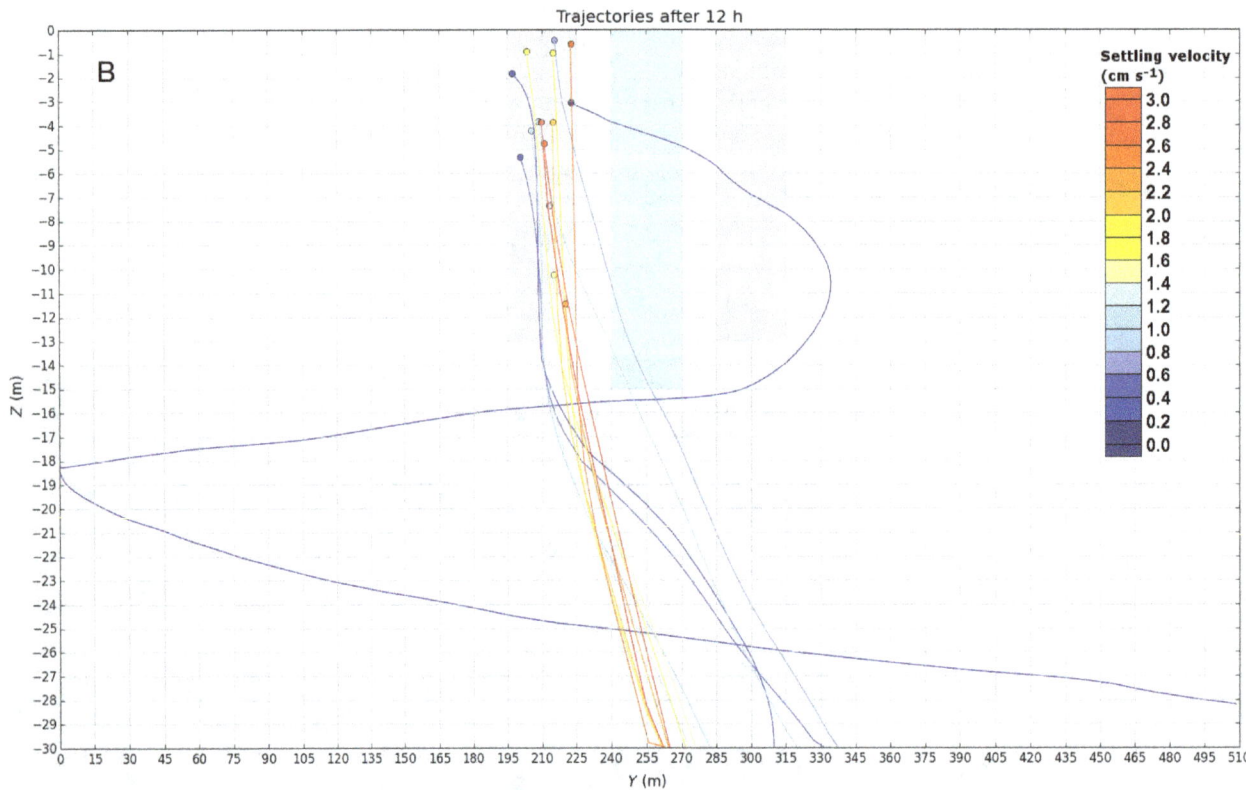

Fig. 8. (A) Plan view (transverse and longitudinal, X and Y, respectively) and (B) vertical profile trajectories after 12 h of a subset of particles with different settling velocities released at random positions in 3 finfish cages (light grey). Mussel longlines are represented in light blue. Note that the vertical and horizontal scales (Z and Y, respectively) in (B) are different for visualization purposes

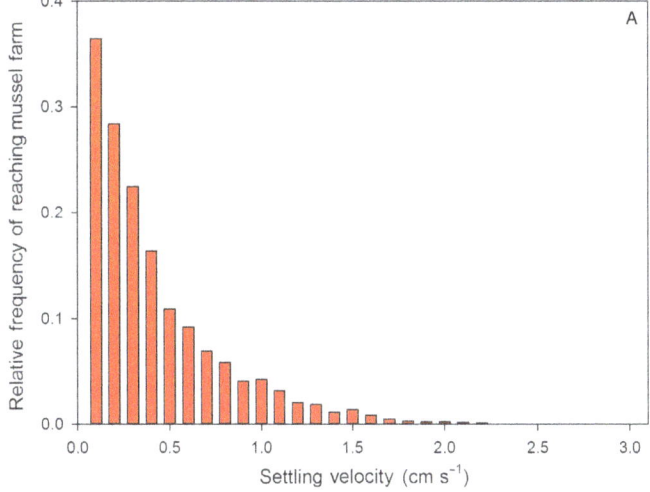

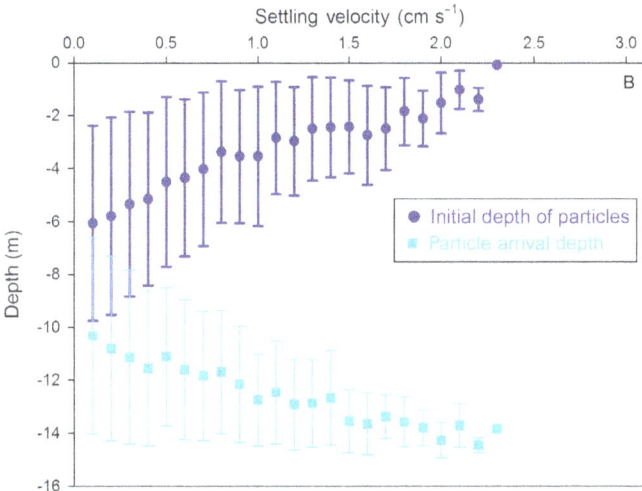

Fig. 9. (A) Probability of reaching the shellfish longlines for particles with different settling velocities released at finfish cages. (B) Mean (±SD) depth at which these particles reach the shellfish longlines (squares) and mean (±SD) initial depth of these particles at finfish cages (dots)

Fate of particles with different settling velocities

One of the key assumptions in the ecosystem model is that each particle type is characterized by averaged parameters and the fluxes are characterized in terms of concentration per time. However, it is well known that FF occur as a range of particle sizes and settling velocities. This conceptual simplification is needed in order to cope with the computational resources required to simulate a complex ecosystem. Nevertheless, this simplification can increase the uncertainty of modelling outcomes, especially when there is sensitivity to one of the parameters related to this simplifi-

cation, i.e. settling velocity. In order to explore the effects of a range of settling velocities on the previous results calculated with averaged values, a different modelling exercise was carried out with focus on the physics of the wastes rather than on the biology. The particle tracking numerical experiment developed in FVCOM allowed a probabilistic analysis of finfish–shellfish connectivity based on the analysis of the trajectories of particles with different settling velocities. In addition, FVCOM allows for the incorporation of the feedback of the aquaculture structures on local currents, that is, flow reduction, which provides a more realistic approach to horizontal fluxes. Fig. 8 presents an example of how FVCOM depicts horizontal and vertical trajectories after 12 h of a subset of particles with different settling velocities released at random positions inside 3 different finfish cages in the 15 m spacing design. This analysis (Fig. 8) provides a visual example of the impact of settling velocity on particle trajectory and probability for waste particles to reach the shellfish farm, showing that this probability increases for particles with low settling velocities while heavy particles tend to sink directly to the bottom in the proximity of the finfish cage.

The trajectories of 129 600 particles (43 200 × 3 experiments) with different settling velocity released in the finfish cages were analysed with the aim of identifying the percentage of the particles that encountered the mussel farm (Fig. 9A). This probability exponentially decays with the settling velocity, reaching null probability at 2.4 cm s^{-1} (Fig. 9A). The results obtained with FVCOM partially disagree with the previous simulations carried out with the hydrodynamic–biogeochemical coupled model, which predicted that some FF (~3 cm s^{-1}, Table 1) could reach the shellfish farm under the same conditions (Fig. 6B). Although the set up of both simulations is conceptually similar, FF are parameterized as independent particles in FVCOM but as a concentration in the coupled model, which implies a different mathematical approach to transport. This can cause the partial disagreement. In addition, FVCOM accounts for flow reduction caused by farming structures, which is not included in the coupled model. Another source of differences is the vertical spatial resolution. The FVCOM model is constructed with 1 m thick layers, but the coupled model integrates 15 m in each layer with the aim of including the whole cage/longline in the same vertical layer in order to save computational resources. Despite the disagreement, both results confirm that heavy particles tend to settle to the bottom before reaching the shellfish farm.

The fine vertical resolution of the FVCOM model also allows for an analysis of the depth at which particles reach the mussel farm. As expected, this depth increases with settling velocity from ~10 m to ~14 m for 0.1 and 2.4 cm s^{-1} particles, respectively (Fig. 9B). The degree of success in reaching the shellfish farm is highly dependent on the depth at which the particle is released (Fig. 9B). While the average initial depth of particles with settling velocity of 0.1 cm s^{-1} that successfully reach the mussel farm is ~6 m, the initial depth of the only particle with a settling velocity of 2.4 cm s^{-1} that reaches the shellfish cage is ~0 m (Fig. 9B). This result matches previous observations and suggests that settling velocity and vertical fluxes dominate IMTA sites, and the connectivity between finfish cages and shellfish structures is highly dependent on positioning, suggesting that only structures that are very close together will be able to be effectively connected.

CONCLUSIONS

The modelling exercises performed in this study focussed on exploring the maximum mitigation efficiency of mussel–finfish IMTA sites, with focus on the solid wastes that potentially cause organic loading. These simulations were performed under ideal conditions that maximize the connectivity between finfish and shellfish cultured areas, e.g. ideal current directionality, and consequently it is expected that the mitigation efficiency estimations are maxima. Despite these ideal conditions that maximize mitigation, the models predicted low mitigation efficiencies, which is a direct consequence of the high settling velocity of finfish aquaculture wastes (faeces and feed), and the consequent dominance of strong vertical fluxes of organic material. A recent study by Bannister et al. (2016) provided a detailed analysis of settling velocity for different size classes of salmon, showing that the settling velocity of faecal particles is higher than 5 cm s^{-1} for more than 60%, and higher than 2.5 cm s^{-1} for more than 80%. Although faecal settling velocity is largely influenced by dietary ingredients (Reid et al. 2009), the results from Bannister et al. (2016) suggest that perhaps our study was very conservative in assuming that the average settling velocity is 3 cm s^{-1}. This updated distribution of settling velocities and the results from our study reinforce the idea that vertical fluxes of organic matter are stronger than horizontal ones in finfish farms, which explain the weak plumes of wastes and limited connectivity

among IMTA structures, critical for mitigation purposes (Reid et al. 2010). The potential for capturing solid wastes improves under specific conditions such as oligotrophic environments, and optimal spatial arrangements, but even in these situations, the net mitigation should be carefully explored because of the additional contribution of ShF to organic loading. In addition, other aspects should be considered to evaluate the viability of IMTA sites. For example, the co-culture design, despite being the one that maximizes the mitigation efficiency, is probably not viable at the commercial scale with the current technology. Other important aspects that would need additional research to evaluate the viability are the effects of the co-cultured species on disease transmission and the effect of chemicals such as medicines or pesticides on the co-cultured species. These interactions are poorly studied and could be critical at IMTA sites. In summary, and similarly to Cubillo et al. (2016), the low potential for mitigation suggests that shellfish suspended culture cannot significantly reduce organic loading. The dominance of vertical fluxes of organic material in finfish aquaculture suggests that organic extractive species in IMTA farms should be located in the benthic environment directly beneath finfish cages in order to maximize the mitigation of accumulated organic wastes. However, it is important to highlight that shellfish could indirectly mitigate organic loading by feeding on phytoplankton populations that could be enhanced due to the remineralization of organic loading and the excretion of nitrogen by fish. These positive effects should be analysed at the watershed scale. Further experimental and modelling studies with focus on the mitigation potential of deposit-feeders such as sea cucumbers will be key in evaluating the future of IMTA.

Acknowledgements. This research was funded by the Natural Sciences and Engineering Research Council of Canada (NSERC) Strategic Research Network titled Canadian Integrated Multi-Trophic Aquaculture Network (CIMTAN). Additional support was provided by an NSERC Industrial Research Chair to J.G.

LITERATURE CITED

Bannister RJ, Johnsen IA, Hansen PK, Kutti T, Asplin L (2016) Near- and far-field dispersal modelling of organic waste from Atlantic salmon aquaculture in fjord systems. ICES J Mar Sci 73:2408–2419

Brager LM, Cranford PJ, Grant J, Robinson SMC (2015) Spatial distribution of suspended particulate wastes at open-water Atlantic salmon and sablefish aquaculture farms in Canada. Aquacult Environ Interact 6:135–149

Brigolin D, Meccia VL, Venier C, Tomassetti P, Porrello S, Pastres R (2014) Modelling biogeochemical fluxes across a Mediterranean fish cage farm. Aquacult Environ Interact 5:71–88

Brooks KM, Mahnken CVW (2003) Interactions of Atlantic salmon in the Pacific Northwest environment II. Organic wastes. Fish Res 62:255–293

Brown JR, Gowen RJ, McLusky DS (1987) The effect of salmon farming on the benthos of a Scottish sea loch. J Exp Mar Biol Ecol 109:39–51

Buschmann A, Troell M, Kautsky N, Kautsky L (1996) Integrated tank cultivation of salmonids and *Gracilaria chilensis*. Hydrobiologia 326-327:75–84

Chang BD, Page FH, Losier RJ, McCurdy EP (2014) Organic enrichment at salmon farms in the Bay of Fundy, Canada: DEPOMOD predictions versus observed sediment sulfide concentrations. Aquacult Environ Interact 5:185–208

Chen C, Liu H, Beardsley RC (2003) An unstructured grid, finite-volume, three-dimensional, primitive equations ocean model: application to coastal ocean and estuaries. J Atmos Ocean Technol 20:159–186

Chen C, Huang H, Beardsley RC, Liu H, Xu Q, Cowles G (2007) A finite volume numerical approach for coastal ocean circulation studies: comparisons with finite difference models. J Geophys Res 112:C03018

Cheshuk BW, Purser GJ, Quintana R (2003) Integrated open-water mussel (*Mytilus planulatus*) and Atlantic salmon (*Salmo salar*) culture in Tasmania, Australia. Aquaculture 218:357–378

Chopin T, Cooper JA, Reid G, Cross S, Moore C (2012) Open-water integrated multi-trophic aquaculture: environmental biomitigation and economic diversification of fed aquaculture by extractive aquaculture. Rev Aquacult 4:209–220

Cowles GW, Lentz SJ, Chen C, Xu Q, Beardsley RC (2008) Comparison of observed and model-computed low frequency circulation and hydrography on the New England Shelf. J Geophys Res 113:C09015

Coyne R, Hiney M, O'Connor B, Kerry J, Cazabon D, Smith P (1994) Concentration and persistence of oxytetracycline in sediments under a marine salmon farm. Aquaculture 123:31–42

Cranford PJ, Reid GK, Robinson SMC (2013) Open water integrated multi-trophic aquaculture: constraints on the effectiveness of mussels as an organic extractive component. Aquacult Environ Interact 4:163–173

Cromey CJ, Nickell TD, Black KD (2002) DEPOMOD—modeling the deposition and biological effects of waste solids from marine cage farms. Aquaculture 214:211–239

Cubillo AM, Ferreira JG, Robinson SMC, Pearce CM, Corner RA, Johansen J (2016) Role of deposit feeders in integrated multi-trophic aquaculture—a model analysis. Aquaculture 453:54–66

Duarte P, Meneses R, Hawkins AJS, Zhu M, Fang J, Grant J (2003) Mathematical modelling to assess the carrying capacity for multi-species culture within coastal waters. Ecol Model 168:109–143

Fang J, Zhang J, Xiao T, Huang D, Liu S (2016) Integrated multi-trophic aquaculture (IMTA) in Sanggou Bay, China. Aquacult Environ Interact 8:201–205

FAO (Food and Agriculture Organization of the United Nations) (2014) The state of world fisheries and aquaculture. FAO, Rome

Farias-Sánchez JA (1983) Experimental trial on the growth

of mussels, *Mytilus edulis*, on ropes suspended from marine fish cages. MSc thesis, University of Stirling

Fennel W, Neumann T (2004) Introduction to the modelling of marine ecosystems. Elsevier Oceanography Series, 72. Elsevier, Amsterdam

Ferreira JG, Saurel C, Ferreira JM (2012) Cultivation of gilthead bream in monoculture and integrated multi-trophic aquaculture. Analysis of production and environmental effects by means of the FARM model. Aquaculture 358–359:23–34

Ferreira JG, Grant J, Verner-Jeffreys DW, Taylor NGH (2013) Carrying capacity for aquaculture, modeling frameworks for determination of. In: Christou P, Savin R, Costa-Pierce B, Misztal I, Whitelaw B (eds) Sustainable food production. Springer Science+Business Media, New York, NY, p 417–448

Filgueira R, Rosland R, Grant J (2011) A comparison of scope for growth (SFG) and dynamic energy budget (DEB) models applied to the blue mussel (*Mytilus edulis*). J Sea Res 66:403–410

Filgueira R, Grant J, Bacher C, Carreau M (2012) A physical-biogeochemical coupling scheme for modeling marine coastal ecosystems. Ecol Inform 7:71–80

Filgueira R, Comeau LA, Landry T, Grant J, Guyondet T, Mallet A (2013) Bivalve condition index as an indicator of aquaculture intensity: a meta-analysis. Ecol Indic 25: 215–229

Findlay RH, Watling L, Mayer LM (1995) Environmental impact of salmon net-pen culture on marine benthic communities in Maine: a case study. Estuaries 18: 145–179

Foreman MGG, Chandler PC, Stucchi DJ, Garver KA, Guo M, Morrison J, Tuele D (2015) The ability of hydrodynamic models to inform decisions on the siting and management of aquaculture facilities in British Columbia. Canadian Science Advisory Secretariat Research Document 2015/005, Department of Fisheries and Oceans Canada, Ottawa

Gillibrand PA, Gubbins MJ, Greathead C, Davies IM (2002) Scottish executive locational guidelines for fish farming: predicted levels of nutrient enhancement and benthic impact. Scottish Fisheries Research Report Number 63/2002

Gowen RJ, Bradbury NB (1987) The ecological impact of salmonid farming in coastal waters: a review. Oceanogr Mar Biol Annu Rev 25:563–575

Gowen RJ, Brown J, Bradbury NB, McLusky DS (1988) Investigations into benthic enrichment, hypernutrification and eutrophication associated with mariculture in Scottish coastal waters. J Exp Mar Biol Ecol 109:39–51

Grant J, Filgueira R (2011) The application of dynamic modelling to prediction of production carrying capacity in shellfish farming. In: Shumway S (ed) Shellfish aquaculture and the environment. Wiley-Blackwell Science Publishers, Ames, IA, p 135–154

Gryska A, Parsons J, Shumway SE, Geib K, Emery I, Kuenster S (1996) Polyculture of sea scallops suspended from salmon cages. J Shellfish Res 15:481

Handå A, Min H, Wang X, Broch OJ, Reitan KI, Helge R, Olsen Y (2012) Incorporation of fish feed and growth of blue mussels (*Mytilus edulis*) in close proximity to salmon (*Salmo salar*) aquaculture: implications for integrated multi-trophic aquaculture in Norwegian coastal waters. Aquaculture 356–357:328–341

Hargrave BT (2010) Empirical relationships describing benthic impacts of salmon aquaculture. Aquacult Environ Interact 1:33–46

Hawkins AJS, Bayne BL, Bougrier S, Héral M and others (1998) Some general relationships in comparing the feeding physiology of suspension-feeding bivalve molluscs. J Exp Mar Biol Ecol 219:87–103

Hawkins AJS, James MR, Hickman RW, Hatton S, Weatherhead M (1999) Modelling of suspension-feeding and growth in the green-lipped mussel *Perna canalicus* exposed to natural and experimental variations of seston availability in the Marlborough Sounds, New Zealand. Mar Ecol Prog Ser 191:217–232

Herndl GJ, Reinthaler T (2013) Microbial control of the dark end of the biological pump. Nat Geosci 6:718–724

Irisarri J, Cubillo AM, Fernández-Reiriz MJ, Labarta U (2013) Growth variations within a farm of mussel (*Mytilus galloprovincialis*) held near fish cages: importance for the implementation of integrated aquaculture. Aquacult Res 46:1988–2002

Irisarri J, Fernández-Reiriz MJ, Cranford PJ, Labarta U (2014) Effects of seasonal variations in phytoplankton on the bioenergetic responses of mussels (*Mytilus galloprovincialis*) held on a raft in the proximity of red sea bream (*Pagellus bogaraveo*) net-pens. Aquaculture 428-429:41–53

Jiang Z, Wang G, Fang J, Mao Y (2013) Growth and food sources of Pacific oyster *Crassostrea gigas* integrated culture with sea bass *Lateolabrax japonicus* in Ailian Bay, China. Aquacult Int 21:45–52

Jones TO, Iwama GK (1991) Polyculture of the Pacific oyster, *Crassostrea gigas* (Thunberg), with Chinook salmon, *Oncorhynchus tshawytscha*. Aquaculture 92:313–322

Jusup M, Gecek S, Legovic T (2007) Impact of aquacultures on the marine ecosystem: modelling benthic carbon loading over variable depth. Ecol Model 200:459–466

Lander T, Barrington K, Robinson SMC, MacDonald B, Martin J (2004) Dynamics of the blue mussel as an extractive organism in an integrated aquaculture system. Bull Aquacult Assoc Can 104:19–28

Lander TR, Robinson SMC, MacDonald BA, Martin JD (2012) Enhanced growth rates and condition index of blue mussels (*Mytilus edulis*) held at integrated multitrophic aquaculture (IMTA) sites in the Bay of Fundy. J Shellfish Res 31:997–1007

Lander TR, Robinson SMC, MacDonald BA, Martin JD (2013) Characterization of the suspended organic particles released from salmon farms and their potential as a food supply for the suspension feeder, *Mytilus edulis* in integrated multi-trophic aquaculture (IMTA) systems. Aquaculture 406-407:160–171

Lefebvre S, Barille L, Clerc M (2000) Pacific oyster (*Crassostrea gigas*) feeding responses to a fish-farm effluent. Aquaculture 187:185–198

Liutkus M, Robinson SMC, MacDonald C, Reid G (2012) Quantifying the effects of diet and mussel size on the biophysical properties of the blue mussel, *Mytilus* spp., feces egested under simulated IMTA conditions. J Shellfish Res 31:69–77

MacDonald BA, Robinson SMC, Barrington KA (2011) Feeding activity of mussels (*Mytilus edulis*) held in the field at an integrated multi-trophic aquaculture (IMTA) site (*Salmo salar*) and exposed to fish food in the laboratory. Aquaculture 314:244–251

McDonnell AMP, Buesseler KO (2010) Variability in the average sinking velocity of marine particles. Limnol

Oceanogr 55:2085–2096

Melbourne-Thomas J, Johnson CR, Fulton EA (2011) Characterizing sensitivity and uncertainty in a multiscale model of a complex coral reef system. Ecol Model 222: 3320–3334

Navarrete-Mier F, Sanz-Lázaro C, Marín A (2010) Does bivalve mollusc polyculture reduce marine fin fish farming environmental impact? Aquaculture 306:101–107

Nobre AM, Ferreira JG, Nunes JP, Yan X and others (2010) Assessment of coastal management options by means of multilayered ecosystem models. Estuar Coast Shelf Sci 87:43–62

Nunes JP, Ferreira JG, Gazeau F, Lencart-Silva J, Zhang XL, Zhu MY, Fang JG (2003) A model for sustainable management of shellfish polyculture in coastal bays. Aquaculture 219:257–277

Parsons GJ, Shumway SE, Kuenstner S, Gryska A (2002) Polyculture of sea scallops (*Placopecten magellanicus*) suspended from salmon cages. Aquacult Int 10:65–77

Peharda M, Župan I, Bav evi L, Franki A, Klanjš ek T (2007) Growth and condition index of mussel *Mytilus galloprovincialis* in experimental integrated aquaculture. Aquacult Res 38:1714–1720

Pridmore RD, Rutherford JC (1992) Modelling phytoplankton abundance in a small enclosed bay used for salmon farming. Aquacult Fish Manag 23:525–542

Reid GK (2007) Nutrient releases from salmon aquaculture. In: Costa-Pierce B (ed) Nutrient impacts of farmed Atlantic salmon (*Salmo salar*) on pelagic ecosystems and implications for carrying capacity. World Wildlife Fund, Washington, DC, p 7–22

Reid GK, Liutkus M, Robinson SMC, Chopin TR and others (2009) A review of the biophysical properties of salmonid faeces: implications for aquaculture waste dispersal models and integrated multi-trophic aquaculture. Aquacult Res 40:257–273

Reid GK, Liutkus M, Bennett A, Robinson SMC, MacDonald B, Page F (2010) Absorption efficiency of blue mussels (*Mytilus edulis* and *M. trossulus*) feeding on Atlantic salmon (*Salmo salar*) feed and fecal particulates: implications for integrated multi-trophic aquaculture. Aquaculture 299:165–169

Reid GK, Robinson SMC, Chopin T, MacDonald BA (2013a) Dietary proportion of fish culture solids required by shellfish to reduce the net organic load in open-water Integrated Multi-Trophic Aquaculture: a scoping exercise with cocultured Atlantic salmon (*Salmo salar*) and blue mussel (*Mytilus edulis*). J Shellfish Res 32:509–517

Reid GK, Chopin T, Robinson SMC, Azevedo P, Quinton M, Belyea E (2013b) Weight ratios of the kelps, *Alaria esculenta* and *Saccharina latissima*, required to sequester dissolved inorganic nutrients and supply oxygen for Atlantic salmon, *Salmo salar*, in Integrated Multi-Trophic Aquaculture systems. Aquaculture 408-409:34–46

Ren JS, Stenton-Dozey J, Plew DR, Fang J, Gall M (2012) An ecosystem model for optimising production in integrated multitrophic aquaculture systems. Ecol Model 246:34–46

Rensel JE, Bright K, Siegrist Z (2011) Integrated fish–shellfish mariculture in Puget Sound. NOAA National Marine Aquaculture Initiative Award NA080AR4170860. Rensel Associates Aquatic Sciences, in association with American Gold Seafoods and Taylor Shellfish, Arlington, WA

Robinson SMC, Reid GK (2014) Review of the potential near-and far-field effects of the organic extractive component of integrated multi-trophic aquaculture (IMTA) in southwest New Brunswick with emphasis on the blue mussel (*Mytilus edulis*). Canadian Science Advisory Secretariat Research Document 2014/026, Department of Fisheries and Oceans Canada, Ottawa

Sarà G, Zenone A, Tomasello A (2009) Growth of *Mytilus galloprovincialis* (Mollusca, Bivalvia) close to fish farms: a case of integrated multi-trophic aquaculture in the Tyrrhenian Sea. Hydrobiologia 636:129–136

Sarà G, Lo Martire M, Sanflippo M, Pulicanò G and others (2011) Impacts of marine aquaculture at large spatial scales: evidences from N and P catchment loading and phytoplankton biomass. Mar Environ Res 71:317–324

Sarà G, Reid GK, Rinaldi A, Palmeri V, Troell M, Kooijman SALM (2012) Growth and reproductive simulation of candidate shellfish species at fish cages in the southern Mediterranean, dynamic energy budget (DEB) modelling for integrated multi-trophic aquaculture. Aquaculture 324–325:259–266

Soto D (2009) Integrated mariculture: a global review. FAO Fish Aquacult Tech Pap 529. FAO, Rome

Stigebrandt A (2011) Carrying capacity: general principles of model construction. Aquacult Res 42:41–50

Stirling HP, Okumus I (1995) Growth and production of mussels (*Mytilus edulis* L.) suspended at salmon cages and shellfish farms in two Scottish sea lochs. Aquaculture 134:193–210

Strain PM, Hargrave BT (2005) Salmon aquaculture, nutrient fluxes and ecosystem processes in southwestern New Brunswick. In: Hargrave BT (ed) Environmental effects of marine finfish aquaculture. Springer, Berlin, p 29–57

Taylor BE, Jamieson G, Carefoot TH (1992) Mussel culture

in British Columbia: the influence of salmon farms on growth of *Mytilus edulis*. Aquaculture 108:51–66

Troell M, Norberg J (1998) Modelling of suspended solids in an integrated salmon-mussel culture. Ecol Model 110: 65–77

Tsagaraki TM, Petihakis G, Tsiaras K, Triantafyllou G and others (2011) Beyond the cage: ecosystem modeling for impact evaluation in aquaculture. Ecol Model 222: 2512–2523

United Nations, Department of Economic and Social Affairs, Population Division (2013) World population prospects: the 2012 revision, highlights and advance tables. ESA/P/WP. 228. Available at esa.un.org/unpd (accessed on 13 June 2013)

Wallace JC (1980) Growth rates of different populations of the edible mussel, *Mytilus edulis*, in north Norway. Aquaculture 19:303–311

Weldrick CK, Jelinski DE (2016) Resource subsidies from multi-trophic aquaculture affect isotopic niche width in wild blue mussels (*Mytilus edulis*). J Mar Syst 157: 118–123

Wildish DJ, Hughes-Clarke JE, Pohle GW, Hargrave BT, Mayer LM (2004) Acoustic detection of organic enrichment in sediments at a salmon farm is confirmed by independent groundtruthing methods. Mar Ecol Prog Ser 267:99–105

Winter JE (1976) A critical review on some aspects of filter-feeding in lamellibranchiate bivalves. Haliotis 7:71–87

Wu Y, Chaffey J, Law B, Greenberg DA, Drozdowski A, Page F, Haigh S (2014) A three-dimensional hydrodynamic model for aquaculture: a case study in the Bay of Fundy. Aquacult Environ Interact 5:235–248

16

Influence of mariculture on the distribution of dissolved inorganic selenium in Sanggou Bay, Northern China

Yan Chang[1,2,*]**, Jing Zhang**[2]**, Jianguo Qu**[2]**, Zengjie Jiang**[3]**, Ruifeng Zhang**[2]

[1]School of Ecological and Environmental Sciences, East China Normal University, Shanghai 200062, PR China
[2]State Key Laboratory of Estuarine and Coastal Research, East China Normal University, Shanghai 200062, PR China
[3]Key Laboratory for Sustainable Utilization of Marine Fisheries Resources, Ministry of Agriculture,
Yellow Sea Fisheries Research Institute, Chinese Academy of Fishery Sciences, Qingdao 266071, PR China

ABSTRACT: Selenium is known as a 'double-edged sword' element on account of its dual beneficial and toxic effects on organisms, depending on its concentration and chemical form. Dissolved inorganic selenium (DISe) concentration in the water column and selenium content in biological species were investigated in a typical aquacultural area in Sanggou Bay, China. In addition to sampling within Sanggou Bay, the main sources of DISe into Sanggou Bay were sampled to estimate selenium transport from different sources. Results showed that DISe and selenite [Se(IV)] concentrations averaged, respectively, 0.69 nmol l^{-1} and 0.28 nmol l^{-1}, with ranges 0.21 to 1.36 nmol l^{-1} and 0.07 to 0.58 nmol l^{-1}, in the surface water of Sanggou Bay. The DISe in Sanggou Bay remained well below the toxic levels. The DISe and Se(IV) concentrations varied temporally, with lows in summer and highs in spring and autumn. Concentrations showed strong horizontal gradients from the coast to offshore areas within the bay, as significantly influenced by the intensive and widespread seaweeds and bivalves aquaculture activity in the bay. The highest selenium content (mean ± SD) was observed in scallops ($3.6 ± 0.7$ µg g^{-1}), followed by oyster ($1.6 ± 0.4$ µg g^{-1}), phytoplankton ($0.9 ± 0.3$ µg g^{-1}), *Gracilaria lemaneiformis* ($0.063 ± 0.008$ µg g^{-1}) and kelp ($0.032 ± 0.005$ µg g^{-1}). The main source of DISe in Sanggou Bay was water exchange with the Yellow Sea, whereas the most important sink was biological activity, which removed $53 ± 12\%$ of the incoming selenium from bay waters.

KEY WORDS: Dissolved inorganic selenium · Aquaculture · Seaweed · Bivalve · Sanggou Bay

INTRODUCTION

Selenium is known to be a 'double-edged sword' element, having one of the narrowest ranges of beneficent effects of all elements, varying between dietary deficiency (<40 µg d^{-1}) and toxicity (>400 µg d^{-1}) (Price et al. 1987, Fernández-Martínez & Charlet 2009). Selenium is an essential trace element required in the diets of many organisms for normal growth and physiological functions (Lin & Shiau 2005, Lobanov et al. 2009). It serves as a component of the enzyme glutathione peroxidase to protect cell membranes against oxidative damage (Rotruck et al. 1973). However, excess dietary selenium behaves as an analogue to sulfur, erroneously replacing sulfur atoms in proteins; this leads to distortion of the structure and eventual dysfunction of enzymes and proteins (Simmons & Wallschläger 2005). The biogeochemical cycle of selenium in aquatic systems has attracted considerable attention in recent decades (Cutter & Bruland 1984, Cutter & Cutter 1995, 2001, Abdel-Moati 1998, Yao & Zhang 2005).

Similar to other trace elements, the chemical form has an important influence on the fate of the sele-

*Corresponding author: cyyc2010@126.com

nium. To understand the biological function of selenium in aquatic organisms, it is necessary to know the levels of different selenium species in the water column. The behavior of selenium in natural waters is complicated by the presence of several oxidation states (−II, IV, VI) and organic species (Conde & Sanz Alaejos 1997). Selenite [Se(IV)] and selenate [Se(VI)] are depleted in surface water and enriched in deep water, and exhibit a nutrient-type profile consistent with other bioactive trace elements in the ocean (Cutter & Bruland 1984, Cutter & Cutter 1995, 2001).

China is one of the largest marine shellfish and seaweed producers in the world (Zhang et al. 2009). Sanggou Bay, located in Shandong Province, is an important aquacultural production area in China (Guo et al. 1999) (see Fig. 1), and is mainly used to culture seaweed and bivalves (Fang et al. 1996). Sanggou Bay has been the focus of research for ~20 yr, and extensive studies have been conducted on hydrodynamic characteristics (Zhao et al. 1996), sediment chemistry (Cai et al. 2003), nutrients (Liu et al. 2004), heavy metals (Jiang et al. 2008), ecosystem services (Zheng et al. 2009), and the sustainable management of aquaculture (Zhang et al. 2009, Shi et al. 2011). Until now, however, little research has been conducted on selenium in the aquacultural areas. Studies have indicated that selenium can stimulate the growth of seaweed (Fries 1982, Horne 1991) and that, through metabolism, seaweed can accumulate over 50 times more inorganic selenium in Se(IV) enriched culture medium than in seawater (Yan et al. 2004). In addition, the selenium assimilation efficiency of bivalves can be as high as 70 to 95% when feeding on the cytoplasm of the prey alga (Wang & Fisher 1996, Reinfelder et al. 1997); these levels may be threatening to upper trophic level birds and fish (Lemly 1995). Further studies are necessary to investigate the distributions of selenium species in aquaculture areas like Sanggou Bay, and to understand the mechanisms controlling the speciation of selenium in natural waters more generally.

The goal of this research was to investigate the distribution of dissolved inorganic selenium (DISe), selenite [Se(IV)] and selenate [Se(VI)] in Sanggou Bay, and in the river and groundwater along the coastline of the bay, in order to (1) determine the distribution of inorganic selenium species under different aquaculture conditions, and (2) estimate the input and output fluxes of selenium species. The results of this study will improve understanding of how the mariculture of seaweed and bivalves affects selenium biogeochemistry in aquacultural areas.

MATERIALS AND METHODS

Study area

Sanggou Bay is situated on the eastern tip of Shandong Peninsula to the northwest of the Yellow Sea (37° 01' to 37° 09' N, 122° 24' to 122° 35' E), with a total area of approximately 144 km^2 and mean depth of 7.5 m (Zhang et al. 2009). The bay has been used for aquaculture with seaweed and bivalves in different regions for >30 yr. Monoculture of seaweeds, including kelp *Saccharina japonica* and *Gracilaria lemaneiformis*, occurs mainly near the mouth of the bay (hereafter 'S-region') from December to May (winter and spring) and from June to November (summer and autumn), respectively. Monoculture of bivalves (in the 'B-region'), including scallops (*Chlamys farreri*) and oysters (*Crassostrea gigas*), is mainly located near the end of the bay, while the middle part of the bay is occupied by seaweed–bivalve polyculture ('SB-region') (Fig. 1). The annual production of kelp and *G. lemaneiformis* is approximately 84.5 × 10^3 and 25.4 × 10^3 t dry weight, respectively, while the annual production of bivalves is approximately 75 × 10^3 t (Rongcheng Fisheries Technology Extension Station 2012). The rivers that enter the bay include the Gu (Chinese name: Guhe), Yatou (Yatouhe), Sanggan (Sangganhe), and Shili (Shilihe) rivers, as well as several other smaller creeks. The annual discharge of these seasonal rivers is from 0.17 × 10^9 to 0.23 × 10^9 m^3 (Editorial Board of Annals of Bays in China 1991).

Sample collection

Samples were collected from marine areas in and adjacent to Sanggou Bay and from major rivers and groundwater along the coastline of the bay. Marine samples were collected during cruises in Sanggou Bay from 22 to 26 April (spring), 21 to 29 July (summer), and 17 to 20 October (autumn) of 2013, to investigate selenium levels in different seasons and various aquacultural activities. During each cruise, approximately 21 marine stations (Stns SG1 to SG19) were sampled (Fig. 1). Surface water was collected upstream and to the side of the boat while the boat moved forward, using a plastic pole sampler 3 to 4 m in length with an acid-cleaned polyethylene bottle attached to the end. Near-bottom water samples were taken with a 5 l organic glass hydrophore. Sediment cores (about 0 to 4 cm depth) were collected using a box core (15 × 15 cm) at Stns SG5 and SG6 in July 2013. Porewaters were extracted and filtered

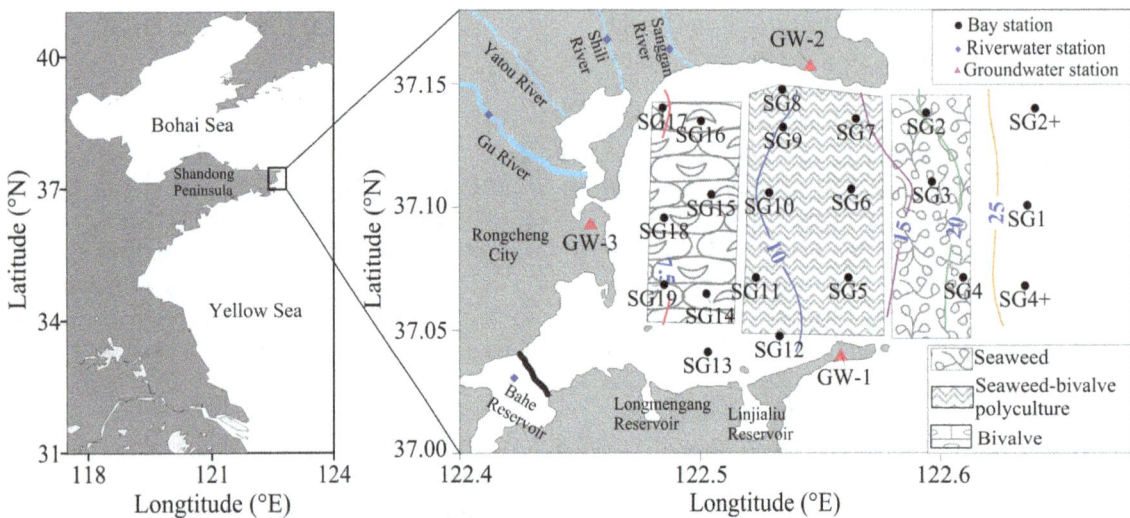

Fig. 1. Locations of field sampling and observation sites in Sanggou Bay in April, July and October 2013

from the sediment cores using 19.21.23F Rhizon CSS soil moisture samplers (Liu et al. 2011) and were frozen until analysis. River water and groundwater sampling was undertaken in April and July 2013 (Fig. 1). On each occasion, a total of 5 river water samples (from the Gu River [2 stations], Shili River, Yatou River, and Bahe Reservoir; Fig. 1) were collected using a 5 l clean plastic bucket from relatively fast-moving areas located away from urban regions. Three groundwater samples (from Stns GW-1, GW-2 and GW-3; Fig. 1) were collected from wells along the coastline of Sanggou Bay. Water temperature and salinity were measured *in situ* using an YSI Professional Plus meter at the time of sample collection. Water samples were filtered in the laboratory within 8 h of collection through precleaned 0.40 µm Nuclepore filters on a class 100 clean bench. The filtrates were placed in acid-cleaned polyethylene bottles and kept frozen until analysis.

Phytoplankton were collected by net tows using a net with a mesh size of 70 µm in April 2015. After the trawl, the plankton were filtered through a 200 µm mesh to remove the zooplankton and then were stored in plastic bottles previously decontaminated with dilute HCl solution. These bottles were kept from the sun and heat in an insulated box containing ice throughout the sampling process and transported to the laboratory. Phytoplankton were collected on the weighed 0.40 µm Nuclepore filters and freeze dried until analysis. Kelp, *G. lemaneiformis*, scallops and oysters samples were also collected in the bay during April, July, July and October 2013, respectively, stored in zip-closure plastic bags and kept in an insulated box containing ice throughout the sampling process. The seaweeds were washed with distilled water

to remove salts and small invertebrates and freeze dried. The muscle was collected from bivalves and freeze dried. Dried samples were ground into powder to pass through an 80 mesh (180 µm) sieve before analysis.

Analytical methods

Measurement of dissolved selenium concentrations

The analytical techniques for Se(IV) and Se(VI) by hydride generation combined with sector field inductively coupled plasma mass spectrometry (HG-ICP-MS) (Element 2™ ICP-MS, Thermol) have been described elsewhere (Zhang & Combs 1996, Chang et al. 2014). Briefly, Se(IV) at an acidity of 2 mol l^{-1} HCl was reacted with NaBH$_4$ to produce hydrogen selenide and was then quantified using HG-ICP-MS. Se(VI) was quantitatively reduced to Se(IV) by heating a sample acidified with 3 mol l^{-1} HCl to 97°C for 75 min, then quickly cooling the sample to room temperature using an ice-water bath, and finally following the steps for Se(IV) determination to yield the concentration of dissolved inorganic selenium (DISe). The reduction recovery ranged from 95 to 103%. This reduction method avoided the problematic variation in Se(VI) reduction behavior with different matrices, and kept the reduction rate at nearly 100% for a longer period of time than previous methods (e.g. Cutter et al. 1978, Yao & Zhang 2003). The Se(VI) concentration was calculated as the difference between DISe and Se(IV). The detection limits for Se(IV) and Se(VI) were 0.025 and 0.030 nmol l^{-1}, respectively. The measurement precisions for Se(IV) and Se(VI) in river water were 3.4 and 3.9%, respectively, and those in seawater were 3.1 and 3.4%, respectively. The spiked standard Se(IV) or Se(VI) recovery ranged from 97 to 103%. The accuracy of the methods was tested with standard solutions, Se(IV) 50031-94 and Se(VI) GBW10032, and showed differences within −3.0 and 0.7%, respectively. The concentration of chlorophyll *a* (chl *a*) was measured using an ACLW-RS chlorophyll turbidity temperature sensor.

Measurement of selenium content of biological tissues

For total selenium content determination, complete digestion of the biological species tissues was performed with a microwave digestion system (MARSX-press, CEM). Samples (about 0.2 g) of dry tissue were soaked in 6 ml concentrated HNO_3. The digestion program was as follows: the sample was heated to 100°C for 10 min, held for 5 min, and then heated to 150°C for 5 min, held for 5 min, and finally heated to 180°C within 5 min and held for 45 min. After cooling, the solution was evaporated at 150°C to dryness using a heating block within about 2.5 h. Subsequently, the residue was dissolved in 5 ml 4 mol l^{-1} HCl, and heated in 110°C using the heating block to reduce Se(VI) to Se(IV) for 45 min. After cooling, samples were added 15 ml H_2O and diluted with 1 mol l^{-1} HCl to 50.0 ml in volumetric flask. The selenium concentration was determined by HG-ICP-MS. The biological standard reference materials GBW010024 (sea scallop), GBW010025 (spiral algae) and GBW010050 (prawn) were measured, with results of 1.49 ± 0.05, 0.23 ± 0.008 and 5.12 ± 0.05 µg g^{-1} (2σ, n = 6), respectively, with relative error −0.6, +3.8, and +0.3%, respectively, confirming the accuracy of the method.

Data statistics and analysis

The statistics software package Statistical Package for the Social Sciences version 16.0 (SPSS) was used for all data analyses. Differences were tested for significance using 1-way and 2-way analysis of variance (ANOVA), and $p < 0.05$ was taken to indicate significant difference. Mean values are presented with standard deviation throughout.

Flux estimates

A steady-state box model based on the Land-Ocean Interactions in the Coastal Zone (LOICZ) Biogeo-chemical Modeling Guidelines was used to construct a DISe budget for Sanggou Bay from water budgets and non-conservative distribution of DISe, which were in turn constrained by the salt balance under steady-state conditions (Gordon et al. 1996).

The vertical diffusional flux of dissolved selenium from the bottom sediment was estimated using a modified form of Fick's first law (Meseck & Cutter 2012):

$$J = \theta^m D_o \left(\frac{\Delta Se}{\Delta z}\right) \qquad (1)$$

where J is the diffusional flux, θ is the porosity, m has a value of 3 for surface sediments (Ullman & Aller 1982), D_0 is the effective diffusion coefficient (−4.87 × 10^{10} m^2 s^{-1}) (Meseck & Cutter 2012), and $\Delta Se/\Delta z$ is the observed concentration gradient of porewater selenium. The selenium concentration at $z = 0$ m (Se_0) in water from the near-bottom of the core was used as the initial point in the concentration gradient. A negative value of J indicates that dissolved selenium is fluxing out of the sediments, while a positive J results from dissolved selenium fluxing into the sediments. The value of θ in Sanggou Bay was 0.7 (Ning et al. 2016, this Theme Section).

RESULTS

Hydrographic properties in Sanggou Bay

The water temperature in the Sanggou Bay displayed a significant horizontal gradient, decreasing from the coast to offshore in spring and summer, with a reversed gradient occurring in autumn (Fig. 2a–c). Mean water temperatures ranged from 7.7 to 20.7°C between seasons, reflecting a remarkable seasonal variation (Table 1). The salinity increased from the coast to offshore (Fig. 2d–f) as a result of water exchange with the Yellow Sea. The mean salinity varied slightly, from 28.3 to 31.4, with lows in July and October due to rainfall and freshwater discharge (Table 1). The average concentrations of phytoplankton biomass, measured as chl *a*, varied between 0.83 µg l^{-1} in spring and autumn, and 6.9 µg l^{-1} in summer (Table 1).

Seasonal variations of inorganic selenium species

Concentrations of DISe ranged from 0.21 to 1.36 nmol l^{-1} for all surface water samples in the bay, with a mean of 0.69 nmol l^{-1} (Table 1). The critical selenium limit in water is classified as 126 nmol l^{-1} in China,

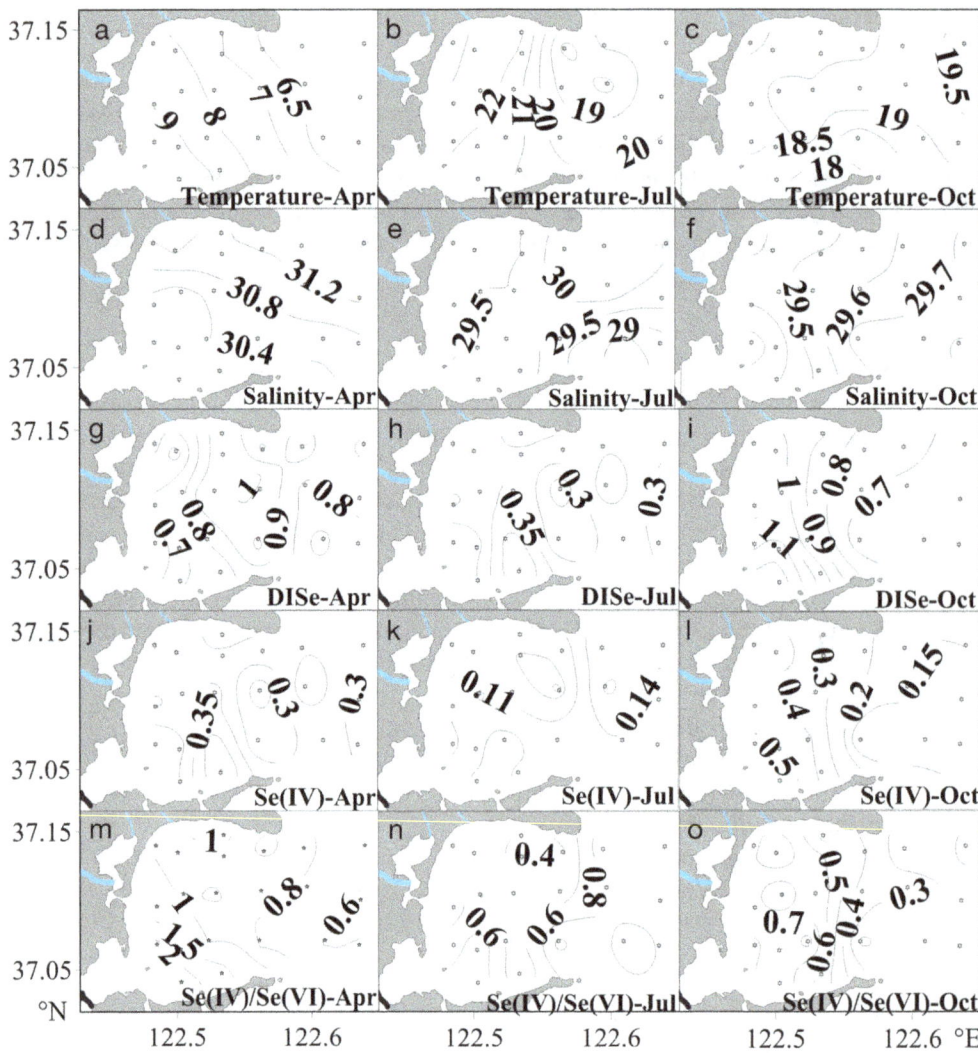

Fig. 2. (a–c) Temperature, (d–f) salinity, (g–i) dissolved inorganic selenium (DISe) concentration, (j–l) Se(IV) concentration and (m–o) Se(IV)/Se(VI) ratio distributions at the surface of Sanggou Bay in April, July, and October 2013 (left, centre and right columns, respectively)

Table 1. Mean values (with ranges in parentheses) of temperature, salinity, chl *a*, dissolved inorganic selenium (DISe) and Se(IV) concentration and Se(IV)/Se(VI) ratio in the surface water of Sanggou Bay in April, July and October 2013

	Spring (April)	Summer (July)	Autumn (October)
Temperature (°C)	7.7 (6.0–9.9)	20.7 (17.7–24.5)	18.5 (16.6–19.6)
Salinity	30.7 (30.1–31.4)	29.6 (28.3–30.5)	29.5 (29.3–29.8)
Chl *a* (μg l^{-1})	0.83 (0.59–2.3)	6.9 (0.86–20)	0.83 (0.36–2.0)
DISe (nmol l^{-1})	0.79 (0.47–1.02)	0.33 (0.21–0.46)	0.89 (0.61–1.36)
Se(IV) (nmol l^{-1})	0.39 (0.24–0.51)	0.12 (0.07–0.18)	0.30 (0.10–0.58)
Se(IV)/Se(VI) ratio	1.07 (0.59–2.30)	0.64 (0.30–1.06)	0.51 (0.18–0.85)

(State Environmental Protection Administration of China 2002); the dissolved selenium in Sanggou Bay remained 2 orders of magnitude below toxic levels.

The minimum mean concentrations of DISe occurred during summer (0.33 nmol l^{-1}), while the maximum occurred during autumn (0.89 nmol l^{-1}) (Table 1). The minimum mean concentrations of Se(IV) also occurred during summer (0.12 nmol l^{-1}), but the maximum occurred during spring (0.39 nmol l^{-1}) (Table 1). One-way ANOVA showed that the concentrations of DISe and Se(IV) were not significantly different between spring and autumn ($p > 0.05$), but that the values in summer were significantly lower than those in spring and autumn ($p < 0.0001$). It is clear from Table 1 that mean concentrations of DISe and Se(IV) showed similar seasonal patterns, with low values in summer and high values in spring and autumn. The mean Se(IV)/Se(VI) ratios for spring, summer, and autumn were 1.07, 0.64, and 0.51, respectively (Table 1). One-way ANOVA showed that the Se(IV)/Se(VI) ratio was not significantly different between summer and autumn (p = 0.25), whereas values in spring were significantly higher than those in summer and autumn (p < 0.0001). The Se(IV)/Se(VI) ratio indicated that Se(VI) was the dominant species of inorganic selenium in the bay during summer and autumn, but Se(IV) was the dominant species in the large proportion along the coast of the bay during spring.

Horizontal distributions of inorganic selenium species

The horizontal distributions of DISe, Se(IV), and the Se(IV)/Se(VI) ratio in surface waters of Sanggou bay show similar features, such as a strong horizontal gradient from the coast to offshore (Fig. 2g–o). The concentrations of DISe along the coast during spring were lower than concentrations in the rest of the bay (Fig. 2g). While DISe showed a strong zonal distribution along the coast during autumn, decreasing offshore from 1.2 to 0.6 nmol l^{-1}, it was rather evenly distributed in summer (Fig. 2h,i). The distribution of Se(IV) in the bay exhibited a similar pattern to DISe in spring, summer, and autumn (Fig. 2j–l). The Se(IV)/Se(VI) ratio had a zonal distribution along the coast during spring and autumn, decreasing offshore from 2 to 0.6 and 0.7 to 0.3, respectively (Fig. 2m,o); however, during summer, the ratio was higher along the coast and at the mouth of the bay (0.8) as compared with in the central region (0.4) (Fig. 2n).

A 2-way ANOVA was conducted that examined the effect of season and space (i.e. S-region, B-region and SB-region, representing the 3 main types of aquaculture in Sanggou Bay) on Se(IV) and Se(VI) distribution. Both season and space significantly affected Se(IV) and Se(VI) concentrations (p = 0.0001). There was also significant interaction between season and space (p = 0.0001). Se(IV) concentrations in the S-region and B-region were significantly less than those in the SB-region in spring (p = 0.046) (Fig. 3a). In contrast, Se(VI) concentrations during spring in the B-region were significantly lower than those in the S-region and SB-region (p = 0.007) (Fig. 3b). During summer, Se(IV) concentrations in the S-region were slightly higher than those in the SB-region and B-region (p = 0.015) (Fig. 3a). There was no significant difference in Se(VI) concentrations between different regions during summer (p = 0.235). In autumn, Se(IV) concentrations in the bay increased in the following order: S-region, SB-region and B-region (p < 0.0001) (Fig. 3a). Se(VI) concentrations shared the same pattern (p < 0.02), with low values in the S-region and high values in the B-region (Fig. 3b).

Riverine input of selenium to Sanggou Bay

As shown in Table 2, the DISe and Se(IV) concentrations in 4 riverine waters were not significantly different between spring and summer (p > 0.2). Overall, DISe concentrations in riverine water ranged from 0.69 to 1.5 nmol l^{-1} with a mean of 1.0 nmol l^{-1}; these were slightly higher than those observed in the water column in Sanggou Bay (0.68 ± 0.29 nmol l^{-1}) during this study (Tables 1 & 2). The mean concentration of Se(IV) was 0.14 nmol l^{-1} in both spring and

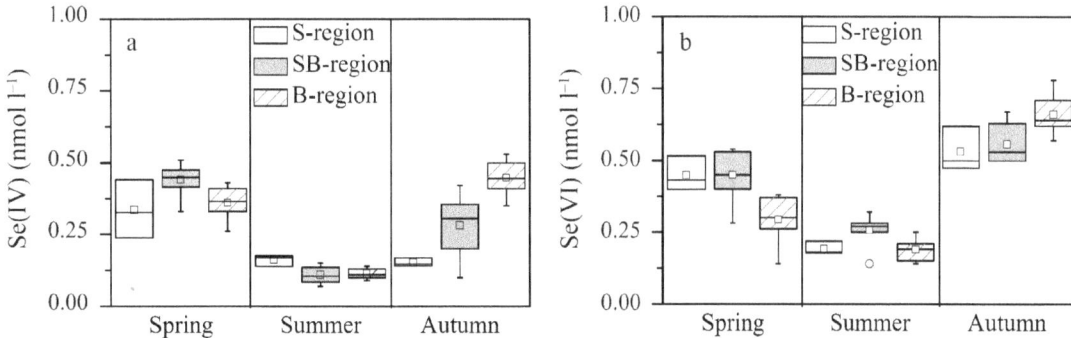

Fig. 3. Box plots of (a) Se(IV) and (b) Se(VI) concentrations in seaweed monoculture (S-region), seaweed-bivalve polyculture (SB-region) and bivalve monoculture (B-region) in spring, summer and autumn of 2013. The ends of the box and the ends of the whiskers, and the line across the box represent the 25th and 75th percentiles, the 1st and 99th percentiles and the median, respectively; the open square inside the box indicates the mean value. A circle represents an outlier

Table 2. Concentrations of dissolved inorganic selenium (DISe) and Se(IV) and Se(IV)/Se(VI) ratio in major rivers that enter into Sanggou Bay and in groundwater (GW) along the coastline of Sanggou Bay, measured in April and July 2013

Water source	Sampling location / station number	DISe (nmol l^{-1})	Se(IV) (nmol l^{-1})	Se(IV)/Se(VI) ratio
Spring (April 2013)				
River	Gu River	1.21	0.18	0.17
	Sanggan River	0.69	0.17	0.21
	Shili River	1.12	0.08	0.18
	Bahe Reservoir	0.91	0.12	0.10
Groundwater	GW-1	34.9	1.36	0.04
	GW-2	3.14	0.60	0.24
	GW-3	3.49	0.26	0.08
Summer (July 2013)				
River	Gu River	1.49	0.11	0.08
	Sanggan River	0.84	0.21	0.18
	Shili River	1.03	0.12	0.25
	Bahe Reservoir	1.16	0.13	0.11
Groundwater	GW-1	27.5	2.76	0.11
	GW-2	5.45	0.49	0.10
	GW-3	3.54	0.89	0.34

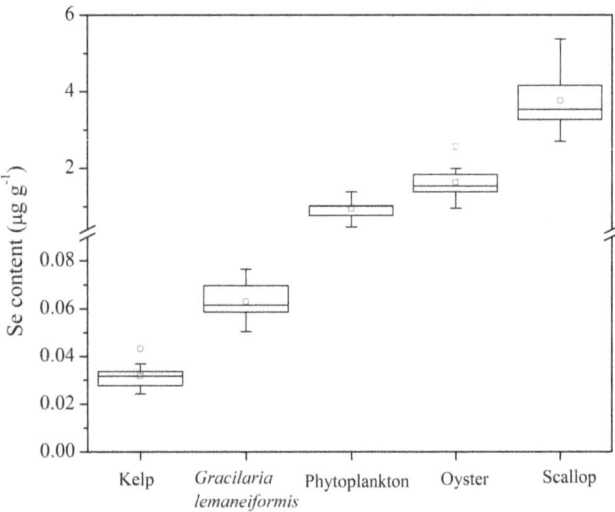

Fig. 4. Selenium (Se) content in kelp (n = 11), *G. lemaneiformis* (n = 9), phytoplankton (n = 6), oysters (n = 9) and scallops (n = 9). Box plot details as in Fig. 3

Groundwater input of selenium to Sanggou Bay

The salinity values of groundwater samples were <0.1, indicating that they are freshwater. The DISe and Se(IV) concentrations in groundwater showed little temporal variation and were not significantly different between spring and summer (p > 0.1). DISe was >25 nmol l^{-1} in the sample from Stn GW-1, while it was <6 nmol l^{-1} in samples from Stns GW-2 and GW-3 (Table 2). The low concentrations of ^{226}Ra in sample GW-1 (Wang et al. 2014) indicates the low activity of rock–water interactions. Moreover, the dissolved inorganic arsenic concentration was nearly 10 times higher in sample GW-1 than samples GW-2 and GW-3 (Li et al. 2014). Stn GW-1 was excluded from the calculation of average selenium concentration. The mean DISe and Se(IV) concentrations were 3.91 ± 1.05 nmol l^{-1} and 0.56 ± 0.26 nmol l^{-1}, respectively. The mean Se(IV)/Se(VI) ratio was 0.15, indicating that Se(VI) was the predominant species in the groundwater. The mean DISe concentration in groundwater (3.91 nmol l^{-1}) was nearly 4 times higher than that in riverine water (1.0 nmol l^{-1}). Compared with surface water, groundwater usually contains higher content due to greater contact time for rock–water interactions (Fordyce 2013).

Selenium content in aquaculture species

The lowest content of Se was present in kelps (0.032 ± 0.005 µg g^{-1}), followed by *G. lemaneiformis* (0.063 ± 0.008 µg g^{-1}) (Fig. 4) these values were in the range of selenium content (0.01 to 0.6 µg g^{-1}) reported elsewhere in seaweed (Liu et al. 1987, Maher et al. 1992, Barwick & Maher 2003). These seaweeds accumulated selenium to concentrations 3 to 4 orders of magnitude above the ambient concentration in the seawater. The Se content for phytoplankton (0.9 ± 0.3 µg g^{-1}) was 10 to 30 times higher than for seaweeds (Fig. 4), and the value was within the range of those previously published for marine phytoplankton (0.5 to 4.5 µg g^{-1}) (Liu et al. 1987, Baines & Fisher 2001, Sherrard et al. 2004). The highest mean Se contents were observed in bivalves, i.e. scallops (3.6 ± 0.7 µg g^{-1}) and oysters (1.6 ± 0.4 µg g^{-1}); these values were consistent with

summer; this concentration was comparable to the one observed in the water column in Sanggou Bay during summer, but was lower than those observed during spring and autumn (Tables 1 & 2). The mean Se(IV)/Se(VI) ratio was 0.16, with a range of 0.08 to 0.25, indicating that Se(VI) was the major inorganic species in riverine water. The DISe concentrations in the Sanggan River during spring and summer were lower than those in the Gu River, the Shili River and Bahe Reservoir; however, concentrations of DISe and Se(IV) varied only slightly among rivers.

the range in bivalves reported elsewhere (0.24 to 4.6 µg g^{-1}) (Liu et al. 1987, Baldwin & Maher 1997, He & Wang 2013).

DISCUSSION

Influence of phytoplankton on selenium distribution

As shown in Table 1, Se(IV) concentrations were lower in summer and higher in spring, while the opposite was true for chl *a* concentrations. Studies have indicated that Se(IV) and Se(VI) can both be assimilated by phytoplankton, with Se(IV) being the preferred species for phytoplankton uptake (Apte et al. 1986, Vandermeulen & Foda 1988, Baines & Fisher 2001). Moreover, as illustrated in Fig. 2m,o, Se(IV)/Se(VI) ratios along the coast decreased towards offshore. In the coastal regions of the bay, the ratio was >0.5 during all 3 seasons, while the ratio in the freshwater end member (river and groundwater) was normally <0.2 (Table 2). There has been a paucity of investigations on selenium species in the Yellow Sea; the mean Se(IV)/Se(VI) ratio for Bohai Sea is 0.45 (Yao & Zhang 2005) and the value for East China Sea was 0.32 (Y. Chang et al. unpubl.). The relatively high Se(IV)/Se(VI) ratios in the bay compared to rivers and surrounding marine basins suggest that either Se(IV) is produced or Se(VI) is preferentially consumed in the bay. Both anions can be assimilated into biomass, but phytoplankton usually has a higher affinity for Se(IV) than for Se(VI) (Fig. 5). Therefore, preferential uptake of Se(VI)

probably cannot explain the observed pattern. However, the organic selenium is later released into the water column where it oxidizes to Se(IV) (Cutter & Bruland 1984), as dissolved oxygen is high in the bay. In contrast, the rate constant of the oxidation from Se(IV) to Se(VI) was 8.7×10^{-4} yr^{-1}, which means it takes >1000 yr to oxidize Se(IV) to Se(VI) (Cutter & Bruland 1984); therefore, this process can be ignored. This may also be the reason why Se(IV)/Se(VI) ratios are elevated at the mouth of the bay during summer, as large amounts of algae would release organic selenium, which would then be oxidized to Se(IV) (Fig. 2n).

Influence of mariculture species on selenium distribution

The distribution of selenium was greatly affected by the mariculture species, as shown in Fig. 3. Laboratory studies have indicated that both Se(IV) and Se(VI) can increase the growth of macroalgae, with Se(IV) taken up more readily than Se(VI) (Fries 1982, Horne 1991). Kelp can bioaccumulate >50 times more inorganic selenium in a Se(IV) enriched culture medium than in seawater (Yan et al. 2004). The kelp in the bay is generally cultivated in November and is harvested in late May. Thus, due to the fast growth of kelp, Se(IV) would be preferentially taken up by the seaweed (Fries 1982). This would explain the relatively low concentrations of Se(IV) in the kelp monoculture region (S-region) in spring (Fig. 3a). Utilization of Se(VI) is more limited compared with Se(IV) (Fries 1982), resulting in the high levels of Se(VI) present in the S-region during spring (Fig. 3b). Moreover, the elevated Se(VI) may be caused by the Yellow Sea input. The *G. lemaneiformis* monoculture is planted after the harvest of kelp in late May. The assimilation of selenium by *G. lemaneiformis* is similar to that of kelp (Fries 1982).

Bivalves mainly accumulate selenium from particulate sources by ingestion and assimilation, while passive uptake from dissolved phases is negligible (Wang & Fisher 1996, Griscom & Fisher 2004, Luoma & Presser 2009). Bivalves have been observed to accumulate ingested selenium to concentrations markedly higher than those present in the algal diet, due to high assimilation rates of cytosolic selenium (Wang & Fisher 1996, Reinfelder et al. 1997). After ingestion, bivalves excrete selenium as dissolved phases probably in the forms of inorganic selenium, including Se(IV) and Se(VI), and organic selenium (Wang & Fisher 1996), and the organic selenium can then be

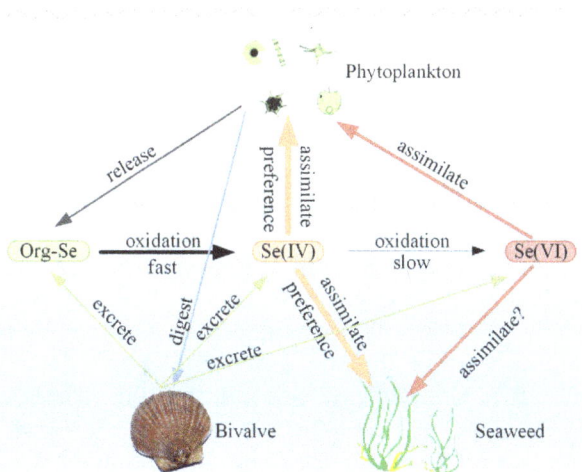

Fig. 5. A conceptual diagram of the biological effects of phytoplankton, seaweed and bivalves on selenium speciation (based on Vandermeulen & Foda 1988, Besser et al. 1994, Hu et al. 1997, Baines & Fisher 2001, present study)

oxidized to Se(IV) (Cutter & Bruland 1984). These processes probably resulted in the high Se(IV) and Se(VI) concentrations in the bivalve monoculture region (B-region) during autumn (Fig. 3a,b). Another possible reason for elevated Se(VI) concentrations in B-region was the input from river or groundwater.

The intensive kelp and bivalve aquaculture activities occurring over large areas of Sanggou Bay have a significant influence on the distribution of selenium in the bay. A conceptual diagram of the biological effects of phytoplankton, seaweed and bivalves on the selenium species in Sanggou Bay is shown in Fig. 5. Phytoplankton and seaweeds (kelp and *G. lemaneiformis*) preferentially assimilate Se(IV) and convert it to organic selenium (Fries 1982, Vandermeulen & Foda 1988, Horne 1991, Besser et al. 1994, Hu et al. 1997, Baines & Fisher 2001). Dissolved organic selenium regenerated from biogenic particles (as phytoplankton cells die and/or bivalves excrete) is quickly oxidized to Se(IV) in the oxygenated water (Cutter & Bruland 1984, Wang & Fisher 1996, Luoma & Presser 2009). However, neither Se(IV) nor organic selenium are reconverted to Se(VI), as these reactions have a half reaction time of hundreds of years (Cutter & Bruland 1984). Thus, the seaweeds assimilate both Se(IV) and Se(VI), resulting in low levels in the water column, while the bivalves assimilate particulate selenium but excrete dissolved selenium, replenishing selenium levels in the water column.

DISe budget for Sanggou Bay

Inputs of selenium to Sanggou Bay

The riverine input of DISe into Sanggou Bay (Y_Q) can reach 15 ± 4 kg yr^{-1} (see Fig. 6) by multiplying the mean DISe concentrations with the annual mean river water discharge.

The submarine groundwater discharge into the bay during summer is $(9.45$ to $11.20) \times 10^9$ m^3 yr^{-1}, as determined by calculation based on the non-conservative inventory of ^{226}Ra and ^{228}Ra (Wang et al. 2014). As submarine groundwater discharge includes recycled seawater (75 to 90 %) as well as fresh groundwater (Moore 1996), the hypothesis of groundwater discharge over the whole bay—instead of just along the shoreline—overestimates the discharge volume. Therefore, we assumed that 5 % of the submarine groundwater discharge represented a best estimate of the fresh groundwater discharge, resulting in a value of 0.47×10^9 m^3 yr^{-1}. The annual

input of DISe from fresh groundwater (Y_G) into the bay was then estimated by multiplying the mean DISe concentrations of groundwater and annual fresh groundwater discharge, giving a value of 146 ± 39 kg yr^{-1}.

The diffusional flux of DISe from the sediment was calculated for the bay using Eq. (1). The DISe concentrations in near-bottom water and porewater for Stn SG5 were 0.32 nmol l^{-1} and 0.45 nmol l^{-1}, respectively, and DISe concentrations in near-bottom water and porewater for Stn SG6 were 0.3 and 0.41 nmol l^{-1}, respectively. Calculations demonstrate that there can be a -0.68 and a -0.58 nmol m^{-2} yr^{-1} flux of DISe between the sediment and water column at Stns SG5 and SG6, respectively. The negative values indicate flux of DISe out of the sediment. The diffusional flux from the sediment to the water column in the bay was estimated by averaging the diffusional flux from each of the different regions. The DISe flux from the sediments (Y_S) was $(7.2 \pm 0.8) \times 10^{-3}$ kg yr^{-1}.

The selenium concentration in rainwater ranges from 1.3 to 2.6 nmol l^{-1} in China (Zhu & Tan 1988). A rainwater sample was collected during summer in the Bay with a selenium concentration of 1.7 nmol l^{-1}, which is within the range of rainwater values in China (Zhu & Tan 1988). Thus, the annual wet deposition of DISe was estimated to be 17.5 kg yr^{-1}, by multiplying the selenium concentration in rainwater with the amount of annual rainfall. The amount of selenium in dry deposition is not known for this region; therefore, the value for the East China Sea (soluble dry deposition of 0.27 ± 0.48 µg m^{-2} d^{-1}) (Hsu et al. 2010) has been adopted to estimate the annual dry deposition of selenium into the bay. Accordingly, the total atmospheric input (wet and dry deposition) of selenium into the bay (Y_P) was 32 ± 25 kg yr^{-1}.

Biological utilization of selenium

The elemental Se:C ratio for phytoplankton in seawater was $(8.5 \pm 3.0) \times 10^{-6}$, which were within the range of those previously pubished for marine phytoplankton (Liu et al. 1987, Baines & Fisher 2001, Sherrard et al. 2004), and the carbon fixed by phytoplankton in Sanggou Bay is 9.5×10^6 kg yr^{-1} (Jiang et al. 2015). Utilization of selenium by phytoplankton in Sanggou Bay was estimated to be -81 ± 29 kg yr^{-1} (negative values indicate removal of selenium from the bay through assimilation by biological organisms; see Fig. 6) by multiplying the carbon fixed by phytoplankton and the elemental Se/C ratio for phytoplankton in Sanggou Bay.

Intensive aquaculture activities have a large influence on selenium levels in Sanggou Bay. The amount of selenium removed from the bay by aquaculture activities was calculated by multiplying aquacultural production in the bay (Rongcheng Fisheries Technology Extension Station 2012) and the selenium content in the cultured species, including kelp, *G. lemaneiformis*, scallops, and oysters. Thus, the amount of selenium fixed by kelp, *G. lemaneiformis*, scallops and oysters was 2.7 ± 0.4, 1.6 ± 0.2, 54 ± 11, and 98 ± 54 kg yr^{-1}, respectively. The highest selenium utilization was by scallops and oysters (151 ± 30 kg yr^{-1}), followed by phytoplankton (81 ± 29 kg yr^{-1}) and seaweed (4.31 ± 0.5 kg yr^{-1}), and the total selenium utilization was estimated to be about 236 ± 42 kg yr^{-1} (see Fig. 6).

Selenium budget for Sanggou Bay

The steady-state box model, illustrated in Fig. 6, calculates the water and salt budgets, and then estimates the mass balance of DISe in Sanggou Bay, including exchange with the Yellow Sea. Freshwater inputs from river discharge (V_Q), groundwater discharge (V_G) and precipitation (V_P) are 0.19 × 10^9 (Editorial Board of Annals of Bays in China 1991), 0.47 × 10^9

(Wang et al. 2014) and 0.13 × 10^9 m^3 yr^{-1} (Shandong province Rongcheng City the Local Chronicles Compilation Committee 1999), respectively. These inputs are reduced by evaporation (V_E), which is 0.15 × 10^9 m^3 yr^{-1} (Shandong province Rongcheng City the Local Chronicles Compilation Committee 1999). From the water mass balance, net water exchange (V_R) is from Sanggou Bay to the Yellow Sea, with a residual flow of −0.64 × 10^9 m^3 yr^{-1} (positive values indicate transport into the bay, negative values export from the bay to the Yellow Sea). Using a salinity of 0 for freshwater input, and salinities of 30 and 32 for Sanggou Bay and the Yellow Sea (Lin et al. 2005), respectively, the water exchange flow from the Yellow Sea to Sanggou Bay (V_X) is 9.96 × 10^9 m^3 yr^{-1}, based on the salt balance in the bay. When calculating the net DISe transport from the Yellow Sea to Sanggou Bay, the mean the DISe concentration (1 ± 0.44 nmol l^{-1}) of Bohai Sea (Yao & Zhang 2005) and the East China Sea (Y. Chang et al. unpubl. data) was used as the value for the Yellow Sea, where DISe data are lacking. The net transport (Y_R) from Sanggou Bay to the Yellow Sea is −42 ± 12 kg yr^{-1} (positive values indicate transport into the bay, negative values export from the bay to the Yellow Sea), and the exchange (Y_X) between Yellow Sea and Sanggou Bay is 253 ± 71 kg yr^{-1}. The data obtained in this study allow for the calculation of selenium budgets in Sanggou Bay (Fig. 6). Atmospheric dry and wet depositions (Y_P), riverine input (Y_Q), groundwater influx (Y_G), exchange between Yellow Sea and Sanggou Bay (Y_X), net transport from the Yellow Sea to Sanggou Bay (Y_R), sediment diffusion (Y_S), and biological fluxes are all shown in Fig. 6. The exchange between Sanggou Bay and the Yellow Sea (Y_X) was the major source of selenium to Sanggou Bay, contributing 57 ± 19% of total DISe inputs. Groundwater discharge (Y_G) accounted for 33 ± 10% of the total DISe input, making it the second largest source of selenium into the bay. However, the exchange of DISe flux (0.002 ± 0.0004%) between the sediment and the water column was negligible compared with other sources. The sediment−water exchange of selenium is also a negligible source in San Francisco Bay, as indicated by stable isotope ratios (Johnson et al. 2000) and sediment porewater values (Meseck & Cutter 2012). The net transport of DISe from Sanggou Bay to the Yellow Sea accounted for 10 ± 3% of the DISe export.

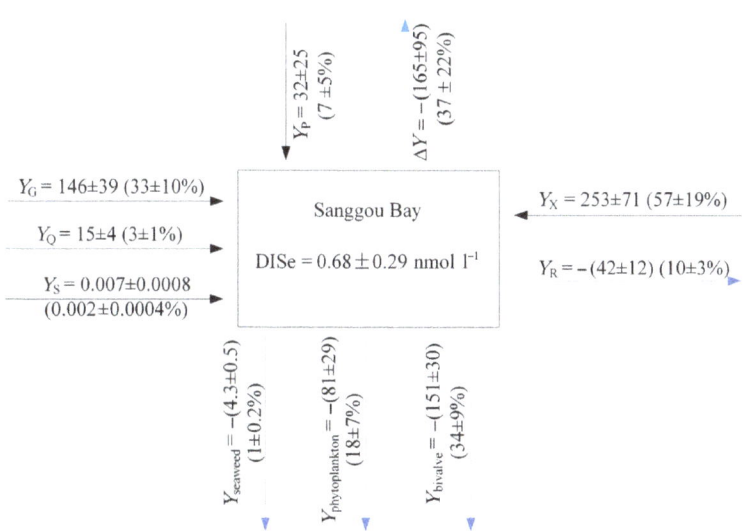

Fig. 6. Selenium budget in Sanggou Bay, showing inputs and outputs (Y) as absolute values (kg yr^{-1}; mean ± SD) and as percentages of total input. Positive values of Y indicate transport into Sanggou Bay; negative values indicate export of dissolved inorganic selenium (DISe) from Sanggou Bay or assimilation by biological organisms. Y_P: atmospheric deposition; Y_Q: riverine input; Y_G: groundwater input; Y_X: exchange between Sanggou Bay and Yellow Sea; Y_R: net transport from Yellow Sea to Sanggou Bay; Y_S: sediment/diffusion; ΔY: net internal sink

The amount of selenium utilized by bivalves was 151 ± 30 kg yr^{-1}. This amount was $34 \pm 9\%$ of the total DISe input (Fig. 6), making bivalves the most important selenium sink. Phytoplankton assimilation was another sink of selenium, using $18 \pm 7\%$ of the total DISe input, while the seaweed assimilated only $1 \pm 0.2\%$ of the total DISe input (Fig. 6). Thus, biological activity removed nearly $53 \pm 12\%$ of the DISe out of the water column and was the major sink of DISe in Sanggou Bay. This means with the harvesting of marine products, nearly half of the selenium was removed out of the bay.

The input of DISe was 445 ± 85 kg yr^{-1}, and output of DISe was 280 ± 42 kg yr^{-1} in the bay. There was a net imbalance between the input and output of dissolved selenium, however. The selenium budgets in Sanggou Bay are only approximations, which depend on the accuracy of the freshwater, water exchange fluxes between Sanggou Bay and the Yellow Sea. If these flux estimates are valid, however, there is a net internal sink of -165 ± 95 kg yr^{-1} ($37 \pm 22\%$) in the bay (Fig. 6). Studies have shown that algae can form dimethylselenide and dimethyldiselenide, which can be volatilized and released into the atmosphere (Ansede & Yoch 1997). The selenium flux of dimethylselenide to the atmosphere was estimated to be 60 to 260 kg yr^{-1} in Gironde Estuary (Ansede & Yoch 1997), and methylation was estimated to account for 10 to 30% of the selenium removed from San Francisco Bay (Hansen et al. 1998). Thus, internal sinks of DISe in Sanggou Bay might include emission from the water column into the atmosphere as volatile selenium, dissimilatory reduction of Se(VI) and Se(IV) to inorganic reduced phases (Se0, Se-II) (Stüeken et al. 2015), and/or transformation to other organic and particulate forms of selenium. To better constrain the uncertainty of budget calculations, more observations are required to understand the biogeochemical process of selenium in Sanggou Bay.

Sensitivity analysis

Box model sensitivity analysis were exampled by changing model parameters by 10% in order to evaluate model response. The sensitivity was quantified by calculating a normalized sensitivity defined as the percentage change in a variable produced by a percentage change in the parameter.

Table 3. Normalized parameter sensitivity of the net internal sink of dissolved inorganic selenium (DISe) (ΔY) for each parameter

Parameter	Treatment (%)	Normalized sensitivity (%)
River discharge (V_Q)	10	4.7
Groundwater discharge (V_G)	10	18
Precipitation (V_P)	10	3.6
Evaporation (V_E)	10	−3.0
DISe concentration in Yellow Sea	10	47
DISe concentration in Sanggou Bay	10	−33
DISe concentration in rivers	10	0.9
DISe concentration in groundwater	10	8.8
DISe concentration in rainwater	10	1.1
Soluble dry deposition	10	0.9
DISe flux in sediments (Y_S)	10	1.1
Se content in phytoplankton	10	−4.9
Se content in kelp	10	−0.2
Se content in G. lemaneiformis	10	−0.1
Se content in oysters	10	−5.9
Se content in scallops	10	−3.3
Carbon fixed by phytoplankton	10	−4.9
Production of kelp	10	−0.2
Production of G. lemaneiformis	10	−0.1
Production of oysters	10	−5.9
Production of scallops	10	−3.3

Ten percentage changes in parameters of DISe concentration for Yellow Sea, DISe concentration for Sanggou Bay and groundwater discharge result in changes in net internal sink (ΔY) yield of 47, −32 and 18%, respectively, while changes in other parameters result in changes in ΔY yield of <10% (Table 3). There has been a paucity of investigations on selenium species in the Yellow Sea; the DISe concentration in Yellow Sea which was used to calculate the net DISe transport from the Yellow Sea to Sanggou Bay (Y_X) was obtained by averaging the values of Bohai Sea (Yao & Zhang 2005) and the East China Sea (Y. Chang et al. unpubl. data). The sensitivity analysis showed that DISe concentration in the Yellow Sea was the most critical parameter and therefore in cases where selenium data are available, it would be worth using a more accurate parameterization of DISe concentration in the Yellow Sea. The DISe concentrations in Sanggou Bay presented significantly variation between seasons and the relative annual mean variation was 42%, which is greater than the 10% changes for sensitivity analysis. Therefore, in future selenium budget calculation, it may be possible to reduce uncertainty in calculating the budget in seasonal levels. Moreover, the box model's sensitivity to changes in groundwater discharge suggested that any improvement in estimates of groundwater discharge is likely to improve the model accuracy. Finally, the budgets were relatively insensitive to riverine input (Y_G), atmospheric deposition (Y_P), sediment diffusion (Y_S), and to biological fluxes. This suggested that inaccuracies in riverine, atmospheric deposition, sediment diffusion and biological data sets have relatively minor impacts on selenium budget calculation, especially in comparison with inaccuracies associated with other model inputs.

CONCLUSION

Distributions of dissolved inorganic selenium species observed in Sanggou Bay provide relevant information that can be linked to the dynamics and biological reactions that take place in the region. Average concentrations of DISe and Se(IV) in surface waters of the bay were 0.67 and 0.28 nmol l^{-1}, respectively, with ranges of 0.21–1.36 and 0.07–0.58 nmol l^{-1}, respectively. The average Se(IV)/Se(VI) ratio was 0.74, indicating that Se(VI) was the predominate inorganic selenium species in large proportion of the bay. DISe concentrations in Sanggou Bay remained 2 orders of magnitude below the critical selenium limit for water in China. The highest selenium content was observed in scallops (3.6 ± 0.7 µg g^{-1}), followed by oyster (1.6 ± 0.4 µg g^{-1}), phytoplankton (0.9 ± 0.3 µg g^{-1}), *G. lemaneiformis* (0.063 ± 0.008 µg g^{-1}) and kelp (0.032 ± 0.005 µg g^{-1}).

The DISe and Se(IV) concentrations were low in summer and high in spring and autumn. The distribution of DISe and Se(IV) in the bay showed strong horizontal gradients from the coast to offshore. The intensive seaweed and bivalve aquaculture present over large areas had a strong influence on selenium distribution. The Se(IV) concentrations in the seaweed monoculture region were low in spring due to the uptake by the seaweed. While the Se(IV) and Se(VI) concentrations in the bivalve monoculture region were high in autumn, probably caused by the bivalves assimilating particulate selenium but excreting dissolved selenium, thereby replenishing selenium levels in the water column.

A simple budget for DISe in Sanggou Bay was estimated in this study. The major source of DISe into Sanggou Bay was water exchange with the Yellow Sea. Groundwater discharge was the second largest source of selenium into the bay. However, intensive bivalve aquaculture removed 34 ± 9% of the DISe input, making it the most important sink.

Acknowledgements. This study was supported by the National Basic Research Program of China (No. 2011CB 409801). The authors thank Chudao and Xunshan Fisheries Corporations for the help in the field work and providing laboratory. We thank colleagues from Yellow Sea Fisheries Research Institute, East China Normal University and Ocean University of China for the assistance in field observations. Anonymous reviewers and the editor are especially acknowledged for their constructive suggestions, which greatly improved the manuscript.

LITERATURE CITED

Abdel-Moati M (1998) Speciation of selenium in a Nile Delta Lagoon and SE Mediterranean Sea mixing zone. Estuar Coast Shelf Sci 46:621–628

ä Ansede JH, Yoch DC (1997) Comparison of selenium and sulfur volatilization by dimethylsulfoniopropionatelyase (DMSP) in two marine bacteria and estuarine sediments. Microb Ecol 23:315–324

ä Apte S, Howard A, Morris R, McCartney M (1986) Arsenic, antimony and selenium speciation during a spring phytoplankton bloom in a closed experimental ecosystem. Mar Chem 20:119–130

ä Baines SB, Fisher NS (2001) Interspecific differences in the bioconcentration of selenite by phytoplankton and their ecological implications. Mar Ecol Prog Ser 213:1–12

ä Baldwin S, Maher W (1997) Spatial and temporal variation of selenium concentration in five species of intertidal molluscs from Jervis Bay, Australia. Mar Environ Res 44: 243–262

ä Barwick M, Maher W (2003) Biotransference and biomagnification of selenium copper, cadmium, zinc, arsenic and lead in a temperate seagrass ecosystem from Lake Macquarie Estuary, NSW, Australia. Mar Environ Res 56:471–502

ä Besser JM, Huckins JN, Clark RC (1994) Separation of selenium species released from Se-exposed algae. Chemosphere 29:771–780

Cai LS, Fang JG, Liang XM (2003) Natural sedimentation in large-scale aquaculture areas of Sungo Bay, North China Sea. J Fish Sci China 10:305–310 (in Chinese with English Abstract)

Chang Y, Qu JG, Zhang RF, Zhang J (2014) Determination of inorganic selenium speciation in natural water by sector field inductively coupled plasma mass spectrometry combined with hydride generation. Chin J Anal Chem 42:753–758 (in Chinese with English Abstract)

ä Conde JE, Sanz Alaejos M (1997) Selenium concentrations in natural and environmental waters. Chem Rev 97: 1979–2004

ä Cutter GA (1978) Species determination of selenium in natural waters. Anal Chim Acta 98:59–66

ä Cutter GA, Bruland KW (1984) The marine biogeochemistry of selenium: a re-evaluation. Limnol Oceanogr 29: 1179–1192

ä Cutter GA, Cutter LS (1995) Behavior of dissolved antimony, arsenic, and selenium in the Atlantic Ocean. Mar Chem 49:295–306

ä Cutter GA, Cutter LS (2001) Sources and cycling of selenium in the western and equatorial Atlantic Ocean. Deep-Sea Res II 48:2917–2931

Editorial Board of Annals of Bays in China (1991) Annals of bays in China. Ocean Press, Beijing (in Chinese)

Fang JG, Kuang SH, Sun HL, Li F, Zhang AJ, Wang XZ, Tang TY (1996) Mariculture status and optimising measurements for the culture of scallop *Chlamys farreri* and kelp *Laminaria japonica* in Sanggou Bay. Mar Fish Res

17:95–102 (in Chinese with English Abstract)

ä Fernández-Martínez A, Charlet L (2009) Selenium environmental cycling and bioavailability: a structural chemist point of view. Rev Environ Sci Biotechnol 8:81–110

Fordyce FM (2013) Selenium deficiency and toxicity in the environment. In: Selinus O (ed) Essentials of medical geology. Springer, Amsterdam, p 375–416

ä Fries L (1982) Selenium stimulates growth of marine macroalgae in axenic culture. J Phycol 18:328–331

Gordon DC, Boudreau P, Mann K, Ong J and others (1996) LOICZ biogeochemical modelling guidelines, Vol 5. LOICZ Core Project, Netherlands Institute for Sea Research, Texel

ä Griscom SB, Fisher NS (2004) Bioavailability of sediment-bound metals to marine bivalve molluscs: an overview. Estuaries 27:826–838

Guo XM, Ford SE, Zhang FS (1999) Mollusca aquaculture in China. J Shellfish Res 18:19–31

ä Hansen D, Duda PJ, Zayed A, Terry N (1998) Selenium removal by constructed wetlands: role of biological volatilization. Environ Sci Technol 32:591–597

ä He M, Wang WX (2013) Bioaccessibility of 12 trace elements in marine molluscs. Food Chem Toxicol 55:627–636

ä Horne AJ (1991) Selenium detoxification in wetlands by permanent flooding: I. Effects on a macroalga, an epiphytic herbivore, and an invertebrate predator in the long-term mesocosm experiment at Kesterson Reservoir, California. Water Air Soil Pollut 57:43–52

ä Hsu SC, Wong GTF, Gong GC, Shiah FK, and others (2010) Sources, solubility, and dry deposition of aerosol trace elements over the East China Sea. Mar Chem 120: 116–127

ä Hu MH, Yang YP, Martin JM, Yin K, Harrison PJ (1997) Preferential uptake of Se(IV) over Se(VI) and the production of dissolved organic Se by marine phytoplankton. Mar Environ Res 44:225–231

Jiang ZJ, Fang JG, Zhang JH, Mao YZ, Wang W (2008) Distribution features and evaluation on potential ecological risk of heavy metals in surface sediments of Sungo Bay. J Agro-Environ Sci 27:0301–0305 (in Chinese with English Abstract)

ä Jiang ZJ, Li J, Qiao XD, Wang GH, Bian DP and others (2015) The budget of dissolved inorganic carbon in the shellfish and seaweed integrated mariculture area of Sanggou Bay, Shandong, China. Aquaculture 446:167–174

ä Johnson TM, Bullen TD, Zawislanski PT (2000) Selenium stable isotope ratios as indicators of sources and cycling of selenium: results from the northern reach of San Francisco Bay. Environ Sci Technol 34:2075–2079

ä Lemly AD (1995) A protocol for aquatic hazard assessment of selenium. Ecotoxicol Environ Saf 32:280–288

Li L, Ren JL, Liu SM, Jiang ZJ, Du JZ, Fang JG (2014) Distribution, seasonal variation and influence factors of dissolved inorganic arsenic in the Sanggou Bay. Environ Sci 35:2705–2713 (in Chinese with English Abstract)

ä Lin CL, Ning XR, Su JL, Lin Y, Xu B (2005) Environmental changes and the responses of the ecosystems of the Yellow Sea during 1976–2000. J Mar Syst 55:223–234

ä Lin YH, Shiau SY (2005) Dietary selenium requirements of juvenile grouper, Epinephelus malabaricus. Aquaculture 250:356–363

ä Liu DL, Yang YP, Hu MH, and others (1987) Selenium content of marine food chain organisms from the coast of China. Mar Environ Res 22:151–165

ä Liu H, Fang J, Zhu J, Dong S, and others (2004) Study on

limiting nutrients and phytoplankton at long-line-culture areas in Laizhou Bay and Sanggou Bay, northeastern China. Aquat Conserv 14:551–574

ä Liu SM, Ling WL, Zhang ZN (2011) Inventory of nutrients in the Bohai. Cont Shelf Res 31:1790–1797

ä Lobanov AV, Hatfield DL, Gladyshev VN (2009) Eukaryotic selenoproteins and selenoproteomes. Biochim Biophys Acta 1790:1424–1428

► Luoma SN, Presser TS (2009) Emerging opportunities in management of selenium contamination. Environ Sci Technol 43:8483–8487

Maher W, Baldwin S, Deaker M, Lrving M (1992) Characteristics of selenium in Australian marine biota. Appl Organomet Chem 6:103–112

► Meseck S, Cutter G (2012) Selenium behavior in San Francisco Bay sediments. Estuaries Coasts 35:646–657

► Moore WS (1996) Large groundwater inputs to coastal waters revealed by ^{226}Ra enrichments. Nature 380:612–614

Ning Z, Liu S, Zhang G, Ning X and others (2016) Impacts of an integrated multi-trophic aquaculture system on benthic nutrient fluxes: a case study in Sanggou Bay, China. Aquacult Environ Interact 8:221–232

► Price NM, Thompson PA, Harrison PJ (1987) Selenium: an essential element for growth of the coastal marine diatom Thalassiosira pseudonana (bacillariophyceae). J Phycol 23:1–9

► Reinfelder J, Wang W, Luoma S, Fisher N (1997) Assimilation efficiencies and turnover rates of trace elements in marine bivalves: a comparison of oysters, clams and mussels. Mar Biol 129:443–452

Rongcheng Fisheries Technology Extension Station (2012) Marine fishery statistics data. Fishery Technology Station of Rongcheng City, Rongcheng

► Rotruck JT, Pope A, Ganther H, Swanson A, Hafeman DG, Hoekstra W (1973) Selenium: biochemical role as a component of glutathione peroxidase. Science 179:588–590

Shandong Flourish the Local Chronicles Compilation Committee (1999) Records Rongcheng. Qilu, Jinan (in Chinese)

ä Sherrard JC, Hunter KA, Boyd PW (2004) Selenium speciation in subantarctic and subtropical waters east of New Zealand: trends and temporal variations. Deep-Sea Res I 51:491–506

► Shi J, Wei H, Zhao L, Yuan Y, Fang J, Zhang J (2011) A physical-biological coupled aquaculture model for a suspended aquaculture area of China. Aquaculture 318: 412–424

► Simmons DB, Wallschläger D (2005) A critical review of the biogeochemistry and ecotoxicology of selenium in lotic and lentic environments. Environ Toxicol Chem 24: 1331–1343

State Environmental Protection Administration of China (2002) Environmental quality standards for surface water (GB 3838-2002). China Environmental Science Press, Beijing (in Chinese)

► Stüeken EE, Buick R, Bekker A Catling D and others (2015) The evolution of the global selenium cycle: secular trends in Se isotopes and abundances. Geochim Cosmochim Acta 162:109–125

► Ullman WJ, Aller RC (1982) Diffusion coefficients in nearshore marine sediments. Limnol Oceanogr 27:552–556

► Vandermeulen J, Foda A (1988) Cycling of selenite and selenate in marine phytoplankton. Mar Biol 98:115–123

► Wang WX, Fisher NS (1996) Assimilation of trace elements and carbon by the mussel mytilus edulis: effects of food composition. Limnol Oceanogr 41:197–207

▶ Wang XL, Du JZ, Ji T, Wen TY, Liu SM, Zhang J (2014) An estimation of nutrient fluxes via submarine groundwater discharge into the Sanggou Bay — a typical multi-species culture ecosystem in China. Mar Chem 167:113–122

▶ Yan X, Zheng L, Chen H, Lin W, Zhang W (2004) Enriched accumulation and biotransformation of selenium in the edible seaweed *Laminaria japonica*. J Agric Food Chem 52:6460–6464

▶ Yao QZ, Zhang J (2003) Salt effect on the determination of inorganic selenium in natural water and its modified method. J Ocean Univ Qingdao 33:765–880 (in Chinese with English abstract)

Yao QZ, Zhang J (2005) The behavior of dissolved inorganic selenium in the Bohai Sea. Estuar Coast Shelf Sci 63:333–347

▶ Zhang LS, Combs SM (1996) Using the installed spray chamber as a gas–liquid separator for the determination

of germanium, arsenic, selenium, tin, antimony, tellurium and bismuth by hydride generation inductively coupled plasma mass spectrometry. J Anal At Spectrom 11:1043–1048

▶ Zhang J, Hansen PK, Fang J, Wang W, Jiang Z (2009) Assessment of the local environmental impact of intensive marine shellfish and seaweed farming — application of the MOM system in the Sungo Bay, China. Aquaculture 287:304–310

Zhao J, Zhou S, Sun Y, Fang J (1996) Research on Sanggou Bay aquaculture hydro-environment. Mar Fish Res 17:68–79 (in Chinese with English Abstract)

▶ Zheng W, Shi H, Chen S, Zhu M (2009) Benefit and cost analysis of mariculture based on ecosystem services. Ecol Econ 68:1626–1632

Zhu F, Tan J (1988) Selenium, iodine and fluorine in rainwater and dustfall in China. Acta Sci Circumst 8:428–437 (in Chinese with English Abstract)

17

Impacts of an integrated multi-trophic aquaculture system on benthic nutrient fluxes

Zhiming Ning[1], Sumei Liu[1,2,*], Guoling Zhang[1], Xiaoyan Ning[1], Ruihuan Li[1,5], Zengjie Jiang[3], Jianguang Fang[3], Jing Zhang[4]

[1]Key Laboratory of Marine Chemistry Theory and Technology MOE, Ocean University of China/ Qingdao Collaborative Innovation Center of Marine Science and Technology, Qingdao 266100, PR China

[2]Laboratory of Marine Ecology and Environmental Science, Qingdao National Laboratory for Marine Science and Technology, Qingdao 266100, PR China

[3]Key Laboratory of Sustainable Utilization of Marine Fisheries Resources, Ministry of Agriculture, Yellow Sea Fisheries Research Institute, Chinese Academy of Fishery Sciences, Qingdao 266071, PR China

[4]State Key Laboratory of Estuarine and Coastal Research, East China Normal University, Shanghai 200062, PR China

[5]*Present address:* State Key Laboratory of Tropical Oceanography, South China Sea Institute of Oceanology, Chinese Academy of Sciences, Guangzhou 510301, PR China

ABSTRACT: Benthic nutrient fluxes in an integrated multi-trophic aquaculture (IMTA) bay— Sanggou Bay, China—were measured in June and September 2012. The benthic nutrient fluxes and total organic carbon (TOC) of sediment in this IMTA system were significantly lower than in monoculture bays. This was due to the efficient recycling of organic matter in the IMTA system, as revealed by historical data of annual production, dissolved inorganic nitrogen (DIN) concentration in seawater and TOC in sediment. Benthic nutrient fluxes in the IMTA system were mainly controlled by seawater temperature, dissolved oxygen (DO) and nutrient concentrations, which were strongly related to aquaculture activities. In June, the early growth phase of cultured finfish and bivalves contributed little to biodeposition, and benthic nutrient fluxes tended to be from the sediment to the seawater and contributed to algal growth. In September, the active growth of finfish and bivalves resulted in high concentrations of nutrients in the seawater and TOC in the sediment; 64 % of the nitrogen and 25 % of the phosphorus metabolized by bivalves were transferred from the seawater to the sediment.

KEY WORDS: Benthic nutrient fluxes · Pore water · Core incubation · Integrated multi-trophic aquaculture · IMTA · Sanggou Bay

INTRODUCTION

World fisheries and aquaculture production has grown rapidly to meet increasing market demand (FAO 2012), and consequently ecosystem biodiversity, productivity and health of marine organisms have been negatively affected. An approach termed 'integrated multi-trophic aquaculture' (IMTA, Fig. 1) was proposed to mitigate these environmental pressures (Tang & Fang 2012) and was implemented in shallow coastal bays including the Bay of Fundy (Canada), and Sanggou Bay and Ailian Bay (China) (Troell et al. 2009, Tang & Fang 2012, Chopin 2013). In terms of production and economic performance, the clear benefits of employing IMTA as opposed to monoculture have been reported (Tang & Fang 2012). However, no sufficient evaluation of the environmental effects of IMTA in comparison to monoculture has been made. Despite the increasing recognition that nutrients are fundamental to the food web in aquaculture eco-

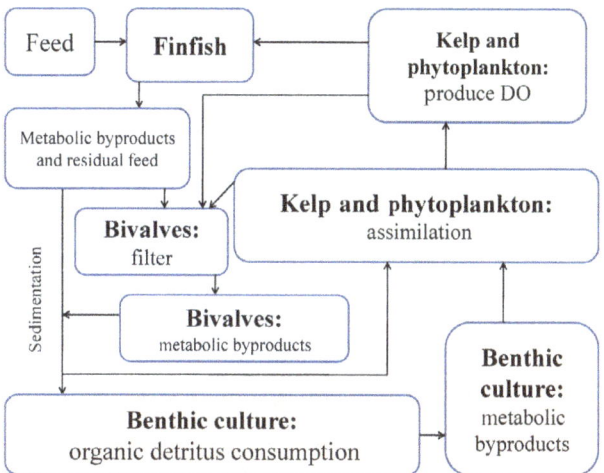

Fig. 1. Diagrammatic representation of the integrated multi-trophic aquaculture (IMTA) system in Sanggou Bay, China, modified from Tang & Fang (2012). DO: dissolved oxygen

systems, information about the internal nutrient cycles in IMTA systems is still unavailable (Sequeira et al. 2008, Troell et al. 2009, Tang & Fang 2012, Chopin 2013). Benthic nutrient regeneration is a significant source of nutrients for primary production in coastal waters (Liu et al. 2003, Sundbäck et al. 2003, Lee et al. 2011). Conversely, nutrients can be stored in the sediments via burial and denitrification (Aller et al. 1985, Song et al. 2013). Hence, an accurate account of nutrient fluxes across the sediment–water interface and the roles of these processes in IMTA systems are of significance to fisheries management.

Many studies have focused on seawater conditions, nutrient uptake efficiency of bivalves, and aquaculture capacity and impacts in Sanggou Bay (Nunes et al. 2003, Mao et al. 2006, Zhang et al. 2009, Lu et al. 2015), but knowledge of the benthic nutrient fluxes in the IMTA system and comparisons of the environmental impacts of IMTA and monoculture are insufficient (Zhang et al. 2006). The aim of this study was to investigate the impacts of aquaculture on benthic nutrient fluxes in the IMTA system, and sedimentary mineralization processes based on nutrient data in pore water, to evaluate the environmental effects of IMTA with respect to benthic nutrient fluxes.

MATERIALS AND METHODS

Study area

Sanggou Bay is a typical IMTA bay located on the western margin of the Yellow Sea (Fig. 2). It is semi-enclosed, with a mean depth of 7.5 m, a total area of 144 km², and a mean salinity of 31 (Zhang et al. 2009). Kelp is cultivated mainly outside the mouth of the bay; bivalves are near the end of the bay. Polyculture of kelp and bivalves occurs centrally between the former 2, and sea cage culture of finfish occurs along the southwest coast. The annual production of kelp, finfish, scallop and oyster were 84 500, 535, 15 000 and 60 000 t in 2012 (the statistical data from the Rongcheng Fishery Technology Extension Sta-

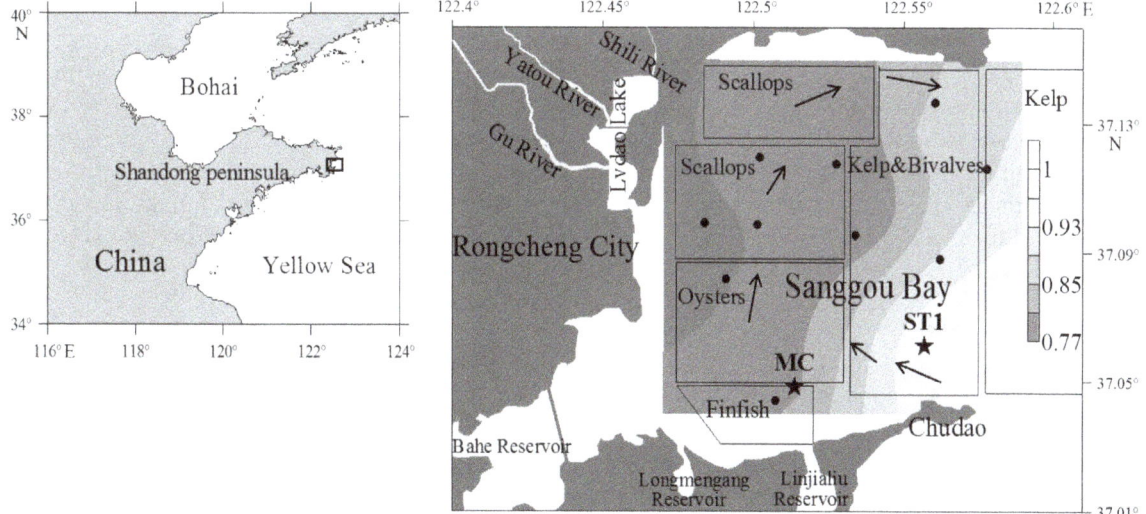

Fig. 2. Aquaculture areas (rectangles, cultured organisms indicated) and study sites in Sanggou Bay, China. ★: stations used for core incubation; ●: stations used for surface sediment sampling. Contours indicate dissolved oxygen saturation levels in bottom seawater in September 2012. Arrows represent current direction at one time of the tidal cycle modified from Bacher et al. (2003)

tion 2012). In an IMTA system (Fig. 1) the bivalves filter suspended particulate matter, including the feces
of finfish and phytoplankton; kelp assimilates nutrients from metabolic byproducts generated by the
bivalves and finfish, and provides dissolved oxygen
(DO) to finfish and bivalves; benthic animals are able
to utilize phytoplankton and sedimentary organic
detritus from aquaculture occurring in the water column, facilitating maximum nutrient recovery efficiency (Tang & Fang 2012, Chopin, 2013). The sediments are predominantly composed of clayey silt
(Zhang et al. 2006).

Seawater and sediment sampling

Field observations were carried out in Sanggou Bay
in 2012, 1–2 June and 24–27 September. Surface sediments for analysis of total organic carbon (TOC) and
porosity were collected from 12 stations (Fig. 2), and 2
stations located in different aquaculture conditions
(polyculture vs. fish culture) were chosen for pore water extraction and core incubation to investigate benthic nutrient fluxes. Diffusion fluxes were derived
from the nutrient profiles in original (i.e. at sampling
of cores and before incubation) pore water obtained
in the field; incubation fluxes were directly measured
from core incubation, and sedimentary mineralization
processes were evaluated based on nutrient data in
pore water before and after incubation.

At each station, bottom seawater was collected
using a Plexiglas sampler; sediments were collected
using a box-sampler; 2 sediment cores were obtained
with Plexiglas tubes (i.d. = 7 cm) and sectioned at 1 or
2 cm intervals within 0.5 h. The resulting sediment
sections from one core were put into plastic bag and
then frozen at −20°C for later analysis, and sections
from the other core were used for pore water extraction (i.e. original pore water). Pore water was
extracted and filtered with Rhizon soil moisture samplers (19.21.23F Rhizon CSS) to vacuum tubes (Song
et al. 2013) and then frozen at −20°C.

Core incubation

Each core (i.d. = 5 cm) was sealed with a gas-tight
lid attached and was pre-incubated in the dark at
room temperature (21°C in June and 24°C in September 2012) for 8–12 h in the presence of bottom water
recirculated using a peristaltic pump (Song et al.
2015). During the following incubation period the
seawater was mixed using a magnetic stirrer turning

a Teflon-coated magnetic stir bar at 60 rpm. At each
sampling time, seawater from triplicate cores was
sampled for measurement of DO and nutrients, and a
sample was taken from the black bucket as a blank.
Seawater for nutrient analysis was filtered with a
0.45 μm pore-size syringe filter (Song et al. 2013), and
the filtrate was frozen at −20°C. At the first and last
sampling time of incubation, sediment cores were
sectioned at 2 cm intervals for pore water extraction
(i.e. pore water before and after incubation).

Physical and chemical analysis

Each frozen sediment sample was freeze dried (AL-
PHA 1–4 LD plus freeze dryer; Martin Christ). The
water content of the sediment was calculated by determining the weight difference before and after
freeze-drying (Song et al. 2013), and porosity was calculated with Berner's equation (Berner 1971). The total organic carbon (TOC) content of sediment was determined using a CHNOS Elemental Analyzer (Vario
EL III, Elemental Analyzer) following removal of the
carbonate fraction via reaction with $4 \ mol \ l^{-1}$ HCl; this
procedure had a precision <6% CV (Liu et al. 2010).

Temperature and salinity were measured by a
multi-parameter instrument (Multi 350i/SET, WTW
GmbH). DO concentration in seawater was measured using the Winkler titration method with a precision better than 0.5% CV (Song et al. 2015). Nutrient
concentrations were determined using an autoanalyzer (AutoAnalyzer 3, SEAL Analytical). The measurement precisions for the NO_3^-, NO_2^-, NH_4^+, PO_4^{3-},
$Si(OH)_4$, total dissolved nitrogen (TDN) and total dissolved phosphorus (TDP) analyses were 1, 1, 2, 1, 0.2,
3 and 5% CV, respectively. Dissolved organic phosphorus (DOP) concentration was calculated as TDP
concentration minus PO_4^{3-} concentration, and dissolved organic nitrogen (DON) concentration was
calculated as TDN concentration minus dissolved
inorganic nitrogen (DIN; sum of the NO_3^-, NO_2^- and
NH_4^+) concentration.

Flux calculations and statistical analysis

Diffusion fluxes were derived from the nutrient
profiles in pore water using Fick's first law of diffusion (Berner 1980, Liu et al. 2003):

$$F = -\phi D_s (\partial C / \partial x)$$

where F is the diffusion flux in mmol m^{-2} d^{-1}, ϕ is the
porosity of the surface sediment, D_s is the whole

Table 1. Biogeochemical properties of the bottom seawater and the sediment at Stns MC and ST1, during June and September 2012. TOC: total organic carbon; S: salinity; DO: dissolved oxygen; DON: dissolved organic nitrogen; DOP: dissolved organic phosphorus

Date	Stn	Water depth (m)	Porosity in sediment	TOC (%) in sediment	Bottom seawater								
					Temp. (°C)	S	DO satura-tion (%)	NH_4^+ (µM)	NO_x^- (µM)	DON (µM)	PO_4^{3-} (µM)	DOP (µM)	$Si(OH)_4$ (µM)
Jun 1	MC	9.2	0.70	0.35	17.7	31.1	96.5	1.50	0.19	12.76	0.08	0.21	3.87
Jun 2	ST1	13.8	0.72	0.40	13.8	31.1	97.4	3.23	1.09	21.38	0.27	0.25	3.49
Sep 24	MC	7.8	0.84	0.68	25.0	30.0	79.2	4.14	5.73	28.83	1.19	0.21	24.45
Sep 27	ST1	11.0	0.80	0.62	23.9	29.9	97.5	5.04	5.49	29.59	0.72	0.26	16.31

sediment diffusion coefficient and $\partial C / \partial x$ is the concentration gradient close to sediment–water interface.

Incubation fluxes, which are a direct measure of net solute fluxes across the sediment–water interface, were calculated from the slope of concentrations versus time (Song et al. 2015).

Standard deviation of the linear rate was derived from the slope standard deviation given by the regression statistic; Pearson correlation was applied to discuss the correlation analysis. Statistical significance was judged using the criterion $p < 0.05$. Incubation fluxes were corrected to the *in situ* temperature using the Arrhenius equation (Aller et al. 1985, Song et al. 2015). In the present study, a positive flux (efflux) value represents a flux into the overlying water from the sediment, and a negative flux (influx) value represents a flux into the sediment from the overlying water.

RESULTS

Sediment and bottom seawater parameters

The TOC in surface sediments at both stations in September were approximately twice the level measured in June; porosities had a similar trend to that of TOC and were higher in September than in June but the values were similar at the different stations (Table 1). The bottom seawater temperature in September was higher than in June and was lower at Stn ST1 than at Stn MC because the water was depth greater at the former station. The salinity at both stations in September was slightly lower than in June. The nutrient concentrations in September were higher than in June. The DO concentrations showed saturated conditions at both stations in June, whereas in September the bottom seawater DO concentration was below saturation in the finfish and bivalve culture areas (Fig. 2).

Benthic fluxes from core incubations and their stoichiometric ratios

The DO content decreased linearly over time during incubations, and the linear slopes of the DO–time plots were similar in the various seasons (Fig. 3), although the TOC content was greater in September than in June. However, the higher *in situ* temperature in September resulted in greater DO influxes than in June (Fig. 4). The DO influx at Stn MC was higher than at Stn ST1 in June but was lower at Stn MC than at Stn ST1 in September.

In June, nutrients were released from the sediment to the seawater (the exception was PO_4^{3-}, which was transferred from seawater into the sediment), and the magnitudes of benthic nutrient flux at 2 stations were similar (Fig. 5). In September: NH_4^+ was largely released at Stn MC, but no NH_4^+ flux was detected at Stn ST1 (Fig. 5a); NO_3^- was largely released at Stn ST1 but was transferred to sediment at Stn MC; NO_2^- was transferred to sediment at Stn ST1 but was released at Stn MC; DON and TDN were transferred to sediment at both stations (Fig. 5b); PO_4^{3-} was transferred to the sediment at both stations, particularly at Stn MC; DOP was strongly released at Stn MC, while DOP and TDP were transferred to sediments at Stn ST1 (Fig. 5c), and the $Si(OH)_4$ efflux was less at Stn MC than at Stn ST1 and was lower in September than in June (Fig. 5d). The O_2:DIN flux ratio was higher in September than in June, and the DIN:PO_4^{3-} flux ratio was lower in September than in June, while the $Si(OH)_4$:DIN flux ratio was higher in September than in June at Stn ST1 but was lower in September than in June at Stn MC (Table 2).

Diffusion fluxes and nutrient profiles in pore water

The concentrations of NH_4^+, NO_x^- ($NO_2^- + NO_3^-$), PO_4^{3-} and $Si(OH)_4$ in pore water were measured

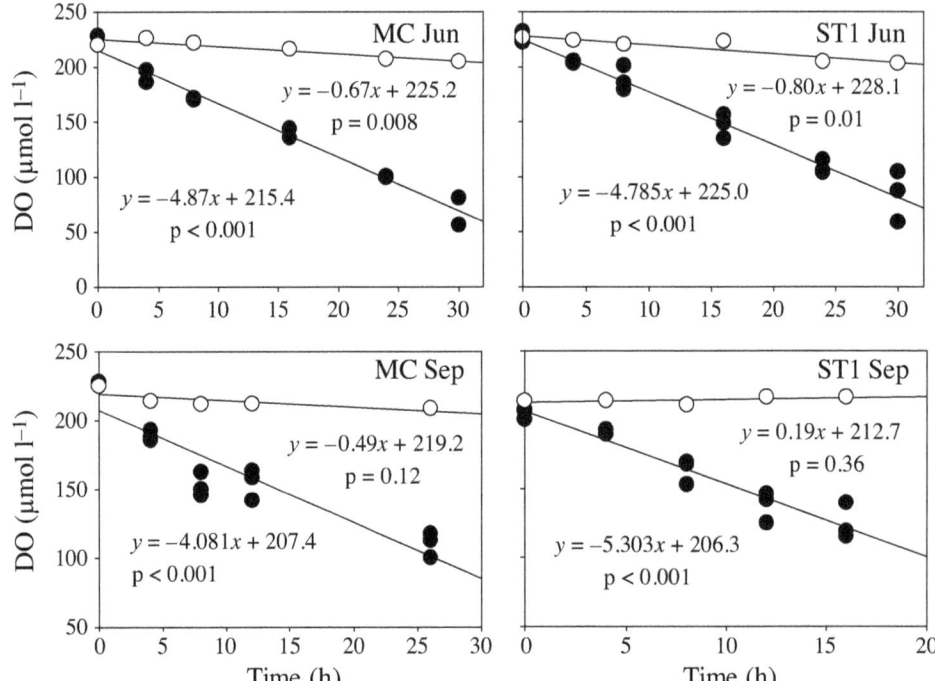

Fig. 3. Time course of dissolved oxygen (DO) concentration during incubation at room temperature at Stns MC and ST1 in June and September 2012. O: DO concentration in the control bucket; ●: DO concentration in the water overlying the sediment

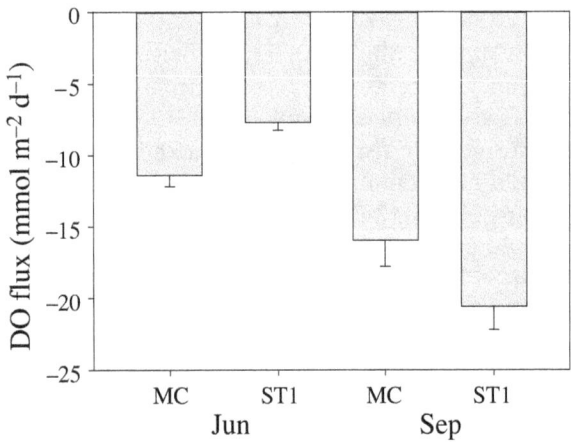

Fig. 4. Temperature-calibrated incubation fluxes of dissolved oxygen (DO) at Stns MC and ST1, during June and September 2012. Error bars show SD

when the core sediments were sampled (original) and before and after incubation (Fig. 6). The nutrient concentrations generally increased with sediment depth; the exception was the NO_x^- concentration. The nutrient diffusion effluxes were supposed to be greater in September than in June as porosities of sediment were higher in September than in June, but the result was opposite. The average diffusion fluxes of DO, NH_4^+, NO_x^-, PO_4^{3-} and $Si(OH)_4$ were 1650, 1405, 7, 14 and 932 μmol m^{-2} d^{-1}, respectively, in June and were 6470, 718, −59, 4 and 818 μmol m^{-2} d^{-1}, respectively, in September.

The nutrient profiles of NH_4^+ were substantially greater after incubation, especially at Stn MC, but there was no difference in NH_4^+ concentrations before and after incubation at Stn ST1 in September; NO_x^- was depleted in deep pore water and increased in surface pore water after incubation, but in September the NO_x^- in surface pore water at Stn MC decreased after incubation; there were minor variations in the PO_4^{3-} profiles for surface pore water, but in deep pore water a significant release of PO_4^{3-} was observed after incubation; the differences in $Si(OH)_4$ concentration before and after incubation were less in September than in June.

DISCUSSION

Environmental factors controlling benthic fluxes

A most important use of DO flux is in the indirect estimation of the total benthic organic carbon mineralization rate (CO_2 flux), which is based on the Redfield ratio; the reported ratio between DO flux and CO_2 flux varies from 0.8 to 1.2, and a O_2:C ratio of 1:1 was used in the present study since this ratio has been widely used for studies involving shallow waters (Glud 2008, Song et al. 2015). The quantity and quality of organic matter, temperature, DO concentration and macrofauna abundance have been suggested to be factors controlling benthic DO fluxes

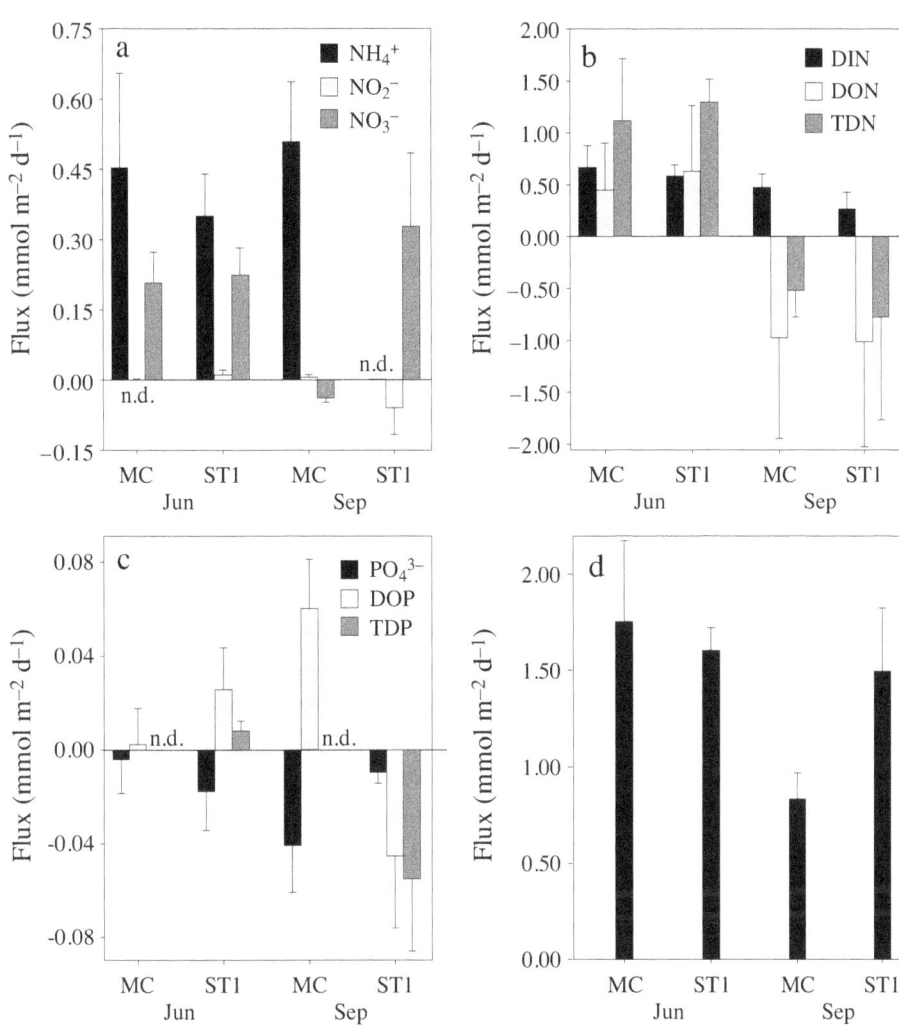

Fig. 5. Incubation fluxes of (a,b) dissolved nitrogen, (c) phosphorus and (d) $Si(OH)_4$ at Stns MC and ST1 in June and September 2012. Positive flux values denote fluxes into the overlying water from the sediment; negative flux values indicate fluxes into the sediment from overlying water. DIN: dissolved inorganic nitrogen; DON: dissolved organic nitrogen; TDN: total dissolved nitrogen; DOP: dissolved organic phosphorus; TDP: total dissolved phosphorus. n.d.: not detectable. Error bars show SD

(Cowan & Boynton 1996). Benthic DO fluxes were similar under similar incubation temperatures (Fig. 3), although the TOC values were higher in September than in June. The positive correlation between calibrated DO influx (F_{DO}) and seawater temperature (T) ($F_{DO} = -0.99T + 0.36$, $R^2 = 0.79$) indicated that temperature rather than TOC is one factor controlling CO_2 fluxes in Sanggou Bay sediment. This was consistent with another IMTA bay, i.e. Ailian Bay, China, in that the contribution rates of biodeposits by the shellfish

Table 2. Stoichiometric ratios of benthic fluxes at Stns MC and ST1, during June and September 2012. PO_4^{3-} fluxes were diffusion fluxes. DIN: dissolved inorganic nitrogen

Date	Stn	O_2:DIN	DIN:PO_4^{3-}	$Si(OH)_4$:DIN
Jun 1	MC	17	40	3
Jun 2	ST1	13	60	3
Sep 24	MC	33	34	2
Sep 27	ST1	76	17	6

and kelp to the sediments in the IMTA area were very low (Ren et al. 2014), but benthic DO fluxes were positively correlated to TOC sedimentation in monoculture areas (Carlsson et al. 2012). Moreover, the low DO saturation level at Stn MC in September resulted in a lesser DO influx than that of Stn ST1 (Fig. 4), suggesting that DO in bottom seawater is also one factor controlling CO_2 fluxes in Sanggou Bay. Benthic CO_2 fluxes removed 12 and 6% of C input via sedimentation in June and September, respectively, but other parts of the sedimentary matter were mainly transported by horizontal fluxes including bioturbation and resuspension (Fig. 7). In Jiaozhou Bay (China) the polychaete bioturbation resulted in a 25% greater DO flux than that in the absence of bioturbation (Zhang et al. 2006). Hung et al. (2013) also reported resuspension may have contributed 27–93% of the POC flux in the East China Sea. However, the sedimentation fluxes may have been overestimated, as it is possible that their values included materials transported horizontally and re-

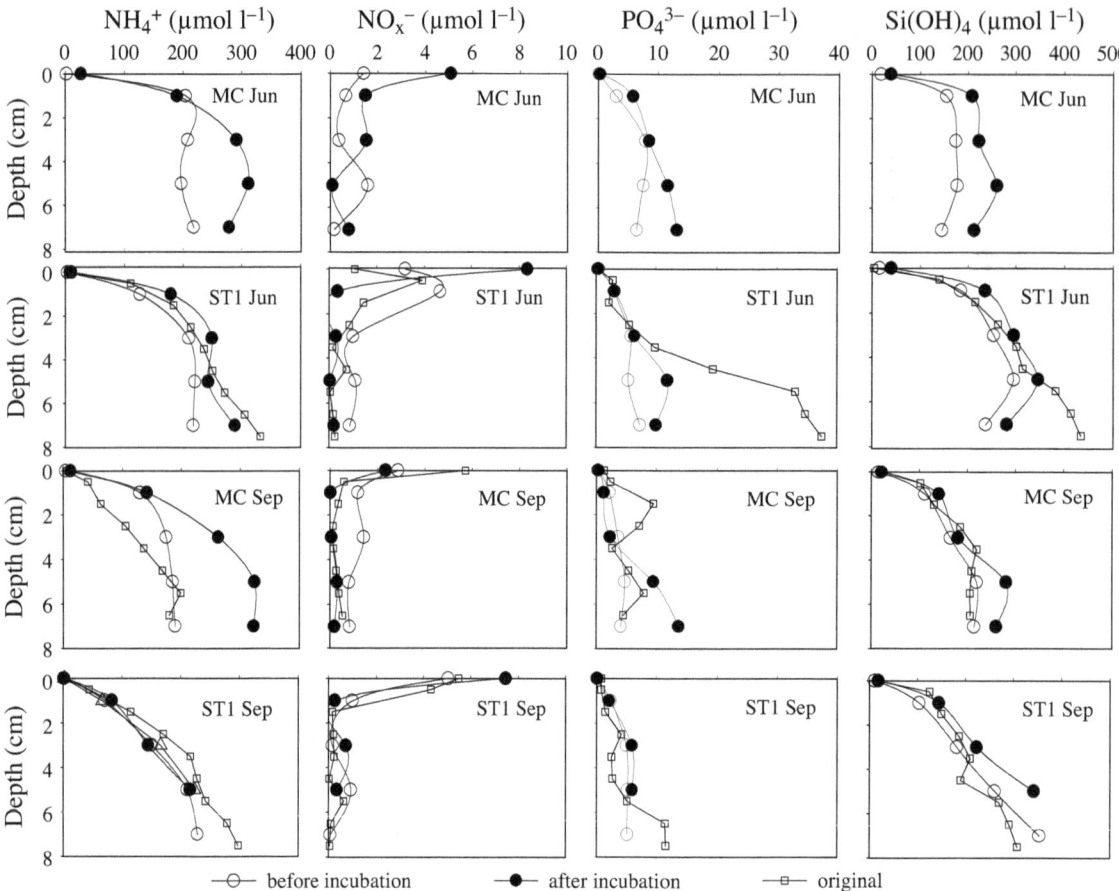

Fig. 6. Nutrient profiles in pore water when the core sediments were sampled (original) and before and after incubation. MC and ST1 are the sampling stations; sampling was done in June and September 2012. Core sediment for original pore water was not sampled at Stn MC in June

mobilized particulates (Hatcher et al. 1994); therefore, these estimated values need confirmation.

Benthic N fluxes are affected by the microbial activities including nitrification and denitrification (Jansen et al. 2012). The elevated O_2:DIN flux ratio (Table 2) was much higher than the Redfield ratio (i.e. 6.6), which suggests that substantial coupled nitrification–denitrification (Cowan & Boynton 1996) occurred in September. The denitrification rate in Sanggou Bay was 0.19–0.37 mmol m^{-2} d^{-1} (Z. Ning et al. unpubl. data), but the high O_2:(DIN+N_2) flux ratio (10–29) (which still exceeded the Redfield ratio) indicated that 30–77% of the mineralized NH_4^+ was retained in the Sanggou Bay sediment. The porosity of the sediment should positively relate to the benthic nutrient diffusion flux (Berner 1980); nevertheless, benthic nutrient fluxes were greater in June than in September, although porosity was higher in September than in June (Fig. 7). Although grain size was not determined in this study, grain sizes at different stations should be similar, since the porosities were sim-

ilar at the 2 stations (Table 1). Hence, neither porosity nor grain size were the main factors controlling benthic nutrient fluxes in the IMTA system.

Fluxes of PO_4^{3-} depend on the PO_4^{3-} production rate, the adsorption–desorption equilibrium in the sediment, and the thickness of the diffusion boundary layer at the sediment–water interface (Sundby et al. 1992). Adsorption of PO_4^{3-} by MnO_2/FeOOH (Woulds et al. 2009) may explain why PO_4^{3-} was transferred to the sediment at both stations (Fig. 5c). Although the N loss by coupled nitrification–denitrification and NH_4^+ adsorption onto clay minerals contributed to the low DIN efflux, the high DIN:PO_4^{3-} flux ratios (Table 2) indicated the degree to which PO_4^{3-} is retained by adsorption in Sanggou Bay. Hence, PO_4^{3-} sorption widely occurred in monoculture (Hyun et al. 2013) and IMTA areas. The DOP fluxes were mainly affected by aquaculture activities (see 'Aquaculture activities and benthic nutrient fluxes in different seasons').

The benthic $Si(OH)_4$ fluxes in Sanggou Bay were higher than the nitrogen and phosphorus fluxes

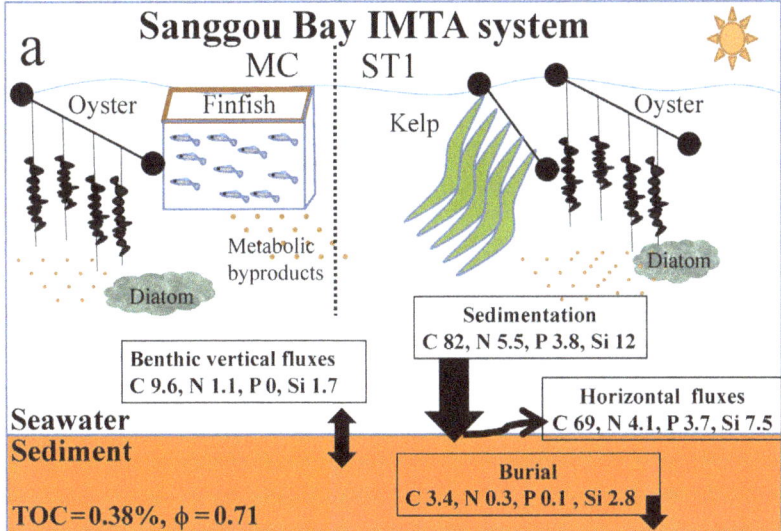

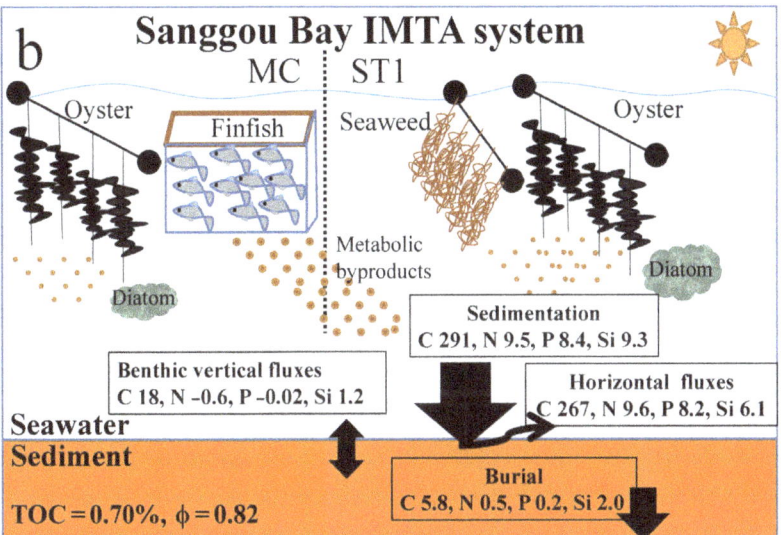

Fig. 7. Sedimentary cycles of C, N, P, Si in (a) June and (b) September 2012 in the Sanggou Bay integrated multi-trophic aquaculture (IMTA) system. Benthic vertical fluxes were measured from core incubation; C fluxes were calculated from the Redfield ratio using dissolved oxygen fluxes. Sedimentation fluxes measured by sediment traps were sourced from Cai et al. (2003); burial fluxes were sourced from Song et al. (2012); horizontal fluxes including resuspension and bioturbation were calculated by difference. All units of fluxes in the boxes are mmol m^{-2} d^{-1}. IMTA: integrated multi-trophic aquaculture; TOC: total organic carbon; ϕ: porosity of the surface sediment. MC and ST1 are the sampling stations

(Fig. 5). Biogenic silica reaches the sediment surface mainly in the form of skeletons or skeletal fragments of silica-secreting microorganisms (Zabel et al. 1998), and dissolution of sedimentary biogenic silica dominates the dissolved silicate content of pore water (Aller et al. 1985, Liu et al. 2003). Diatoms were predominant in the phytoplankton community in Sanggou Bay (Yuan et al. 2014). Consequently, tempera-

ture and the biomass of diatoms in seawater are the main factors controlling benthic Si(OH)$_4$ fluxes. The seawater temperature was higher in September than in June, and therefore the Si(OH)$_4$ fluxes were expected to be higher in September but were found to be higher in June (Fig. 5d). The higher Si(OH)$_4$ concentration in seawater in September (Table 1) was related to a lesser biomass of diatoms in the seawater, because the abundance of phytoplankton was tightly controlled by filter-feeding oysters (Hyun et al. 2013); therefore, heavy grazing by oysters may result in the reduction of the Si(OH)$_4$ flux at Stn MC in September. In comparison to other monoculture areas, competition with co-cultivated kelp resulted in lower diatom biomass in the IMTA system (Yuan et al. 2014), resulting in lower benthic Si(OH)$_4$ fluxes in the IMTA than in monoculture (Table 3).

Aquaculture activities and benthic nutrient fluxes in different seasons

In June the concentrations of nutrients in seawater were quite low because the kelp *Saccharina japonica* assimilated substantial nutrients in spring (Shi et al. 2011), and the metabolic byproducts of finfish and oysters in the early growth stages produced low levels of nutrients in seawater (Fig. 7a). In September the seaweed *Gracilaria lemaneiformis* replaced kelp, and finfish and oysters were in active growth stages and generated large quantities of metabolic byproducts (Fig. 7b). The maximum metabolic rates from Pacific oyster were recorded in July and August (Mao et al. 2006), and decomposition resulted in high nutrient concentrations in the seawater. In addition to assimilation by kelp, Si(OH)$_4$ concentration was tightly related to the biomass of diatoms, as diatoms were predominant in the phytoplankton community in Sanggou Bay (Yuan et al. 2014). Hence, ratios of Si(OH)$_4$:DIN concentrations were higher in September than in June, especially at Stn MC due to heavy grazing by oysters

(Hyun et al. 2013). When discussing the impacts of aquaculture on benthic nutrient fluxes, it is important to clarify the sources of brodeposits by the marine organisms to the sediments using sediment traps or natural isotopic tracers, etc. However, TOC was not a directly controlling factor of benthic fluxes in an IMTA system as discussed in 'Environmental factors controlling benthic fluxes'; therefore, the sources of biodeposits by the marine organims to the sediment were not an object of this study.

In June, the decrease in nutrient concentrations in seawater enlarges the concentration gradient in the sediment–water interface, which may result in larger diffusion effluxes (Berner 1980). Hence, all nutrients are released from the sediments to the seawater except PO_4^{3-}, and the effluxes in June were greater than in September (Fig. 5). The benthic effluxes of DIN and $Si(OH)_4$ contributed 4 and 11 %, respectively, of gross primary productivity (GPP) (including the GPP of kelp). DON can be assimilated by seagrass and macroalgae (Vonk et al. 2008). Assuming DIN and DON released from the sediment was completely consumed by phytoplankton and kelp, the benthic TDN efflux contributed 8 % of GPP. The benthic nutrient contributions to GPP were much smaller than that in the Mandovi Estuary (Pratihary et al. 2009), on the west coast of Sweden (Sundbäck et al. 2003) and in Jinhae Bay (Lee et al. 2011), since substantial cultivation of kelp made the highest contribution to GPP in the IMTA system. If only the GPP of kelp is taken into account, the benthic effluxes of DIN and $Si(OH)_4$ contributed 7 and 18 % of algal N and Si demands. The low contribution of benthic mineralization may be due to efficient recycling of organic matter in the IMTA system, which will be discussed in 'Benthic nutrient fluxes in different aquaculture modes'. The fact that benthic PO_4^{3--} fluxes made no contribution to GPP in Sanggou Bay is consistent with the finding of Hatcher et al. (1994) that suspended mussel culture had little impact on sediment phosphorus dynamics in Upper South Cove (Nova Scotia, Canada). The sedimentation flux of carbon was 82 mmol m^{-2} d^{-1} in June (Cai et al. 2003), which was much lower than that in September, and therefore the TOC in sediment remained at a low level (0.30%). With respect to nutrient feedback in pore water (Fig. 6), large amounts of DIN and $Si(OH)_4$ were generated after incubation, suggesting large potential DIN and $Si(OH)_4$ effluxes, while the generated PO_4^{3-} was not released to the seawater; the decrease in surface PO_4^{3-} after incubation was probably caused by adsorption by Mn/Fe oxides (Woulds et al. 2009). However, this was offset by the release of DOP.

Table 3. Comparison of benthic fluxes in Sanggou Bay, China, with other regions. Ranges or means ± SD. TOC: total organic carbon; DO: dissolved oxygen. CO_2 fluxes represent total benthic organic carbon mineralization rate, calculated from the Redfield ratio (C:O_2 = 1:1) using DO fluxes. DON: dissolved organic nitrogen. nd: no data

| Sites | TOC in sediment (%) | Fluxes (mmol m^{-2} d^{-1}) | | | | | | | References |
		DO	CO_2	NH_4^+	NO_x^-	DON	PO_4^{3-}	$Si(OH)_4$	
Aquaculture bays									
Sanggou Bay	0.35 to 0.68	−21 to −7.7	7.7 to 21	0 to 0.51	−0.030 to 0.27	−1.01 to 0.63	−0.04 to 0	0.83 to 1.76	Present study
Stn MC	nd	−13 ± 1.0	13 ± 1.0	0.48 ± 0.12	0.090 ± 0.030	−0.26 ± 0.16	−0.02 ± 0.01	1.29 ± 0.22	Present study
Stn ST1	nd	−14 ± 0.86	14 ± 0.86	0.35 ± 0.04	0.25 ± 0.084	−0.19 ± 0.44	−0.01 ± 0.01	1.55 ± 0.18	Present study
Tolo Harbour	nd	−39 to −15	15 to 39	3.3 to 5.9	0.010 to 0.025	0.64 to 1.5	0.070 to 0.098	nd	Chau (2002)
Horsens Fjord	8	−300 to 0	0 to 300	−0.7 to 12	−5 to 0.2	nd	0 to 5	nd	Christensen et al. (2000)
Río San Pedro creek	1.44 to 2.67	−79 to −16	16 to 79	3.4 to 21.5	−5.0 to 5.6	nd	0.2 to 2.4	0.7 to 10.2	Ferrón et al. (2009)
Upper South Cove	7.13	−50 to 10	0 to 50	0 to 30	−2 to 3	−20 to 30	−3 to −2	nd	Hatcher et al. (1994)
Jinhae Bay	1.97 to 4.15	−328 to −58	58 to 328	6 to 41	−5.4 to 0.37	nd	0.90 to 3.0	15 to 45	Lee et al. (2011), Hyun et al. (2013)
Non-aquaculture areas									
East China Sea	0.2 to 0.5	nd	3 to 13	−0.10 to 0.54	−0.04 to 0.02	−1.2 to 0.15	−0.04 to 0	0.55 to 2.6	Qi et al. (2006), Hung et al. (2013)
Yellow Sea	nd	nd	nd	−1.1 to 0	−0.44 to 0.02	−0.42 to 1.3	−0.02 to 0	0.65 to 2.9	Qi et al. (2006)

In September, an intense biodeposition resulted in high levels of TOC accumulation in sediment and high DO and DON influxes (Hatcher et al. 1994). Based on the metabolic rates of NH_4^+ (57 t N) and PO_4^{3-} (11 t P) from the Pacific oyster (Mao et al. 2006) and benthic influxes of TDN and TDP in September, sediment may be able to take up 64% of the N and 25% of the P metabolized by oysters. With respect to nutrient feedback in pore water (Fig. 6), at Stn MC the NH_4^+ level in pore water was significantly increased and the NO_x^- was depleted after incubation, which is consistent with high levels of NH_4^+ efflux and NO_x^- influx at high biodeposition sites (Gilbert et al. 1997, Christensen et al. 2000). When NO_x^- is depleted, MnO_2/FeOOH were reduced and the adsorption of PO_4^{3-} substantially decreased, which explains why a marked increase in the PO_4^{3-} concentration was observed in deep pore water after incubation at Stn MC. At Stn ST1 there was no obvious increase in the NH_4^+ concentration in pore water after incubation, probably because of the removal of N by coupled nitrification–denitrification or adsorption (discussed in 'Environmental factors controlling benthic fluxes').

Benthic nutrient fluxes in different aquaculture modes

Increased biodeposits produced by the actively growing animals can result in a substantial increase in the organic content of sediment (Hatcher et al. 1994, Christensen et al. 2000, Ferrón et al. 2009, Lee et al. 2011); the mineralization of sedimentation can release substantial nutrients from the sediment to the seawater, which may result in the deterioration of seawater quality (Chau 2002). Hence, the TOC in the sediment and benthic effluxes of nutrients in traditional aquaculture areas were extremely high (Table 3). Monoculture was implemented in Sanggou Bay in the 1970s; the extremely high TOC in the sediment and the low DIN concentration in the seawater may have resulted in great benthic nutrient fluxes in this monoculture period (Fig. 8). Since 1980 the introduction of polyculture in Sanggou Bay has resulted in the reduction of TOC in the sediment (Song et al. 2012). And the high DIN concentration in the seawater indicated that substantial organic matter was recycling in the seawater during the polyculture period. The efficient recycling of organic matter and nutrients explains why the TOC of sediment and

the benthic effluxes in Sanggou Bay were significantly less than in other monoculture areas. During the polyculture period, the annual gross yield of seafood increased especially in the 2000s, and the proportion of different species changed continuously so that the optimal aquatic environment was obtained (Zhang et al. 2009). Once the IMTA was widely implemented in Sanggou Bay, the DIN concentration dropped to a moderate level; the TOC of sediment and the benthic effluxes in Sanggou Bay are comparable with that in non-aquaculture areas such as the East China Sea (Table 3), though substantial aquaculture activities have been implemented in Sanggou Bay.

In Sanggou Bay, the benthic mineralization rates (CO_2 fluxes) at the 2 different stations were similar, but the benthic nutrient fluxes were different, which reflected the impacts of different aquaculture modes (Table 3). In September, DO was at near saturation levels at Stn ST1 (polyculture area of kelp and oyster) but below saturation at Stn MC (the fish culture area, and near the oyster area) (Fig. 2; contours of DO saturation); the lower DO level at Stn MC led to an increase in the NH_4^+ efflux and a decrease in the NO_x^- efflux. Hyun et al. (2013) reported DO concentrations less than saturation in bottom waters at an oyster farm, presumably because of the combination of DO consumption at the sediment–water interface and the dense suspended culture that limits seawater exchange and the replenishment of DO. Conversely, at Stn ST1, DO provided by kelp helps to maintain the DO saturation level. Hence, greater NO_x^- efflux was observed at Stn ST1 than at Stn MC. The influxes of DON and PO_4^{3-} were higher at Stn MC

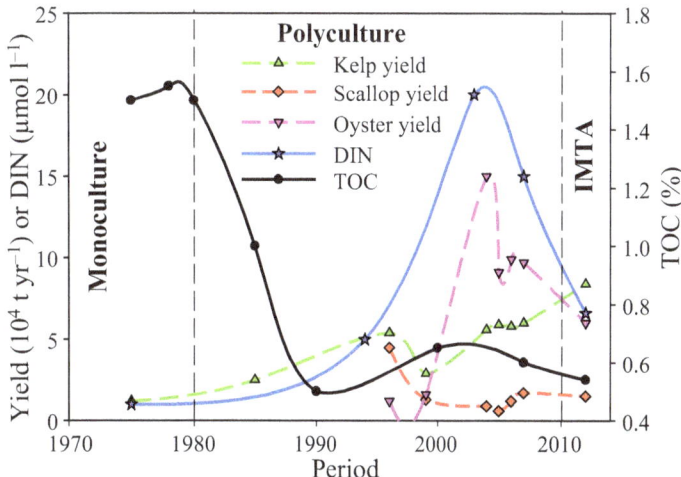

Fig. 8. Historical data of yield, dissolved inorganic nitrogen (DIN) concentration in seawater and total organic carbon (TOC) in sediment, which were sourced from Zhang et al. (2009), Li et al. (2016, this Theme Section) and Song et al. (2012), respectively. IMTA: integrated multi-trophic aquaculture

than at Stn ST1, probably due to the greater metabolic rates from bivalves in the oyster culture area than in the polyculture area of kelp and oyster (Cai et al. 2003). More filtration of diatoms by bivalves in the oyster culture than in the kelp and oyster polyculture area may explain why $Si(OH)_4$ efflux at Stn MC was lower than at Stn ST1.

In summary, the benthic nutrient fluxes were significantly lower in the IMTA system than in other monoculture areas and were impacted by DO levels at different culture stations rather than by sedimentary TOC generated from aquaculture species. Seasonal variations in benthic fluxes were controlled by temperature and nutrient concentrations related to aquaculture.

Acknowledgements. This study was supported financially by the Ministry of Science & Technology of China (2011 CB409802) and the National Science Foundation of China (40925017). Additional financial support was provided by SKLEC/ECNU for TOC determination. We sincerely thank Senlin Wang, Director of the Chudao Fisheries Corporation, for his cooperation, and thank Guodong Song, Xuming Kang, Shuhang Dong and Mingshuang Sun for their help in the field. We are grateful to the 3 anonymous reviewers and the editor for their comments.

LITERATURE CITED

ä Aller RC, Mackin JE, Ullman WJ, Wang CH and others (1985) Early chemical diagenesis, sediment–water solute exchange, and storage of reactive organic matter near the mouth of the Changjiang, East China Sea. Cont Shelf Res 4:227–251

ä Bacher C, Grant J, Hawkins AJS, Fang JG, Zhu MY, Besnard M (2003) Modelling the effect of food depletion on scallop growth in Sungo Bay (China). Aquat Living Resour 16:10–24

Berner RA (1971) Principles of chemical sedimentology. McGraw-Hill, New York, NY

Berner RA (1980) Early diagenesis: a theoretical approach. Princeton University Press, Princeton, NJ

Cai LS, Fang JG, Liang XM (2003) Natural sedimentation in large-scale aquaculture areas of Sungo Bay, north China Sea. J Fish Sci China 10:305–310 (in Chinese with English Abstract)

Carlsson MS, Engström P, Lindahl O, Ljungqvist L, Petersen JK, Svanberg L, Holmer M (2012) Effects of mussel farms on the benthic nitrogen cycle on the Swedish west coast. Aquacult Environ Interact 2:177–191

ä Chau KW (2002) Field measurements of SOD and sediment nutrient fluxes in a land-locked embayment in Hong Kong. Adv Environ Res 6:135–142

Chopin T (2013) Aquaculture, integrated multi-trophic (IMTA). In: Christou P, Savin R, Costa-Pierce B, Misztal I, Whitelaw B (eds) Sustainable food production. Springer, New York, NY, p 184–205

ä Christensen PB, Rysgaard S, Sloth NP, Dalsgaard T, Schwærter S (2000) Sediment mineralization, nutrient fluxes, denitrification and dissimilatory nitrate reduction

to ammonium in an estuarine fjord with sea cage trout farms. Aquat Microb Ecol 21:73–84

ä Cowan JLW, Boynton WR (1996) Sediment–water oxygen and nutrient exchanges along the longitudinal axis of Chesapeake Bay: seasonal patterns, controlling factors and ecological significance. Estuaries 19:562–580

FAO (Food and Agriculture Organization of the United Nations) (2012) The state of world fisheries and aquaculture. FAO, Rome

ä Ferrón S, Ortega T, Forja JM (2009) Benthic fluxes in a tidal salt marsh creek affected by fish farm activities: Río San Pedro (Bay of Cádiz, SW Spain). Mar Chem 113:50–62

ä Gilbert F, Souchu P, Bianchi M, Bonin P (1997) Influence of shellfish farming activities on nitrification, nitrate reduction to ammonium and denitrification at the water–sediment interface of the Thau lagoon, France. Mar Ecol Prog Ser 151:143–153

ä Glud RN (2008) Oxygen dynamics of marine sediments. Mar Biol Res 4:243–289

ä Hatcher A, Grant J, Schofield B (1994) Effects of suspended mussel culture (*Mytilus* spp.) on sedimentation, benthic respiration and sediment nutrient dynamics in a coastal bay. Mar Ecol Prog Ser 115:219–235

ä Hung CC, Tseng CW, Gong GC, Chen KS, Chen MH, Hsu SC (2013) Fluxes of particulate organic carbon in the East China Sea in summer. Biogeosciences 10:6469–6484

ä Hyun JH, Kim YT, Mok JS, Lee JS, An SU, Lee WC, Jung RH (2013) Impacts of long-line aquaculture of Pacific oysters (*Crassostrea gigas*) on sulfate reduction and diffusive nutrient flux in the coastal sediments of Jinhae–Tongyeong, Korea. Mar Pollut Bull 74:187–198

Jansen HM, Verdegem MCJ, Strand Ø, Smaal AC (2012) Seasonal variation in mineralization rates (C-N-P-Si) of mussel *Mytilus edulis* biodeposits. Mar Biol 159:1567–1580

ä Lee JS, Kim YT, Shin KH, Hyun JH, Kim SY (2011) Benthic nutrient fluxes at longline sea squirt and oyster aquaculture farms and their role in coastal ecosystems. Aquacult Int 19:931–944

Li R, Liu S, Zhang J, Jiang Z, Fang J (2016) Sources and export of nutrients associated with integrated multi-trophic aquaculture in Sanggou Bay, China. Aquacult Environ Interact 8:285–309

ä Liu SM, Zhang J, Jiang WS (2003) Pore water nutrient regeneration in shallow coastal Bohai Sea, China. J Oceanogr 59:377–385

ä Liu SM, Zhu BD, Zhang J, Wu Y and others (2010) Environmental change in Jiaozhou Bay recorded by nutrient components in sediments. Mar Pollut Bull 60:1591–1599

ä Lu JC, Huang LF, Luo YR, Xiao T, Jiang ZJ, Wu LN (2015) Effects of freshwater input and mariculture (bivalves and macroalgae) on spatial distribution of nanoflagellates in Sungo Bay, China. Aquacult Environ Interact 6:191–203

ä Mao YZ, Zhou Y, Yang HS, Wang RC (2006) Seasonal variation in metabolism of cultured Pacific oyster, *Crassostrea gigas*, in Sanggou Bay, China. Aquaculture 253:322–333

ä Nunes JP, Ferreira JG, Gazeau F, Lencart-Silva J, Zhang XL, Zhu MY, Fang JG (2003) A model for sustainable management of shellfish polyculture in coastal bays. Aquaculture 219:257–277

ä Pratihary AK, Naqvi SWA, Naik H, Thorat BR, Narvenkar G, Manjunatha BR, Rao VP (2009) Benthic fluxes in a tropical estuary and their role in the ecosystem. Estuar Coast Shelf Sci 85:387–398

Qi XH, Liu SM, Zhang J (2006) Sediment–water fluxes of nutrients in the Yellow Sea and the East China Sea. Mar

Sci 30:9–15 (in Chinese with English Abstract)

▶ Ren L, Zhang J, Fang J, Tang Q, Zhang M, Du M (2014) Impact of shellfish biodeposits and rotten seaweed on the sediments of Ailian Bay, China. Aquacult Int 22:811–819

▶ Sequeira A, Ferreira JG, Hawkins AJS, Nobre A and others (2008) Trade-offs between shellfish aquaculture and benthic biodiversity: a modelling approach for sustainable management. Aquaculture 274:313–328

▶ Shi J, Wei H, Zhao L, Yuan Y, Fang JG, Zhang JH (2011) A physical–biological coupled aquaculture model for a suspended aquaculture area of China. Aquaculture 318: 412–424

Song XL, Yang Q, Sun Y, Yin H, Jiang SL (2012) Study of sedimentary section records of organic matter in Sanggou Bay over the last 200 years. Acta Oceanol Sin 34: 120–126 (in Chinese with English Abstract)

▶ Song GD, Liu SM, Marchant H, Kuypers MMM, Lavik G (2013) Anammox, denitrification and dissimilatory nitrate reduction to ammonium in the East China Sea sediment. Biogeosciences 10:6851–6864

▶ Song GD, Liu SM, Zhu ZY, Zhai WD, Zhu CJ, Zhang J (2015) Sediment oxygen consumption and benthic organic carbon mineralization on the continental shelves of the East China Sea and the Yellow Sea. Deep-Sea Res II, doi: 10.1016/j.dsr2.2015.04.012

▶ Sundbäck K, Miles A, Hulth S, Pihl L, Engström P, Selander E, Svenson A (2003) Importance of benthic nutrient regeneration during initiation of macroalgal blooms in shallow bays. Mar Ecol Prog Ser 246:115–126

▶ Sundby B, Gobeil C, Silverberg N, Mucci A (1992) The phosphorus cycle in coastal marine sediments. Limnol Oceanogr 37:1129–1145

Tang QS, Fang JG (2012) Review of climate change effects in the Yellow Sea large marine ecosystem and adaptive actions in ecosystem based management. In: Sherman K, McGovern G (eds) Frontline observations on climate change and sustainability of large marine ecosystems. UNDP, New York, NY, p 170–187

Troell M, Joyce A, Chopin T, Neori A, Buschmann AH, Fang JG (2009) Ecological engineering in aquaculture–potential for integrated multi-trophic aquaculture (IMTA) in marine offshore systems. Aquaculture 297:1–9

▶ Vonk JA, Middelburg JJ, Stapel J, Bouma TJ (2008) Dissolved organic nitrogen uptake by seagrasses. Limnol Oceanogr 53:542–548

▶ Woulds C, Schwartz MC, Brand T, Cowie GL, Law G, Mowbray SR (2009) Porewater nutrient concentrations and benthic nutrient fluxes across the Pakistan margin OMZ. Deep-Sea Res II 56:333–346

▶ Yuan ML, Zhang CX, Jiang ZJ, Guo SJ, Sun J (2014) Seasonal variations in phytoplankton community structure in the Sanggou, Ailian, and Lidao Bays. J Ocean Univ China 13:1012–1024

▶ Zabel M, Dahmke A, Schulz HD (1998) Regional distribution of diffusive phosphate and silicate fluxes through the sediment–water interface: the eastern South Atlantic. Deep-Sea Res I 45:277–300

Zhang XL, Zhu MY, Chen S, Grant J, Martin JLM (2006) Study on sediment oxygen consumption rate in the Sanggou Bay and Jiaozhou Bay. Adv Mar Sci 24:91–96 (in Chinese with English Abstract)

▶ Zhang JH, Hansen PK, Fang JG, Wang W, Jiang ZJ (2009) Assessment of the local environmental impact of intensive marine shellfish and seaweed farming — application of the MOM system in Sungo Bay, China. Aquaculture 287:304–310

PERMISSIONS

All chapters in this book were first published in AEI, by Inter-Research; hereby published with permission under the Creative Commons Attribution License or equivalent. Every chapter published in this book has been scrutinized by our experts. Their significance has been extensively debated. The topics covered herein carry significant findings which will fuel the growth of the discipline. They may even be implemented as practical applications or may be referred to as a beginning point for another development.

The contributors of this book come from diverse backgrounds, making this book a truly international effort. This book will bring forth new frontiers with its revolutionizing research information and detailed analysis of the nascent developments around the world.

We would like to thank all the contributing authors for lending their expertise to make the book truly unique. They have played a crucial role in the development of this book. Without their invaluable contributions this book wouldn't have been possible. They have made vital efforts to compile up to date information on the varied aspects of this subject to make this book a valuable addition to the collection of many professionals and students.

This book was conceptualized with the vision of imparting up-to-date information and advanced data in this field. To ensure the same, a matchless editorial board was set up. Every individual on the board went through rigorous rounds of assessment to prove their worth. After which they invested a large part of their time researching and compiling the most relevant data for our readers.

The editorial board has been involved in producing this book since its inception. They have spent rigorous hours researching and exploring the diverse topics which have resulted in the successful publishing of this book. They have passed on their knowledge of decades through this book. To expedite this challenging task, the publisher supported the team at every step. A small team of assistant editors was also appointed to further simplify the editing procedure and attain best results for the readers.

Apart from the editorial board, the designing team has also invested a significant amount of their time in understanding the subject and creating the most relevant covers. They scrutinized every image to scout for the most suitable representation of the subject and create an appropriate cover for the book.

The publishing team has been an ardent support to the editorial, designing and production team. Their endless efforts to recruit the best for this project, has resulted in the accomplishment of this book. They are a veteran in the field of academics and their pool of knowledge is as vast as their experience in printing. Their expertise and guidance has proved useful at every step. Their uncompromising quality standards have made this book an exceptional effort. Their encouragement from time to time has been an inspiration for everyone.

The publisher and the editorial board hope that this book will prove to be a valuable piece of knowledge for researchers, students, practitioners and scholars across the globe.

LIST OF CONTRIBUTORS

Tariq Mahmood
School of Resources and Environmental Science, East
China Normal University, 3663 North Zhongshan
Road, Shanghai 200062, PR China
National Institute of Oceanography, St 47, Block 1,
Clifton, Karachi 75600, Pakistan

Jianguang Fang and Zengjie Jiang
Key Laboratory of Sustainable Utilization of Marine
Fisheries Resources, Ministry of Agriculture, Yellow
Sea Fisheries Research Institute, Chinese Academy
of Fishery Sciences, Qingdao 266071, PR China

Jing Zhang
State Key Laboratory of Estuarine and Coastal
Research, East China Normal University, 3663
North Zhongshan Road, Shanghai 200062, PR China

Nelson A. Lagos and Samanta Benítez
Centro de Investigación e Innovación para el
Cambio Climático (CiiCC), Facultad de Ciencias,
Universidad Santo Tomás, 8370003 Santiago, Chile

Cristian Duarte
Departamento de Ecología y Biodiversidad, Facultad
de Ecología y Recursos Naturales, Universidad
Andrés Bello, 8370251 Santiago, Chile

Marco A. Lardies
Facultad de Ingeniería & Ciencias y Facultad
de Artes Liberales, Universidad Adolfo Ibáñez,
7941169 Santiago, Chile

Bernardo R. Broitman
Centro de Estudios Avanzados en Zonas Áridas
(CEAZA), Universidad Católica del Norte, Larrondo
1281, 1781421 Coquimbo, Chile

Christian Tapia and Pamela Tapia
Cultivos Invertec Ostimar S.A., Tongoy, 1780000
Coquimbo, Chile

Steve Widdicombe
Plymouth Marine Laboratory, Prospect Place, West
Hoe, PL1 3DH Plymouth, UK

Cristian A. Vargas
Laboratorio de Funcionamiento de Ecosistemas
Acuáticos (LAFE), Departamento de Sistemas
Acuáticos, Facultad de Ciencias Ambientales,
Universidad de Concepción, 4070386 Concepción,
Chile

S. Bui
Sustainable Aquaculture Laboratory – Temperate
and Tropical (SALTT), School of BioSciences,
University of Melbourne, Victoria 3010, Australia

T. Dempster
Sustainable Aquaculture Laboratory – Temperate
and Tropical (SALTT), School of BioSciences,
University of Melbourne, Victoria 3010, Australia
Institute of Marine Research, Matredal 5984,
Norway

M. Remen and F. Oppedal
Institute of Marine Research, Matredal 5984,
Norway

**Maximilian de Kantzow, Paul Hick, Joy A. Becker
and Richard J. Whittington**
Faculty of Veterinary Science, The University of
Sydney, 425 Werombi Road, Camden, NSW 2570,
Australia

Tanglin Zhang, Jiashou Liu and Zhongjie Li
State Key Laboratory of Freshwater Ecology and
Biotechnology, Institute of Hydrobiology, Chinese
Academy of Sciences, Wuhan 430072, China

Wei Li
State Key Laboratory of Freshwater Ecology and
Biotechnology, Institute of Hydrobiology, Chinese
Academy of Sciences, Wuhan 430072, China
School of Aquatic and Fishery Sciences, University
of Washington, Box 355020, Seattle, Washington
98195-5020, USA

Huaiyu Ding
Jiangsu Engineering Laboratory for Breeding
of Special Aquatic Organisms, Huaiyin Normal
University, Huaian 223300, China

Fengyin Zhang
College of Life Sciences, Jianghan University, Wuhan 430056, China

Annette Bruhn, Thorsten Johannes Skovbjerg Balsby and Michael Bo Rasmussen
Department of Bioscience, Aarhus University, Vejlsøvej 25, 8600 Silkeborg, Denmark

Ditte Bruunshøj Tørring, Jens Kjerulf Petersen and Paula Canal-Vergés
Danish Shellfish Centre, Institute of Aquatic Resources, Technical University of Denmark, DTU-Aqua, Øroddevej 80, 7900 Nykøbing Mors, Denmark

Marianne Thomsen
Department of Environmental Sciences, Aarhus University, Frederiksborgvej 399, 4000 Roskilde, Denmark

Mette Møller Nielsen
Department of Bioscience, Aarhus University, Vejlsøvej 25, 8600 Silkeborg, Denmark
Danish Shellfish Centre, Institute of Aquatic Resources, Technical University of Denmark, DTU-Aqua, Øroddevej 80, 7900 Nykøbing Mors, Denmark

Martin Mørk Larsen
Department of Bioscience, Aarhus University, Frederiksborgvej 399, 4000 Roskilde, Denmark

Karin Loft Eybye
Division of Life Science & Food Technology, Danish Technological Institute, Kongsvang Allé 29, 8000 Aarhus C, Denmark

Knud Simonsen and Øystein Patursson
Fiskaaling – Aquaculture Research Station of the Faroes, við Áir, 430 Hvalvík, Faroe Islands

Esbern J. Patursson
Fiskaaling – Aquaculture Research Station of the Faroes, við Áir, 430 Hvalvík, Faroe Islands
VKR Centre for Ocean Life, National Institute of Aquatic Resources, Technical University of Denmark, Kavalergaarden 6, 2920 Charlottenlund, Denmark

André W. Visser
VKR Centre for Ocean Life, National Institute of Aquatic Resources, Technical University of Denmark, Kavalergaarden 6, 2920 Charlottenlund, Denmark

Frida Solstorm and Anders Fernö
Institute of Marine Research, 5984 Matredal, Norway
Department of Biology, University of Bergen, 5006 Bergen, Norway

David Solstorm, Frode Oppedal and Lars Helge Stien
Institute of Marine Research, 5984 Matredal, Norway

Rolf Erik Olsen
Institute of Marine Research, 5984 Matredal, Norway
Department of Biology, Norwegian University of Science and Technology, 7491 Trondheim, Norway

Tina Oldham
Aquatic Animal Health Group, Institute for Marine and Antarctic Studies, University of Tasmania, Launceston, Tasmania 7250, Australia

Tim Dempster
Sustainable Aquaculture Laboratory – Temperate and Tropical (SALTT), School of BioSciences, University of Melbourne, Parkville, Victoria 3052, Australia

Jan Olav Fosse and Frode Oppedal
Institute of Marine Research, Matredal 5984, Norway

Daphne Munroe, David Bushek, Patricia Woodruff and Lisa Calvo
Haskin Shellfish Research Laboratory, Rutgers University, 6959 Miller Ave., Port Norris, NJ 08349, USA

Paul M. South
Cawthron Institute, 98 Halifax Street East, Nelson 7010, New Zealand
Institute of Marine Science, University of Auckland, Private bag 92019, Auckland, New Zealand

Oliver Floerl
Cawthron Institute, 98 Halifax Street East, Nelson 7010, New Zealand

Andrew G. Jeffs
Institute of Marine Science, University of Auckland, Private bag 92019, Auckland, New Zealand

Li Ma, Jie Zhu, Qi Chen and Guo-Hua Huang
Hunan Provincial Key Laboratory for Biology and Control of Plant Diseases and Insect Pests, Hunan Agricultural University, Changsha 410128, Hunan, PR China

Wei Li
State Key Laboratory of Freshwater Ecology and Biotechnology, Institute of Hydrobiology, Chinese Academy of Sciences, Wuhan 430072, Hubei, PR China

Konrad Hughen, Tracy J. Mincer, Justin Ossolinski, Laura Weber and Amy Apprill
Department of Marine Chemistry and Geochemistry, Woods Hole Oceanographic Institution, Woods Hole, MA 02543, USA

Cynthia Becker
Department of Marine Chemistry and Geochemistry, Woods Hole Oceanographic Institution, Woods Hole, MA 02543, USA
Department of Biology, Ithaca College, Ithaca, NY 14850, USA

José L. Stech
Remote Sensing Division, National Institute for Space Research, 12227-010 São José dos Campos, SP, Brazil

Carlos A. S. Araújo
Remote Sensing Division, National Institute for Space Research, 12227-010 São José dos Campos, SP, Brazil
Département de Biologie, Chimie et Géographie, Université du Québec à Rimouski, Rimouski, QC G5L 3A1, Canada

Fernanda G. Sampaio
Embrapa Environment, Brazilian Agricultural Research Corporation, 13820-000 Jaguariúna, SP, Brazil

Enner Alcântara
Department of Environmental Engineering, São Paulo State University, 12247-004 São José dos Campos, SP, Brazil

Marcelo P. Curtarelli
Green Economy Center, Reference Center Foundation for Innovative Technologies, 88040-970 Florianópolis, SC, Brazil

Igor Ogashawara
Department of Earth Sciences, Indiana University–Purdue University Indianapolis, Indianapolis, IN 46202, USA

R. Filgueira
Marine Affairs Program, Dalhousie University, 1355 Oxford St., Halifax, NS B3H 1R2, Canada

T. Guyondet
Department of Fisheries and Oceans, Gulf Fisheries Centre, Science Branch, Moncton, NB E1C 9B6, Canada

G. K. Reid
Canadian Integrated Multi-Trophic Aquaculture Network (CIMTAN), University of New Brunswick, Saint John, NB E2L 4L5, Canada Department of Fisheries and Oceans, St. Andrews Biological Station, 531 Brandy Cove Rd., St. Andrews, NB E5B 2L9, Canada

J. Grant
Department of Oceanography, Dalhousie University, Halifax, NS B3H 4R2, Canada

P. J. Cranford
Department of Fisheries and Oceans, Bedford Institute of Oceanography, 1 Challenger Dr., Dartmouth, NS B2Y 4A2, Canada

Yan Chang
School of Ecological and Environmental Sciences, East China Normal University, Shanghai 200062, PR China
State Key Laboratory of Estuarine and Coastal Research, East China Normal University, Shanghai 200062, PR China

Jing Zhang, Jianguo Qu and Ruifeng Zhang
State Key Laboratory of Estuarine and Coastal Research, East China Normal University, Shanghai 200062, PR China

Zengjie Jiang
Key Laboratory for Sustainable Utilization of Marine Fisheries Resources, Ministry of Agriculture, Yellow Sea Fisheries Research Institute, Chinese Academy of Fishery Sciences, Qingdao 266071, PR China

Zhiming Ning, Guoling Zhang and Xiaoyan Ning
Key Laboratory of Marine Chemistry Theory and Technology MOE, Ocean University of China/ Qingdao Collaborative Innovation Center of Marine Science and Technology, Qingdao 266100, PR China

Sumei Liu
Key Laboratory of Marine Chemistry Theory and Technology MOE, Ocean University of China/ Qingdao Collaborative Innovation Center of Marine Science and Technology, Qingdao 266100, PR China Laboratory of Marine Ecology and Environmental Science, Qingdao National Laboratory for Marine Science and Technology, Qingdao 266100, PR China

Ruihuan Li
Key Laboratory of Marine Chemistry Theory and Technology MOE, Ocean University of China/ Qingdao Collaborative Innovation Center of Marine Science and Technology, Qingdao 266100, PR China State Key Laboratory of Tropical Oceanography, South China Sea Institute of Oceanology, Chinese Academy of Sciences, Guangzhou 510301, PR China

Zengjie Jiang and Jianguang Fang
Key Laboratory of Sustainable Utilization of Marine Fisheries Resources, Ministry of Agriculture, Yellow Sea Fisheries Research Institute, Chinese Academy of Fishery Sciences, Qingdao 266071, PR China

Index